Population, Development and Environment

About the Editors

Editor-in-Chief

Prof. (Mrs.) Tara Devi Singh, (b. 1947) is Professor of Geography at Banaras Hindu University. Her areas of interest are Urban and Population Geography. She has published numerous research papers, monograph, edited volume and authored one book on **Indian Ocean** in Hindi for which she is awarded by the Government of India in 2010.

Editors

Dr. A.P. Mishra, (b. 1958) is Associate Professor of Geography at Banaras Hindu University. His specialization includes rural poverty, resource, regional development studies and gender studies. He has completed three projects and published more than 40 research papers, one research monograph and three books.

Dr. Arun K. Singh, (b.1966) is Associate Professor of Geography in MMV, Banaras Hindu University. His area of interest includes Urban Geography, Population Geography, Participatory Planning and Governance. He is running major research project on urban water supply funded by the UGC.

Dr. Archana K. Roy, (b.1975) is Associate Professor of Geography at Banaras Hindu University. Her areas of interest are in population, migration and urbanization studies, women studies and study of deprived and marginalized groups. She authored a book "Distress Migration and 'Left behind' Women".

Dr. Narender Verma, (b.1972) is Associate Professor in the Department of Geography, Banaras Hindu University, His areas of interest includes Physical Geography and Disaster Management. He has attended several international and national conferences and has to his credit six papers at the international level.

Population, Development and Environment

A CONTEMPORARY DEBATE

Editor-in-Chief

Tara Devi Singh

Editors

Anand Prasad Mishra
Arun K. Singh
Archana K. Raj
Narender Verma

CONCEPT PUBLISHING COMPANY PVT. LTD.,
NEW DELHI-110059

ISBN-13: 978-81-8069-916-0

First Published 2013

Published and Printed by

Concept Publishing Company Pvt. Ltd.
A/15-16, Commerical Block, Mohan Garden
New Delhi - 110059 (India)
T : 25351460, 25351794, F : 091-11-25357109
E : publishing@conceptpub.com, W : www.conceptpub.com
Editorial Office : H-13, Bali Nagar, New Delhi-110015, India

Cataloging in Publication Data--*Courtesy:* D.K. Agencies (P) Ltd. <docinfo@dkagencies.com>

Population, development and environment : a contemporary debate / editor-in-chief, Tara Devi Singh ; editors, Anand Prasad Mishra ... [et al.].
p. cm.
Includes bibliographical references and index.
ISBN 9788180699160

1. India—Population—Environmental aspects—Congresses. 2. Sustainable development—India—Congresses. 3. Urbanization—India—Congresses. 4. Rural-urban migration—India—Congresses. 5. Economic development—India—Congresses. 6. India—Environmental conditions—Congresses. I. Singh, Tara Devi, 1947- II. Mishra, Anand Prasad.

DDC 304.60954 23

Dedicated to

Mahamana Pt. Madan Mohan Malviyaji

The Great Educationist, Visionary, Nationalist
and Founder of Banaras Hindu University

On the Occasion of 150th Birth Anniversary
(1861-1946)

Dr. Lalji Singh
Ph.d. D.Sc. (Hon.)
FNA, FASc, FNASc, FNAAS,
FTWAS
Padmashri
Bhatnagar Fellow (CSIR)
Former Director, CCMB,
Hyderabad
Vice-Chancellor

काशी हिन्दू विश्वविद्यालय
BANARAS HINDU UNIVERSITY
(Established by Parliament by Notification No. 225 of 1916)
VARANASI-221 005 (INDIA)
Phones : 91-542-2368938, 2368339
Fax : 91-542-2369100, 2369951
e-mail : vobhu1@gmail.com,
vc_bhu@sify.com
website : www.bhu.ac.in

Foreword

Population, development and environment have been the issues of debate among the academicians since the time of Thomas R. Malthus with the thrust on maintaining the balance between population and natural resources. Discourse on environmental issues and its explicit relationship with population in the context of development could take place in the second half of 20th century with the Stockholm Conference (1972); The United Nations Conference on Environment and Development (1992); International Conference on Population and Development (1994) and United Nations Conference on Human Settlements (Habitat II 1996). Since then efforts have been made to formulate and implement the policies which integrate these issues and look at the things in a holistic manner. However, because of administrative shackles and lack of co-ordiantion and convergences at the implementation level, the desired result is yet to come out. Hence, the idea of bringing out an edited volume entitled "*Population, Development and Environment: A contemporary Debate*" is a welcome and timely effort of Prof. Tara Devi Singh, Department of Geography, BHU and her team members especially because it is coming at the time when the University is celebrating 150th birth anniversary of its founder the great educationist, visionary and freedom fighter Mahamana Pt. Madan Mohan Malaviyaji. I consider this enderavour as a true tribute and modest step towards the fulfilment of the vision of our founder by flagging the issues which touch every section of the society. Title of the book is justified as the papers cover a wide range of spectrum from conceptual aspects of the theme to the policy level interventions involving all the stakeholders of the society in the successful implementation at the grassroots level.

The book adopts a multidisciplinary approach to explore the complex, continuous and eternal relationship between population, development and environment in the Indian context, drawing concepts and methods from several academic disciplines like geography, demography, environment, economics, sociology, botany and social work. This book will be useful for academicians, researchers, NGOs, policy makers, planners and all those who are concerned for the common people, the people who are concerned with population, development and environment.

(Lalji Singh)

Preface and Acknowledgements

Discussion and debate on the interrelationships among population, environment and development has been initiated way back in second half of eighteenth century with the writings of Thomas R. Malthus and his contemporary philosophers and thinkers. But they were concerned mainly with the balance between population and natural resources conceptualized as means of subsistence and livelihood (food and water). Discourse on environmental issues and its relationship with population in the context of development could take place in the mid twentieth century and became apparent at the first session of the Population Commission in 1947. The argument has been between those who view population numbers *per se* the main culprit in increasing pressure on the environment and those who place more blame on economic development, non-sustainable agricultural and industrial practices, and excessive and wasteful consumption. In fact, both population growth and non-sustainable development are cause for concern in India. Though the relationship is complex, population size and growth tend to expand and accelerate these human impacts on the environment.

Last century has witnessed an unprecedented population growth, economic development and environmental change. During 1901-2001 world population grew from 1.6 billion to 6.4 billion and India's population increased from 236 million to 1027 million recording around four times increase in population in both the cases. During the same period world's real gross domestic product (GDP) increased 20 to 40 times (De Long, 1998 quoted from UN Publication 2001), allowing the world not only to sustain a fourfold population increase but also to maintain the quality of life. But population growth and economic growth occurred unevenly over the space because of different level of knowledge base, skill and technological development. The United Nations Conference on Environment and Development (1992) was a milestone in the evolution of an international consensus on the relationships among population, development and environment, based on the concept of sustainable development which is defined as development that "meets the needs of the present without compromising the ability of future generations to meet their own needs". Rio Declaration stated that "to achieve sustainable development and a higher quality of life for all people, states should reduce and eliminate unsustainable patterns of production and consumption and promote appropriate demographic policies". The Programme of Action of the International Conference on Population and Development (1994) noted "population, poverty, patterns of production, consumption and the environment are so closely interconnected that none of them can be considered in isolation". Since then efforts have been made to formulate and implement the new policies which integrate the issues of population, development and environment. However, because of administrative shackles and lack of co-ordination and convergences the desired result is yet to come out.

The idea of bringing out an edited volume entitled "Population, Development and Environment: A Contemporary Debate" lies in the successful organization of the national seminar on the topic "Population, Development and Environment: Issues and Challenges" in the Department of Geography, Banaras Hindu University which received overwhelming response from different corners of the country and attracted thought provoking quality papers from academicians, planners and policy makers. Title of the book seems to be justified because of the fact that papers cover a wide range of spectrum ranging from conceptual aspects of population, development and environment to the policy level interventions involving all the stakeholders in the successful implementation of the policy.

From the conception of the national seminar to bringing out the book in the present form is a combined effort of so many organisations, agencies and individuals, who rendered support in different forms at different points of time. First of all, I extend my deep sense of gratitude to the chief patron (Hon'ble Vice Chancellor, Banaras Hindu University), Prof. V.K. Kumra (Former Head of the Department of

Geography) and Prof. M.B. Singh (Head of the Department of Geography) who had been with us at every time whenever their support was needed, I will fail in my duty if I forget the names of Prof. K.N.P. Raju and Prof. Jagdish Singh (Co-ordinators, SAP) whose untiring efforts and financial support through DRS-SAP put our morale high through the event.

Successful management of any such mega event, requires a lot of financial resources. I am grateful that agencies, oranisations and institutions like UGC, SAP-DRS, DST, Ministry of Earth Sciences, UNFPA, CSIR, PFI, ICMR and Banaras Hindu University, Varanasi provided us the financial support in organizing the event. Thanks goes to all the contributors who have been with us right from paper presentation to its writing, modification and submission. I express my gratitude to all of them for collaborating and cooperating with the editorial board.

I acknowledge the prompt and timely cooperation of my editorial board members (Dr. Anand Prasad Mishra, Dr. Arun K. Singh, Dr. Archana K. Roy and Dr. Narender Verma) at every moment from the conception of idea to giving it a concrete shape and sending it to the publisher.

I feel obliged to Mr. Ashok Kumar Mittal of Concept Publishing Company (P) Ltd., New Delhi who kindly agreed to publish the manuscript in the present form. Last but not the least, I personally extend my thanks to Mr. Abhishek Vishwakarma who made continuous, untiring efforts from typing to setting and formatting the text.

Finally, I extend my thanks to all those who have directly or indirectly helped in bringing out the volume.

I believe the book will be useful for those interested in the study of population, development and environment and policy makers. If there is any shortcoming in the book that is mine. However, views expressed in papers are the views of contributors and editorial board is not responsible for any violation of the copyright.

Banaras Hindu University,
Varanasi-221005
Date: 19.01.2012

Tara Devi Singh
Editor-in-Chief

[illegible] and Prof. M.B. [illegible] Head of the Department of [illegible] who [illegible] at every time whenever their support was needed. [illegible] forget the names of Prof. K.S. [illegible] and Prof. [illegible] Singh (Co-ordinators, SAP) [illegible] and [illegible] support [illegible] through the event.

Successful [illegible] of any such [illegible] requires [illegible] resources. I am grateful that agencies, organisations and institutions [illegible] (Ministry of Earth Sciences) [illegible] and Banaras Hindu University [illegible] the [illegible] the event. Thanks [illegible]

[illegible]

[illegible] (Dr. [illegible]) [illegible] to giving [illegible] and [illegible]

[illegible]

Contents

PART II
MIGRATION AND URBANIZATION

PART III
SOCIO-ECONOMIC DEVELOPMENT

Contributors

A.A. Ansari, Associate Professor of Geography, Faculty of Science, The Maharaja Sayajirao University of Baroda, Sayajigunj, Vadodara-390002, E-mail: ansari_geomsu@yahoo.co.in.

Aalok Ranjan Chaurasia, Director, 'Shyam' Institute, Bhopal, Email:aranjan@shyaminstitute.in

Ajay Kumar, Assistant Professor of Architecture, National Institute of Technology, Patna-800 005, Email: ajay.archi@gmail.com

Anand Prasad Mishra, Associate Professor of Geography, Banaras, Hindu University, Varanasi-221005, email: adeepayan@ gmail.com

Archana K. Roy, Assistant Professor of Geography, Banaras Hindu University, Varanasi-221 005

Arun K. Singh, Associate Professor of Geography (MMV), Banaras Hindu University, Varanasi-221 005, Email: aksmmvbhu@gmail.com

B.L. Nagda, Assistant Director, Population Research Centre, Mohanlal Sukhadia University, Udaipur-313 001, Email: blnagda@yahoo.co.in

Bijay Kumar Das, Assistant Professor of Architecture, National Institute of Technology, Patna-800 005, Email: bijaykdas@yahoo.com

Fulena Rajak, Associate Professor of Architecture, National Institute of Technology, Patna-800 005, Email: fulenarajak@yahoo.com

Jyotirmayee Acharya, Assistant Professor, School of Rural Management, KIIT University, Bhubaneswar. Email: jacharya9@ gmail.com

Kashi Nath Singh, Professor (Retd.), Department of Geography, BHU, Varanasi-221 005.

Mithlesh Gupta, Assistant Professor of Sociology, B.S.R. Govt. Arts College, Alwar-301 001.

M.H. Qureshi, Former Professor of Geography, CSRD, J.N.U., New Delhi-110 067. Presently Chair Professor, Shri A.M. Khwaja Chair, Jamia Millia Islamia, New Delhi-110 025. Email: mhqureshi42@gmail.com.

Mrutuyanjaya Sahu, Academic Research Fellow, Economic and Social Science Area, Indian Institute of Management Bangalore, Bannerghatta Road, Bangalore-76, Email: mrutuyanjaya.sahu@iimb.ernet.in; mrutuyanjoyhcu@gmail.com

Narender Verma, Assistant Professor of Geography, Banaras Hindu University, Varanasi-221 005. Email: nverma1972@gmail.com

Neelam Saraswat, Assistant Professor of Economics, B.S.R. Govt. Arts College, Alwar-301 001. Email-dr.neelamsaraswat@yahoo.com

Neelam Tripathi, Institute of Natural Resources, Massey University, Palmerston North, New Zealand.

Prabhat Kumar Rai, Assistant Professor, School of Earth Sciences and Natural Resource Management, Mizoram University, Tanhril, Aizawl-796 004. Email: prabhatrai24@yahoo.co.in

R.B. Bhagat, Professor and Head, Migration and Urban Studies, Institute for Population Sciences, Mumbai-400088. E-mail: rbbhagat@iips.net

Ram Bali Singh, Professor of Urban Planning and Development, School of Graduate Studies, Ethiopian Civil Service University, P.O. Box 5648, Addis Ababa, Ethiopia. E-mail: singh.rb1941@gmail.com.

Rajani Srivastava, Assistant Professor, Department of Environmental Science and Technology, RGSC, Banaras Hindu University, Varanasi-221 005. E-mail: srivastava_252003@yahoo.com, rajani.srivastava25@gmail.com

Rajesh Kumar Chauhan, Joint Director, Population Research Centre, Department of Economics, University of Lucknow, Lucknow-226007. Email: rajesh_kumar_chauhan@hotmail.com

Rajneesh Kumar, Lecturer in Chemistry, Patna Science College, Patna University, Patna-800 005. Email: rajneesh.scp@gmai.com

Ranjana Bajpai, Associate Professor of Geography, Department of Applied Geography, School of Earth Sciences and Regional Studies, Ravenshaw University, Cuttack-753 003. Email: ranjudel@gmail.com

Rashmi Kant, Institute of Natural Resources, Massey University, Palmerston North, New Zealand.

Ratnawali, Assistant Professoı, Centre for Social Studies, Surat. Email: sinha_ratnawali@yahoo.com info@css.ac.in

Shekhar Mukherjee, Professor (Retd.), Migration and Urban Studies IIPS, Mumbai, NL-6-2-7, Sector 8, Nerul (W), New Mumbai-400 706. Email: spm631@vsnl.net

Shiv Shankar Mishra, Assistant. Professor of Commerce, P.G. College, Malikpura, Ghazipur.

Sibabrata Das, Assistant Professor, Department of Applied Geography, Ravenshaw University, Cuttack-753003. Email: sibabrata2007@gmail.com, sibabrata@ravenshawuniversity.ac.in

Srikrishan Singh, Assistant Professor of Commerce, P.G. College, Malikpura, Ghazipur. Email: gurudhamvns@rediffmail.com

Sudha Srivastava, Professor and Head, Department of Geography, University of Mumbai, Mumbai. Email: srivastavasudha @hotmail.com

Tara Devi Singh, Professor of Geography, Banaras Hindu University, Varanasi-221005. Email: tarad.singh@yahoo.co.in

Upendra Nath Roy, Rural Devlopment, NITTTR, Sector, 26, Chandigarh.

Vandana Shiva, Enviornmentalist, A-60, Hauz Khas, New Delhi-110 016. Emails: vandana@vandanashiva.com; vandana.shiva @gmail.com; vshiva@vsnl.com

Vishal Gupta, Department of Geography, Faculty of Science, The Maharaja Sayajirao University of Baroda, Sayajigunj, Vadodara-390 002. E-mail: gupta_geomsu @yahoo.co.in

Introduction

Tara Devi Singh, Anand Prasad Mishra, Arun K. Singh, Archana K. Roy and *Narendra Verma*

Population and development generally refers to the study of the consequences of population trends on socio-economic development, human welfare and the natural environment. One school of thought adopts "doomsday scenario" which argues that population growth has adverse consequences on human welfare and the natural environment. The belief is the numerous we become, the poorer we will be. Another school of thought considers that population is not a problem rather it is social structure, economic policies and government initiatives and interventions which create, the problem. Some contend that numbers in themselves do not equal to poverty; rather, poorly structured societies and economies foster poverty. These views are two extremes and the lies in between the two arguments.

This debate is not new. There are a number of conventional models of the impact of population size and growth on the economy. Throughout the history, and even to some modern theories, a large and growing population has been considered as a sign of prosperity. The most forceful thought was given by Thomas R. Malthus, in 1798 through his work 'An Essay on the Principle of Population'. He invented the theory of the declining rate of return for each additional unit of labour in an agricultural economy with constant technology. In an overpopulated rural economy with no reserve of cultivatable land, population growth leads to underemployment and an increasing fraction of the population is pushed into abject poverty. Karl Marx rejects the Malthusian theoretical construct. He believed capitalism creates surplus population through physical separation of producers (peasants, artisans, workers) by landowners from their means of production (land, machinery, tools etc.). Poverty occurs, not because of overpopulation, but rather due to lack of access to means of gaining subsistence. Resources are confined to a capitalist or land owning class, potential always exists for poverty and hunger and peasants and working poor have very little bargaining power compared to landlords and capitalists—hence it leads to poverty and hunger.

In last few decades inside Europe, the debate had shifted from food-population balance to balance between nations and possibility of expansion with definitely nationalist and military connections. However, in later half of the twentieth century, geo-political scenario changed. Rapid population growth was taking place in developing countries without much economic development and again the debate shifted back to population and development relationship. The debate was oscillating between populations negatively affecting the development or cause underdevelopment.

There is a broad range of neo Malthusians, ranging from the primarily polemical (such as the Paul Ehrlich of *The Population Bomb*) to demographers (Kingsley Davis, 1967) who believe the current world population is unsustainable. The neo-malthusian concept of carrying capacity, which talked about limit of land and water to sustain human population and supported the strategy of family planning, contraception, abortion and population problem, cannot be handled in short time. The concept of population bomb was brought forward by Paul Ehrlich in the late 1960s. Fast population growth was seen as a threat the word "bomb" referred to the perceived lethal character of the problem. Ehrlich and others continued the basic Malthusian numbers game in which population growth outstrips food production. They moved beyond Malthus in their consideration of many environmental issues and resource shortage. The exemplary research in this tradition was by Ansley Coale and Edgar Hoover (1958) in their book *Population Growth and Economic Development in Low Income Countries*. High levels of fertility create a very youthful age structure with a high ratio of children to working age adults. Social expenditure on school and health due to young age structure diverted funds from capital investment and thus high population growth causes poor socio-economic development. Therefore, governments should intervene to control population.

Contrary to the Malthusian and Neo-Malthusian pessimistic viewpoints, another view put forward was the creative pressure, which argues that solution lies in the problem itself. Large populations will stimulate technological change and productivity. The counter hypothesis to Malthus' pessimistic account, namely Ester Boserup's (1981) thesis that densely settled populations provide the motivation and resources necessary for technological breakthroughs. The pressures of growing populations have probably stimulated large scale migration around the globe, the origins and spread of agriculture, the shortening of the fallow, irrigation, and most other developments that we associate with civilization in traditional preindustrial societies. The most famous advocate of this position is the late Julian Simon, whose book, *The Ultimate Resource* (1981) argued that each person was a potential source of ingenuity and creativity. Societies with larger populations would be more likely to develop because of their larger number of potential scientists, inventors and creative minds.

The consequences of this debate became concrete. While it was often recognized that development had an impact on lowering fertility, but the time necessary for this effect was questioned particularly in the context of countries with high population growth rates. Number of countries formulated policies based primarily on the recognition of the negative consequences of rapid population growth. The feeling of urgency, led to setting up of military style campaign in some cases setting targets led to violation of human rights and abuses. Such abuses provoked to reviewing of programmes and adaptation of social engineering. Now in practice a number of countries promoted activities in population and development rather than choosing one at the expense of others. Although these countries succeeded in reduction of fertility and acceleration of economic development, however, the reductionist strategy led to neglect of other issues like migration, environment, quality of life and thus precipitating difficulties in certain areas.

With emerging concerns for environmental protection and the introduction of quality of life considerations into development, the Malthusian debate was reborn like the phoenix. It now raises the question of limits that at certain levels of population growth and consumption and production patterns are compatible with environmental protection and conservation? Such question is difficult to answer but is useful and legitimate. Earlier the issue of discussion was on population dynamics and factors that influence population trends and its impact on the economy, but later on the dependent variables shifted the topics in purview of other sciences, such as soil erosion, future water supplies or the carbon cycle etc. There has also been a huge debate over the demands of population size and growth on the availability of food supplies, the exhaustion of natural resources, and negative externalities on the environment. Until the last decade, the literature on the effects of population growth on natural resources could be divided into two communities that had little contact with each other. Those who were primarily concerned about the exhaustion of natural resources—let's call them the "pessimists"—could model the limits of sustainability and predict a doomsday scenario when the world's resources would be exhausted. On the other side were the economists and historians—let's call them the "optimists"—who would report that new sources of energy or resources would be discovered or invented. Since the exhaustion of wood for fuel was replaced by charcoal, and in turn, coal was discovered as a substitute for charcoal, followed by oil, natural gas and presumably by atomic and/or solar power, so the doomsday will never arrive.

This Volume

Two hundred years after Thomas R. Malthus (1798) since he wrote his famous treaties on population, the debate continues. Does population growth really lead to any type of calamity or environment disaster? Do Indian tsunami of 2004, major earthquake/tsunami of March 2011 or other havocs going on in different parts of the globe tell anything about Malthusian concept of preventive checks or positive checks? How could it be minimized, mitigated or controlled? How one region of the globe is developing at the cost of other, denying the spirit of justice, equity and equality. The edited volume entitled *Population, Development and Environment: A Contemporary Debate* is virtually an outcome of intense interaction, discussion and debate on population, development and environment among academicians, planners and policy makers held during the national seminar in February 2010 in the Department of Geography, Banaras Hindu

University, Varanasi. The volume contains 26 papers exploring the relationship among the three dimensions, i.e. population, development and environment especially in the Indian context. The chapters are organized into five parts based on the broader topics of the book. The first part presents a theoretical framework of the relationship between the trilogy from different perspectives. The second part revolves around the various aspects of migration, one of the important components of population dynamics and process of population redistribution, i.e. urbanization. It brings together paradoxical issue of distress labour migration and displacement on the one hand and development and urbanization on the other. The third part is webbed around the socio-economic developmental issues. Taking the cases of Eastern UP, Bihar, Orissa, Gujarat and Rajasthan (coincidently all the case studies except Gujarat belong to the BIMARU states) it raises a series of issues like, why certain areas are not picking up development? Even where the economic development has taken place, who have been benefited and who were the sufferers in this process? The fourth part throws light on environmental issues. It starts with the contemporary debate on earth democracy *versus* corporate globalization and later on it examines the impact of economic development on environment and impact of environment on human development. The last part discusses the data sources needed for research in the holistic study of population, development and environment and policy concerns. However, most of the chapters dealing with specific problem suggest various policy measures. But in this section strategies for better implementation of population policy in India for inclusive growth has been discussed.

Overall, the book follows a multidisciplinary approach to explore the complex, continuous and relationship among population, development and environment in the context of India, drawing concepts and methods from several academic disciplines like geography, demography, environment, economics, sociology, botany and social work. This book will be useful for academicians, researchers, NGOs, policy makers, planners and all those who are concerned for the common people and the people who are concerned with population, development and environment. The description of the each part has been given in following paragraphs.

Part I: Population Development and Environment: A Conceptual Framework

Discussing the interrelated issues of the population, development and environment, Weaks in Population: An introduction to concepts and issues (2008) noted the three perspectives—boomsters, doomsters and neo-Marxist—carries a very different set of policy prescriptions and a different range of environmental consequences, so it matters who is right. Boomsters believe that population growth stimulates economic development; doomsters advocate that in trying to not only to feed but also to improve the lives of an ever increasing population, we are polluting the land, affecting the climate, using the critical resources which are finite; and the third perspective, i.e. neo-Marxist is that growth has nothing to do with economic development at all. The economic development where it lags is held back by the dominant countries. Like doomster view, that population cannot grow indefinitely and people will lower their fertility when they see advantage from doing so which means lifting them out of poverty. In this light, this section provides the global overview of the debate on population, development and environment. Kashi Nath Singh in his paper *Population, Environment, Economy and Sustainable Development: Some Interlocking Patterns* initiates the debate with the hypothesis that population, environment, economy and sustainable development are all intertwined into a multiplex, dynamic and tightening interlocking pattern in which population factor occupies the locus and other three are inextricably interwoven with each other. While discussing these issues he argues that technology, being more benign, and used in the widest perspectives, including social and politico-economic engineering, must keep innovating and adapting to these newer needs of the society and ecology with a 'human face'. Economic growth, development and ecology must go together in reciprocally balancing act.

Human beings form the centre of gravity of the triangle whose three edges are environment, technology and development and it forms the core of geographical studies. Geographers have been interested in studying man-environment relationship and its impact on society since the dawn of civilisation. These issues have been dealt by M.H. Qureshi in his paper *Population, Development and Environment:*

Issues and Challenges. He has discussed the issues in historical perspective, highlighted the contemporary challenges and identified the role of present generation geographers in resolving the crisis at various fronts by applying powerful and sophisticated tools in the post modern world. He has talked about the convergence of the experience of the old and energy of the young geographers which could lead to the welfare of the society. He concludes that the population, development and environment have their own dynamic relationship which need to be balanced and be ensured by accommodating the aspirations of the population, by reducing the regional imbalances and by protecting the environment.

R.B. Bhagat's paper on *Population and Environment Relationship: The Context of Climate Change* attempts to present a synthesis of the nature of relationship between the three in the context of climate change which is the result of global warming in the wake of rising greenhouse gases (GHG) because of different anthropogenic interventions in different parts of the world. The rising level of GHG in the atmosphere is attributed to the rising affluence, increasing consumption, technology and rapid population growth. The relationship among these factors is complex and multifarious which has been addressed by the author. There is a need to incorporate methodology of green accounting in our national accounts statistics to change the calculation of national income from Gross National Product to Gross Nature's Product. Finally, he also recapitulates some of the policy issues regulating environment and development at national and international levels.

The multiple dimensional facets of population and development is influenced by several factors like geographical location, level of development and cultural background and needs fresh orientation towards the explanation of nature in development studies. Most of the formulations on the question of population get derivation through colonial mindset. Malthusian view on population is still dominating in most of our policies and academic discourses. However, the ground reality always contradicts and reminds us our tradition to consider people as assets for resource enhancement process and future growth in the form of a dictum '*doodho nahao, pooto fallo*'. Such innovative and path breaking viewpoints have been put forward by Anand Prasad Mishra in his paper entitled *Contextualizing Population in Developing Societies: Third World Perspective*.

Sustainable Development a buzz word in the development discourse includes in itself economic and social development and environmental protection by rational use of natural resources without compromising the ability of future generations to meet their own needs. But this concept is not acceptable to many developing countries since it seems to disregard their aspirations for growth and development. The paper contributed by Ajay Kumar, Bijay Kumar Das, Rajneesh Kumar and Fulena Rajak entitled *Climate Change and Sustainable Development: Issues and Challenges in India* states that challenges associated with sustainable development facing India and other developing countries seem to be gigantic and multidimensional and humans will have to decide the issues like whether the development means material comfort or peace, prosperity and happiness.

Part II: Migration and Urbanization

This section includes the papers related to the two most important issues of contemporary world, i.e. urbanization and migration. This part contains five papers. Unlike West, the migration in India mainly occurs from poor rural "peasant" sector to poor urban "informal" sectors. This is a typical spatial manifestation of poverty, underdevelopment and spatial disorganization of the economy of India, which occurred due to past colonization and its adverse consequences on its space economy. Shekhar Mukherji in his paper *Migration in India: Poverty, Underlying Phenomena, Spatial-Economic Processes and Development Policies* blames the government and academicians for not playing instrumental role in settling the poverty issues at priority basis. He suggests unless we soulfully do so, neither can we develop the correct theoretical perspective nor can we develop right planning prescriptions. His paper deals with the scenario of ongoing trends of migration, unfold the underlying socio-economic-spatial processes of such poverty-induced migration and recommends development policies and strategies for relieving such problems of distress migration and lopsided urbanization in India.

India is a country of paradoxes usually manifested in various forms like regional imbalances, mass illiteracy and limited skill base.

Agriculture the mainstay of Indian economy is contributing to less than 20 per cent of the Gross Domestic Product whereas more than two-third population is engaged in it in various forms. Shrikrishna Singh and Shiv Shankar Mishra in their paper *Globalisation Induced Migration of Labour and their Exploitation* have highlighted that in India migration primarily occurs due to regional disparity and has been accelerated even after adopting New Economic Policy in 1991. The old economy is shrinking and new investment is coming in selected pockets. The widespread mechanization in agriculture and destruction of cottage industries have rendered thousands of people in rural areas jobless and forcing them to work in even at the lower wages and in distressed condition.

In situ rapid urban population growth fuelled by rural influx has been the trademark of leading mega cities of India, creating a "top-heavy" syndrome of Indian urbanisation. This has resulted into physical expansion of the cities, congestion in the core areas and shrinking urban amenities and infrastructure leading to deterioration of the environment quality in most of these cities. These issues have been raised by Tara Devi Singh in her paper entitled *The Mega Cities of India: A Question of their Sustainability* taking example of Mumbai, Delhi and Kolkata. The author has discussed the pattern of urbanisation, urban growth and consequent urban environmental problems. However, because of their contribution in national economy these cities are considered as engine of growth and bound to decide the direction of development. These mega economic engines cannot sustain for longer time unless exogenous model of urban management is adopted, like investment in infrastructure, developing public transport system, improving the quality of life etc.

World is being urbanized day by day accommodating around half of the population in just over three per cent of land surface. Because of unchecked rural urban migration urban resettlement issue has emerged as a major concern as it dislocates the lives and livelihood of thousands of people, especially the poor. This issue has been addressed by Mrutuyanjaya Sahu in his paper *Urban Development and Involuntary Resettlement: An Indian Experience*. He attempts to portray the rehabilitation and resettlement programme for people affected by urbanization in the country and suggests that such programmes should focus on enhancing human capabilities and expanding social opportunities by addressing the social and personal constraints that restrict people's choice. The author advocates for an appropriate planning framework for urban involuntary resettlement.

The paper contributed by Sibabrata Das entitled *Spatial Patterns of Urbanisation and Development in Odisha, India* highlights that the southern and western part of the state is least urbanized and also least developed. Against the assumption that the process of economic liberalization would accelerate rural to urban migration and boost urban growth in post liberalized period in Odisha, the pace of urbanization has slowed down in 1990s compared to 1980s. The state has registered negative net migration, which reflects a weak pull of urban centres within the state. Author pleads for the development of small and medium towns which would act as focal points and accentuate the process of development in these areas.

Part III: Socio-economic Development

This section entitled socio-economic developmental issues covers the regional studies of various parts of India, which try to investigate varying nature of micro realities and ultimately incorporate these elements for suitable strategies to development planning. In this context, Ram Bali Singh has made an attempt in his paper entitled *Development of Eastern Uttar Pradesh: Challenges and Opportunities* to bring out the level of development in eastern U.P. in relation to the state and the country in terms of selected variables and the cumulative development index and inter-district variation in performance level with respect to 11 development indicators as well as the composite index of development worked out by using 36 indicators. Suggestions have been put forth to ameliorate the existing situation and to accelerate the process of development and transformation in the region. This paper could be used as a pointer by the planners and policy makers in determining and deciding the road map for eastern Uttar Pradesh.

Further Rajesh Kumar Chauhan in his paper *Progress of Poverty and Health Related Goals of MDG: Special Reference to Eastern Uttar Pradesh* using DLHS-3 data which measures relative position of eastern UP in the state in achieving the Millennium Development Goals related to poverty and hunger (goal 1) and health (goal 4 and

5), which are set to be achieved by the year 2015. He observed, that as far as health and poverty related goals of MDG are concerned state has tough task ahead. However, the eastern region has displayed a satisfactory performance on rural MMR decline and this experience may need to be replicated in other regions to arrest the high MMR in the state.

In this continuation, Archana K. Roy, Neelam Tripathi, Upendra Nath Roy and Rashmi Kant in their paper *Issues of Population and Development in Bhojpuri Speaking Region of India with Special Reference to Millennium Development Goals (MDGs)* find that Bhojpuri speaking region which spreads across the border of UP and Bihar is characterized by underdevelopment, primitive and subsistence agriculture, lack of infrastructure, deindustrialization and high population concentration. Due to its backwardness, migration of youth has become a way of life in this region. Authors raise the question that despite its prosperous history and high physical and human resource potential why this region is backward. The contributors have tried to extensively elaborate the cultural and political elements which create obstacle in the process of modernization and industrial growth of Bhojpuri speaking region.

Taking the case of one of the most developed states, Gujarat, Ratnawali in her paper *Industrial Growth and Deprivation: A Paradox of Developmental Strategy* talks about paradoxical issue of land acquisition and displacement. She highlights that the poor and marginalized people who have limited voice to advocate their problem are being displaced. She put her argument that acquisition of land and rehabilitation should go side by side. A humane approach is required towards providing necessary living condition and livelihood options to the displaced.

Addressing this issue of gender and development, Neelam Saraswat and Mithlesh Gupta in their paper *Relationship between Women Work Participation and Education in Rajasthan* finds that the work participation and education are negatively related in rural areas whereas it has a positive association in urban setups. There is a need to mainstream women by motivating them not only for higher qualifications, but also for workforce participation.

B.L. Nagda's paper *Demographic and Health Composition of Tribal Population in Rajasthan* deals with the issues of tribal health and its demographic composition in Rajasthan. The author analyses various indigenous medicinal practices in tribal population of Rajasthan, especially the herbal use for birth control. The paper also highlights about the various prevailing knowledge on herbal medicine in public domain which always maintained a trade secret by local tribes. Lastly, the paper also presented significant information about cumulative knowledge that prevailed among the tribes of Rajasthan.

On the contrast, the mineral rich but otherwise poor state of Odisha is marching towards industrialization in an ecologically sensitive area of the state leading to turmoil over land acquisition and revolt by the people in the tribal dominated belt. Large tract of agricultural land and reserved forests are being destroyed in the name of development. Mining and industrial sludge are polluting the water and air and the profits are going to the private parties. This issue has been pertinently raised by Ranjana Bajpai in her paper entitled *Development Induced Displacement: A Case Study of Odisha*. The author tries to enquire into the problems of land acquisition and displacement of people in the context of current industrial development scenario citing cases from different parts and ultimately asking basic question to the state as *Development for Whom*? Even in the economically developed state it is observed that the poor are the most sufferer and bears the heaviest cost of development.

Part IV: Environmental Issues

The impacts of growing population, development projects and exploitation of resources through rapid strides in science and technology have raised the issues of directionless development and environmental degradation leading to the deterioration of quality of human life. Hence, there is an urgent need to conserve the indigenous resources by using them wisely on the one hand and regenerating them to maintain the environmental quality. Though policies and plans are well-drafted, their implementation is poor because of poor coordination at various levels.

The environmentalist, Vandana Shiva in her paper *Earth Democracy, Living Democracy: Rebuilding Security in an Age of Insecurity* debates between the corporate globalization and earth democracy. The laws for maximizing corporate profits are based on privatising the earth, enclosing the commons and externalising the costs of ecological destruction of hazards. While the laws for protecting the rights of Mother Earth are based on respecting the integrity of the earth's ecosystems and ecological process and recovery of the commons and internalising ecological costs. This shift is also an ecological imperative. She adds that the impact of green economy and institutional framework for sustainable development depend in the context they are explained that is the old paradigm of market driven solutions or in the context of the emerging paradigm of the rights of mother earth when the green economy is Gaia's economy, and the institutional framework is the earth democracy, *i.e.* democracy rooted in the Earth. People's needs for food and water can only be met if nature's capacity to provide food and water are protected. Dead soils and dead rivers cannot give food and water. Hence defending the rights of mother earth is, therefore, the most important human rights and struggle for social justice of our times.

The impacts of a growing population, development projects and exploitation of resources through rapid strides in science and technology have given rise to profound changes and innumerable problems linked to the environment. There is an urgent need to conserve the indigenous forests and to increase the area under forests to mitigate the impacts of climate change. Though policies and plans are well-drafted, their implementation leaves a lot to be desired. The paper *Declining Biotic Cover and Strategies for Conservation* written by Sudha Srivastava brings out issues and problems of deforestation and loss of bio-diversity in Mumbai Metropolitan Region. The author suggests for the conservation of the fragile and sensitive ecosystems that have extremely important roles to play in supporting and sustaining development activities of a region. The ruthless use of resources is imprinted in the land use change owing to anthropogenic activities.

The paper *Climate Change, Glacial Recession in the Himalayas and Its Impact on Indian Economy* by Narender Verma deals with impact of global warming on the Himalayan Glaciers and its likely impact on the Indian economy. Author believes that global warming is resulting in the rapid depletion of glaciers all-round the world including the Himalayan glaciers. This recession of the glaciers would have a long term deleterious impact on the ecology and environment of the Himalayas and the socio-economic conditions of the people. The hardest hit would be the rural locals who thrive on the ecology of the Himalayas for their sustenance.

Vishal Gupta and A.A. Ansari in their article *Pattern of Land-Use Changes and Associated Environmental Issues: A Case of Coastal Districts of Kutch, Gujarat* raise that how vociferous growth in population and expansion of human activities has witnessed a sporadic change in the pattern of land use and ruthless exploitation of natural resources which has caused an adverse impact on the surrounding natural environment. They have taken the coastal districts of Kutch region in Gujarat as example to highlight this issue. Further, though there has been an all-round development in Kutch district as a whole but it is highly skewed with development concentrated in a few pockets of the region. In order to prevent further damage to the environment the authors have stressed the need for pursuing environmental friendly policies aimed at sustainable development, environment impact assessment of all projects and restoration of degraded mangrove forests in the study area.

Prabhat Kumar Rai in his paper *Arsenic Pollution and Its Sustainable Eco-Remediation from Aquatic Ecosystems: An Overview* deals with the problems caused by arsenic contamination of ground water and its consequence on health and ecology and its management in aquatic ecosystem. He has then discussed the relative advantages of eco-remediation of Arsenic management, i.e. Phyto-remediation technology over conventional chemical methods of treating arsenic which according to him are more cost effective as compared to the conventional methods. There is a debate on the cause of global warming. One group of scholars say it is caused by excessive emission oft CO_2 and CFC gases while other says earth climate is cyclic in nature and hence the global warming is the result of natural process of heating of the earth, however, the human factor has accelerated this process. Whether climate change is anthropogenic or not will remain an issue of debate but the fact is that the temperature is rising and resulting into melting of the glaciers.

Rajani Srivastava, in her paper *Diversity of Weed Species and Soil Seed Bank in Indian Dryland and Irrigated Agro-ecosystems* studies the case of two rice based (one dryland and other irrigated) ecosystem in India and found that intensive agriculture based on crop monoculture and high inputs of mineral fertilizers and pesticides have had negative impact on species richness and agro ecosystem diversity. The impacts of growing population in an effort of attaining the higher standard of living has resulted in excessive use (or abuse) of the chemicals has also been reflected in terms of pollution and degradation of natural resources.

Part V: Policies and Planning

UNFPA in the background paper 14 on Population, Environment and Poverty Eradication for Sustainable Development—Action Towards Johannesburg 2002 stresses upon that the greatest challenge today is to ensure better environment and sustainable future by dealing with the issues of population and poverty eradication. Thus, it brings a great sense of responsibility to opt for right path of development or choices to secure sustainable future. There is a real challenge to build a new world that rejects the current condition of poverty for many; controls unsustainable consumption and production behaviour; and keeps balance on the socio-cultural, ethical, human, and economic dimensions of lives for the present and future generations. This requires concerted efforts and a long term planning with multiple strategies to build and strengthen institutional capacity in developing countries to mainstream population issues into environmental and sustainable development policies, plans and programmes. Planners and decision makers need to be educated with better research studies clarifying the causes and cumulative effects of population-environment-poverty linkages. In order to explain the inter-webbed relationship between demographic factors (population size, distribution and composition) and environmental change, the mediating factors that influence this relationship (technological, institutional, policy and cultural forces) and climate change and land use change by population dynamics this section deals with two important issues. The first thing is to access and integrate different socio-economic, demographic and environmental data sources by sensitizing future generation through improvement in course curriculum and the second issue is to influence programmes and policies related to population, environment and development.

The various data sources on population, development and

environment are vital tool to achieve the target of sustainable development. The data can help in estimation of stock and flow population, size and structure of population, people's quality of living, population and resource relationship, bearing of human activities on the environment which are critical for the success of development policies and programmes. But access to quality data sources is the real challenge to go for evidence-based planning. Aalok Ranjan Chaurasia in his paper *Data for Development Policy, Planning and Evaluation Sources, Nature and Limitations* discusses about the nature and limitations of data required for planning, monitoring, evaluation and impact assessment of development programmes and policies for improvement in quality of life and impact on environment.

The growth of world population combined with unsustainable consumption patterns with respect to present production puts increasingly severe stress on the life-supporting capacities of our planet. These interactive processes affect the landuse, water, air, energy and other resources. Rapidly growing cities, unless well-managed poses a major environmental threat. The increase in both the number and size of cities calls for greater attention to issues of local government and municipalities. The human dimensions are key elements to consider in this intricate set of relationships and they should be adequately taken into consideration in comprehensive policies for sustainable development. Such policies should address the linkages of demographic trends and factors, resource use, appropriate technology dissemination and development. Population policy should also recognize the role played by human beings in environmental and development concerns. Arun K. Singh in his paper *India's Population Policy; Operational Framework for Implementation* has traced the India's population policy right from the pre-independence age and focused on shifting emphasis from top-down to bottom up culminating into highlighting the role of different stakeholders in ensuring implementation at grassroots level under National Population Policy, 2000. He presented an operational framework which recognizes two sets of interventions; the first meets the specific family welfare needs and the second contributes towards creating enabling environment. In both the cases the key issues are accessibility, availability and acceptability of the services to the people.

Water is scarce but precious resource. Therefore, management of water particularly in agriculture is critical for livelihood of the rural people. However, in the present millennium when water has emerged as gendered resource, it is pertinent to look critically about gender and inclusive growth in the conventional water irrigation management focusing on the participatory discourse. Paper by Jyotirmayee Acharya entitled "*Irrigation Management and Inclusive Growth: AKRSP (India)'s Intervention in Tribal Gujarat*" addresses issue like the contribution of small water harvesting irrigation schemes of Aga Khan Rural Support Programme (AKRSP) intervention in bringing about change in the trend of women's participation in agricultural practices and reduction in poverty and in the life of women. She puts strong advocacy for gender concern in policy on water and land resource management and governance to bring significant difference in life of tribal women of Gujarat.

In conclusion, it can be stressed that if India aims to become a developed nation by 2030, then we need to focus on the problems associated with population growth, pros and cons of development and suitable strategies for environmental conservation and management by involving different sections of society which include state, CBOs, NGOs and other stakeholders with great vigour and multi-prompt and result-oriented strategies. It is hoped that the research paper of this book will strengthen further research initiatives in the field of population, development and environment.

Part I
Population Development and Environment: A Conceptual Framework

1

Population, Environment, Economy and Sustainable Development: Some Interlocking Patterns

Kashi Nath Singh

Introduction

As the topic suggests main hypothesis of the paper is woven around the issue that population, environment and economy, and sustainable development are all intertwined into a multiplex, dynamic and tightening interlocking pattern. While population factor holds the pivotal locus in this syndrome, other three dimensions too are inextricably interwoven within its fold. It is the earth's environment on which all bio-life, including humankind live and on whose resources all depend for subsistence and survival and further development. While all non-human animal or plant world are tightly directly interlocked for the 'given' kind of subsistence on the natural environment and space for movement, in their particular specieted ways and feed systems, humankind carved out its different version of life-system. The humans developed two very significant techniques of life, during their evolutionary journey, e.g. particularly in their cultural upward and civilizational period. Of these two techniques, one was that they never specieted in nature under any lock-up. They unlocked from the nature's direct controlling stronghold on many resources and space and mobility. They became generalists, which would fit them, as it well visualized today in their global ecumenal space, in most of the given environments on the earth. At the same time, they become omnivorous, e.g. subsisting on all kinds of eatables, as vegetarians and meat-eaters, and hence multiplied their food sources several fold. Foods of different kinds are available in all the earth's environments wherever people are found. Second technique was adopting free breeding (reproducing) strategy, and human male or female could reproduce offsprings from any counter-partner within the entire spectrum of the *Homo sapiens*. This strategy led to the increase of human population throughout their *ecumene* and reproduced a whole of different ethnic groups and cultural systems of different colours, and culinaries and belief systems.

Changing Relations with Ecology

In the civilizational period, humans enhanced their level of interaction with Nature and environment in a rather aggressive self-propelled anthropocentric mindset to extract the environmental resources, based on their own perception with enhanced zeal. Ancient Greek scholar, Cicero, put in the bug of anthropocentrism into the human mind, exhorting man to modify, make possible changes, discover and learn the mysteries and musings, and laws and working mechanisms of Nature and their rhythmic patterns and trends in space and time cycles, and possibly even transform the edifice of primal or Original First Nature into the human-desired Second Nature. Exactly that the human systems have been all through doing. To a large extent, except the larger systems of biosphere or other spheres, forms of original Nature have been drastically changed, modified, moulded or transformed. It was assumed that natural systems and living within and depending directly on Nature could not allow procurement of the human-desired resource quantities or types to fufil even their basic needs for food, dwellings or clothing for their increasing size of population. This is what was conceptually made explicit by the nineteenth century economists.

Meanwhile during the long-continuing enhancing evolutionary process, mother nature bequeathed a unique gift in the form of a rare equipment to her prodigal human baby, e.g. the living brain of a size no other primate or animal species ever had. This unique equipment of brain could enable him like a Promethian god to discover fire, which led to multiple developmental dividends to mankind. The human brain developed into a very highly functional size of about 1450 cc by 50,000 years BCE (Leakly & Lewin, 1982). The brain led to the evolution of the *Homo sapiens*, the *genus* of the thinking or 'Wise' kind. The human mind pocketed within the brain, gave him multiple faculties to think, imagine, and calculate or observe and perceive the changes occurring in the environmental setting around to which mankind could systematically adapt in order to survive. The long extinct primates or even series of human ancestral species perished under the crushing onslaught of the environmental changes which they could not ever perceive to adjust to these systems. Large big belly lizard or reptiles, e.g. the Dinosaurs became extinct some 65 million years BCE due to climatic changes. Modern *Homo sapiens* has learnt to remain one point up above the environmental changes. However, the currently going-on anthropogenic pressures on the climate are already known in the form of depletion of ozone layer, glacial melts, and zonal shifts in climatic zonal belts or rising sea-level. Our change-monitoring systems are far from perfect, and the IGBP (Global Change Science) loudly argues for mankind to induce its anthropocentric behaviour into focusing together on geocentric adaptations and retracing its profligate disrupting steps *vis-à-vis* ecological processes and frameworks. Howsoever our natural sciences might boast of their knowledge of Nature's First Principles, they all pertain at best second class proxies; Nature's First Principles are far beyond these sciences (Radhakrishnan, 1966). However, our existing knowledge about the various anthropogenic agencies released into the atmosphere like the greenhouse gases due to our unbridled overuse of fossil fuels in the form of CO_2 or methane gas or others, forest fires, over-devegetation, disturbance of coral reefs, mangrove forests, overuse and increasing pollution-contamination of fresh water resources, wild life endangerment, biodiversity losses, and our dumping of wastes into the lands, water channels, seas and bays, can serve us well to take effective, corrective and restorative measures to stem this worsening rot on ecosystem. Our economy and sociology must fall into this adjustive web with the ecology. In fact, our living and reckless behaviour patterns need our full cooperation with bio-systems. As Oldak (1977) stated, Natural systems alone cannot carry on alone or increasing load any longer on health fully.

There has been an intimate linkage between the low resource expropriation from Nature by mankind and its early natural living trend over the historical period. Hunter-gather-forage societies lived in isolated small bands of about 25 people in a group of being part of the natural order. In general, the people must have maintained a secure and efficient balance with the environment, finely turning their survival needs with increasing experience, learning of the different habits of the prey animals, as well as the growing, fruiting, flowering and ripening cycles of eatable plants. Certainly, some hunting groups hunted out certain large animals like Bisons from North American plains, but in general the balance was maintained in most cases. It was in the Agricultural Revolution, since last 10,000 years, that farming potentials were steadily recognized by mankind to produce their own food, e.g. growing crops for food and raising livestock for dairy and meat. Certainly, farming disrupted the natural order more vehemently. It was followed by the more powerful Industrial Revolution which led to a technological revolution followed by the IT revolution. Exploitation of environmental and natural resources intensified to a very high scale and depth. These anthropocentric self-interests led to an exponential increase in human population, standing by the century to about 7 billion. These developmental Eltonian revolutions and increasing populations have led to multifold increases in human needs and wants and deeply intensively ingrained material accumulative tendencies. For instance, one single factor of enormous and seemingly unsustainable high energy use in the rich countries, and now in the developing world also, particularly in the emerging economies, has become highly disruptive to the ecobalance. All in all, '...humanity's high-energy activities amount to a large scale direct attack on the earth's ecosystems and the critical services they provide' (Ehrlich, 1994) (Box 1.1).

Population Factor

Large size and rather rapid growth rate of human population in the really economically and socially underdeveloped countries is globally aberrant as it retards their economic and social advancement efforts as is so far observed. This leads to worsening poverty, resourcelessness, food insecurity, starvation, disease, lack of social

Box 1.1: Energy factor: increasing use with development

Expansion and expropriation of energy sources of the environment by modern humanity is a great instance of human profligacy against nature's eco-balance and endurance. Whereas, early hominids lived on a bare energy use of 2000 kcal per capita/per day, hunting-gathering society with fire use lifted it to 5000 kcal. However, the amount rose to 12 kcal in elementary cultivation stage, but went up to 26 kcal in advanced cultivation stage of human society. Industrialization led to much higher level of energy consumption, and in early phase of industrial development, England ascended to 77 kcal (Ca 1850). But it rose up very high with intensive industrial times in the next over hundred years. The merging post-industrial society began to use as much as 230 kcal (USA, 1970). A break up of this use in early phase of industrial development, England (Ca 1850) and USA (1970) shows the increasing role of some crucial sub-sectors of industries and food-production, and of transportation. Transportation indeed took up only 14 kcal of total of 77 kcal in England (about 1850), but as high as 63 of the total 230 kcal in the post-industrial society of USA about 1970. Obviously, "it is not only the number of people on Earth that has led to the great increases in energy use, but also the expansion of transportation (i.e. of trade)", particularly in industrial-colonial times and now in post-industrial globalizing economy (O'Sullivan, 2004).

provisions like people's education, healthcare, unemployment, shelterlessness, all leading to a level of sub-human living conditions for the bottom messes. Environmental degradation is the natural consequence as the poor use wasteful, crude technologies of resource take-off direct from the environment. Deforestation, soil erosion, biodiversity loss, and low production naturally follow. This straight-line cause-and-effect relationship is jacketed in the Malthusian and Neo-Malthusian Syndrome, as ending in a series of 'vicious circles' of maladies. The developed world feels that the Aid, FIDs, other kinds of financial assistance, debts, and other efforts hardly make any dent on the South's development, largely because of its fast growing large population size and related issues, project execution policy failure, and also failure on genuine social developmental fronts to tackle poverty issues.

However, the population issue is not only a numbers game (Harrison, 1994a). It is only a part of the problem of resource and environmental degradation issue; while larger load and pressure on the environment emanates from high Northern resource consumption and consumerist living style even though it carries only over 18 per cent of the world population (2008), against about 82 per cent of the population load in the South, giving birth to about 95 per cent of the coming-in world population.

Harrison (1994b) put that (even though very small) population growth in the United States (with high energy use, CO_2 emission, and high meat consuming culture) exerts at present more impact on the global climate than population growth of India. High proteinous food chain with beef and other meat intake is greatly dangerous. US feeds over 70 per cent of its grains to the livestock (primarily cattle to produce meat) which factor contributes to world hunger. Frances Moore Lappe (1979), a food economist, found that 145 million tonnes of grain fed to the livestock (cattle, poultry and pigs) provided only 21 million tonnes of meat available to humans. He notes that an acre of cereals can produce five times more protein than an acre devoted to meat production. In fact, 'Legumes can produce ten times more, and leafy vegetables fifteen times more' (c.f. Rifkin, 1992).

Forcing of large global population, and fulfilment of their basic needs and wants, increasing consumerism and opulent, even hedonistic life style of the rich and their greedy psychomaterialism have led to the profligate behaviour of the humankind to over-exploit and over-tax "the environment beyond the limits of our own (time-tested) adaptability if not that of the earth itself (Leaky & Lewin, 1982). This gross imbalance and maladjustment with the ecology besides, "long-term stability for a single highly complex species (e.g. llama sapiens sapiens), no matter how adaptable, is biologically impossible" (Leaky & Lewin, 1982). Long human tenure as such is highly questionable. The enormity of these forcings has led to highly contentious global debate in regard to the principal role of Northern Over-consumption or Southern large population size and rapid growth, (which is overturned analytically into the phrase), e.g. Northern over-population *versus* Southern poor consumption in the enormous and even irreversible damages to the environment. It was termed as the 80 : 20 ratio, which means 80 per cent global population and a bare

20 per cent of wealth in the South, in contrast to that of 20 per cent population of the North owning 80 per cent of global wealth. The issue was raised forcefully at the UN Conference on Environment and Development (UNCED) in 1972, and led to a fierce debate between the rich and developing countries. UN Population Conference in Bucharest in 1974, intensified the debate, all the more.

Paul Ehrlich remarked that at this level of global disparity:

> 'The main population problem is in wealthy countries. There are in fact too many people. The amount of resources each person consumes and the damage done by the technologies to supply them need to be taken into account as much as the size of the population. In theory, the three factors (e.g. population size, technology factor and resource consumption) should be multiplied together to obtain an accurate measurement of the impact on the planet.'
>
> (Paul Ehrlich, 1994)

The relationship is expressed through I = PAT, in which (I) indicates Impact on the earth environment, equal to population size in numbers (P), multiplied by Affluence (A) or per capita resource consumption, multiplied again by Technologies (T) used to supply each unit of consumption. It means, higher the per capita resource consumption, higher and deeper the impact. Moreover, industrial rich economies thrive on internalizing high global environmental costs into their life-style.

As such, Kyte (1994) noted:

> 'There is no directly proportional link between numbers of people and environmental degradation. There is a clear link, however, to the threats that their (rich peoples' of the industrial countries) life-styles pose.'

Economics rises to be the Dominating Sector in Social Life (Karl Polanyi's Critic: 1944)

Karl Polanyi made a scathing criticism of both Liberalism and

Marxism for their carrying aloft their theory of 'economism' and positioning it into a transhistorical context. He elaborated that in earlier societies, the economy—the arrangement of enduring human kinds livelihood—was embedded in the overall social relations, in fact, subordinate to religion, politics, and other social arrangements. "Economic aspect played a miner role in individual's life. Only in the Nineteenth century economic theory, economic self-interest was lifted to the dominant principle in social life. Market-based societies raised the economic imperative level in shaping social (and in reality, the dominant ruling) in human life. This 'economism' is thus historically contingent, a recent phenomenon of social history articulated by the nineteenth century economic theory" (Polanyi, 1944).

The nineteenth century economists put that economic activity is the process of providing for the material needs of society against the background of scarcity. This notion of scarcity of resources was altogether a novel idea brought into the new economic theory. It is a creation of modern economism against the background of assumed 'parsimony of Nature which sets the stage for the economic problems' (Heilbroner, 1972). Obviously, the concept of parsimoniousness of Nature or the Natural economy as well as that of scarcity are both very recent, historically contingent economic constructs.

Factually speaking, Nature or resources hidden into its enormous system and processes and sectors cannot provide for the needs and wants of modern man, without being activated by human efforts, technologies, or resourceability working on the natural settings. The man's wants, prestige and status in society are all human made aspects. Nevertheless, the humans need not take their problem as a licence to be profligate to disrupt the ecobalance abusively.

Ecology *versus* Economics: Debate on each other's Imperative Significance

Environmental Crisis

Ecology is defined as the scientific study of the relation of plants and living creatures to each other species and to their abiotic surroundings. It is a natural biological system which together with abiotic surroundings does affect and is affected by the humankind and their economy or economic activities. Economics is a social science or principle of production, distribution and consumption of goods and services, with reference to cost and efficiency.

Economy is defined as the entire web of human economic activities and arrangement. Doubtlessly, it is clear that 'the economy exists within a larger setting, e.g. set within the ecology, is surrounded as it were by the ecology, influences the ecology, and is limited by the ecology' (Keyfitz, 1991). As such, as will be clear in the following discussion, that economy is set within ecology which is the key foundation of the life and sustainable development thesis. These two systems, however, put their own imperatives are the dominant set in their role in the environmental crisis. While we discuss the point, we have to evade certain 'fixed' ideas and fallacious natural or social science-led dictums, which were put without paying much attention to the enormously powerful role played by human cultural, edifice and devices in this situation as a powerful intermediary force. Environmentalist O'Sullivan (2004) has been critically successful in explaining such concepts and remarks, which otherwise lead to misunderstanding of the truth. Some scientists stressed 'the similarities between human beings and non-human species', some of them directly put that 'human societies are ecosystems' (Odum, 1971); that 'evolve solely via Darwinian mechanisms', that human behaviour is largely genetically determined (Wilson, 1978; Hardin, 1968); and that 'the numbers of people on earth is the most important human factor driving environmental change' (Messerli *et al.*, 2000). On analytical screening based on hard evidences, O'Sullivan found that human cultural interventions are important role players in human dealing with all these statements. For instance, other "human systems are certainly part of the ecosystem but they are also more than ecosystems". Darwinian mechanisms do also apply to humans as equally to other species, but change in human societies is not "evolution *sensu stricto*" (O'Sullivan, 2004). It does not occur all 'through natural selection, but depends on human choices'. So also, human behaviour is not as strongly conditioned by genetic influence, as the other species: cultural influences and other such factors are more important (Midgley, 1995). It is quite wrong to state that environmental change is driven by human numbers as strongly as by the per capita high levels of resource consumption supplied by high technologies. Human cultural devices are largely absent in other species, such as language, education and health and utility systems, sciences, knowledge creation, skills, technologies to solve problems, social organization, experiences, systematically organized governance and governments, ideologies, morals, ethics, religions, customs and rituals, value systems, marriage systems, traditions, and law and order. All together with others play dominant roles between human societies and economies, on one hand and ecology on the other. Also other fallacies abound. For example, economics, including environmental economics, and even ecological economics, assess nature and environmental in terms of market prices, than in values (O'Sullivan, 2004). All said, however, environmental edifice is very important to influence human systems.

However, in the intensive anthropocentric fondness, man elevates himself to a very great role player in his destiny, even though within the limits of Nature. Europe's great psychoanalyst, Victor E. Frankl exhorts that:

'A human being is not one thing among others, things determine each other, but man is ultimately self-determining. What he becomes—within the limits of endowment and environment—he has made out for himself' (1984). Frankl rejects any kind of "pan-determinism". 'One of the main features of human existence is the capacity to rise above (any biological, psychological or sociological) conditions to grow beyond them. Man is capable of changing the world for the better.' Eurocentric values and anthropocentric central isms put by many European scholars have led to many concepts of overwhelming arrogance of man against nature and ecology. Spinoza's theory of happiness, Kant's doctrine of 'autonomous' man, Friedrich Nietzsche's high moral tone of "man is his own creator and thus has the right to free himself from his pasts, from his beliefs, from all obligations in order to reach out for authentic creativity". Colonial mindsets harped on social Darwinism, and theory of race or lording over Nature or environment, or weaker mankind (Schumeli, 1980). The renaissance humanists exhorted, 'it is man's dignity to stand at the centre of the universe, capable of realizing his possibilities in whatever way he chooses. No stars, no fate, not even gods will of themselves dominate man's capability to self-determination' (Ernst Cassier's assertion worded by Schumeli, 1980). Hegel, Marx and, above all, John Stuart Mill's formulation of human freedom was that of man pursuing "over own good in our own way". This super-dominative or all-dominative mindset of modern industrial men led them to believe that economic setting was to serve man according to the dominant technological human capacity, with unfettered free market mechanism, and removal of natural or human barriers, as Hobbs declared. However, these claims are on the extremistic frame.

Although man is the greatest resource, as Julian Simon (1981) and Nandan Nilkeni (2008) have emphasized, yet the unbridled large and rapidly growing population size in an environment of mass poverty and deteriorating ecological system, without adequate food, health and education, skill or utilities for large masses is a great problem. A mass of uneducated, unhealthy people cannot provide beneficial 'demographic dividend' for higher economic development. Essential social and living provisions have to be well created for the poor masses by drastic poverty reduction and quality health, education and skill formation to bring about a motivated human capital.

The Two Imperatives

An economist states that: "In starkest terms, the ecologist confronts us with an environmental imperative that requires an end to economic growth or at least a sharp curtailment of it as the price of biological survival, and health. In contrast, the economist counters with a socio-economic imperative that requires the combination of economic growth as the price of social survival. Some ecologists see the reduced (economic) growth, as a necessary though not sufficient condition for saving the ecosystem. The economist sees (economic) growth as a necessary, though not sufficient condition for social progress and stability" (Heller, 1973).

In fact, the differences between the two thought systems are deeper and sharper. The economist not only denies to pay the price for the ecological losses (no environmental accounting system in economy), and even asserts that economic growth is not at the root of any such evil, and that it is "one of the prerequisites to success in restoring the environment". Economic growth of the poor will not only enhance their well-being but also help them with resources to clean up their degraded environment and habitats. Thus, Henderson sees this imperative fix: economists see ecology as a subset of economics, while ecologists see economics as a subset of ecology (1973).

Ecologists correctly visualize the high-tech, energy-intensive polluting farming, industry, or other production or transportation systems as fundamentally environment-degrading systems even more than large, poor and fast-growing population (Commoner, 1972; Paul Ehrlich, 1994). The ecologist wants the economist to include in his ledger the full ecological and social cost of diseconomies created, internal or external, and also to internalize all these in his economic activity and costs and prices.

Happily, the agreement between the two is emerging around the point that maximizing production must not lead to maximize ecological depletion, pollution, or toxicity, and the economy must be increasingly linked to maximizing, and conserving, (and even preserving/protecting under the human stewardship) the ecological capital stock on the ledger. The economy must not regard the externalities caused as unaccountable. It is realized that striking balance with ecological wealth is all the more beneficial to the economy. However, depollution investments, price policy (in terms of full economic, social and ecological cost), 'polluter pay principle' or withdrawal of subsidy must not be pushed in haste beyond the economy's functioning (production/supply) tolerance level.

What the economist thinks as rational is to: "seek not total or maximum cleansing/restoration of the environment, but an optimum arising out of a careful matching of the "bads" that we overcome and the "goods" that we forego in the process" (Heller, 1973).

However, 'polluter pay principle' would take care of the effluent fees, or the now largely free use of clean air, water, or land areas for dumping mineral tailings, and industrial and municipal wastes. Yet, without very tight and effective material-to-cost punishment which genuinely hurts the polluters, environmentalists take these devices as a licence to pollute or to exploit the biosphere and public resources up to the level of paying capacity as wrong and removable if not effectively executed by laws or execution. There are several problems asking for the scientifically investigated uncertainty:

- Does the process of economic growth and the rising levels

of economic activity inexorably lead to the degradation of environment? What is the share of problems arising in the environment on its own natural process?

- Does the society in general have a genuinely realistic option, conceptually and realistically, socially and politically acceptable in at least substantially reducing its economic growth level, for the sake of ecological restoration of its health?
- Is there essentially a mismatch in the contention between economic growth (or development) and environmental quality in the sense that while economic growth is a means to social survival ends, ecology being of instrumental value and bestower of social benefits, the most in the developing countries, for the existential purpose of the society, 'the quality of the environment is an end in itself or at best an important component of the quality of existence' (sought after hotly by the rich and the developed countries). Indeed, the environmental quality has become a divided issue—one for the poor for their survival, and the other facet for the rich seeking ideal environment for ecotourism, pleasure spots, natural grandeur, and other positional goods and services.
- If the society is willing to substantially reduce economic growth, it would mean an absolute reduction in material living standards for all, but worst for the poor, whose life will be worsening, given the population growth and distribution share of national income, at the cost of a more livable natural environment.
- Since it is impossible to come to zero population growth or economic growth, there is wide open space for reconciliation between the technological processes and attendant economic growth, and its trade off with ecology and environment.

Technology, being more benign, and used in the widest perspectives, including social and politico-economic engineering, must keep innovating and adapting to these newer needs of the society and ecology with a 'human face'. Economic growth and development and ecology must go together in reciprocally balancing act.

The ecologist further desires economists to upgrade their models and methods to:

> "incorporate more hard data from the physical sciences as well as develop the philosophical overviews necessary to interpret itself (and rediscover) with anthropology, biology, psychology, and the major systems of ethical and moral values" (Henderson, 1973). So also, the economist needs to adjust to novel requirements. Even economist Maynard Keynes thought of spiritual links of economic goals and approaches towards more sublime disciplines. In this spaceship economy, as Jay Forrester stated, the narrow short-term goals of sub-systems of the economy are generally in conflict with long-term goals of optimizing the macro-system. Henderson wishes man to "evolve and adapt his value system to the requirements of the interdependent world society which must operate in equilibrium on a finite planet" (1973).

Development Issues

Development embodies several wings within its corpus: development, economic growth, social development, linked with poverty elimination, and sustainable growth and sustainable development. Some crucial elaboration on these wings seems to stand in order.

Development

Development connotes philosophical as well as popular wish-fulfilment. Philosophically, it connotes to achieve a goalless spiral of constant upward ascendance in a comprehensive conceptual frame (Pretes, 1997). For Marx, it is to realize a sort of *'human species potential'* (although he never defined it). Philosopher Radhakrishnan (1966) and biologist Waddington (1978), said that whatever the future will be, it will be a 'human willed future'. In popular sense, development is to acquire a personally happy, blissful materially fulfilling and spiritually subliming life. Although people widely differ in their wishes, most mankind seem to attain a 'good life'.

Economic Growth

Economic growth is conceptualized as to enhance and expand the social resources, create wealth, to be used to extend the system of physical and social infrastructures, institution-building, eliminate or drastically reduce poverty, inequalities and create other facilities for peoples' well-being. However, generally, economic growth does not in and by itself lead to economic and social development. 'Growth without development or jobless growth' is a great fault-line (Prakash). With lapse of time, it becomes economistic, investing and re-investing the created wealth and resources for its own sake. Research studies across nations find out that while national GNP or GDP and per capita gross national income (GNI) grow, about half to two-third of masses are in poverty or even further immiserized. As such, the concept of economic growth must be accompanied by a proportion of gains towards social development. This aspect is conceptually built-in 'economic growth' strengthened by strong policy measure on social dividend.

Nobel Laureate Arthur Lewis (1970) prompted a seminal paper 'Is economic growth desirable?' His work is in no way 'economistic' but is explicitly advocating for socially equitable distribution of the growth gains to enrich the general people's life system. He strongly advocated the case for growth as 'no growth' means 'no change' in *status quo* and no societal development. He wished for growth dividends, while extending its frontiers, also to be employed in elimination of poverty, underdevelopment, illiteracy, inclusion of women and focus on gender issues, development of physical and social infrastructures, institutions, sports grounds and stadia, health and education provisions, housing for all, and so on. Growth creates more wealth, and even if more wealth may not necessarily enhance happiness, it does create more income facilities for enjoying a good material life; moreover, it enhances the peoples' range of choices, and capability to exert greater and effective control of physical disasters, and other 'odds' in life, such as incidence of chronic illness, giving the best of education to children, and raising their own productivity. Anyway, the 'growth' gains are only instrumental to be socially allocated by committed policy drives of governments, which creates 'enabling environment' for good farming and industries, and other further growth and development systems. It may lead to political freedom, social democracy, social harmony, value choices, and human rights. It does away with the age-old archaic social taboos, superstitions, and loyalty to narrow systems of caste or tribes, and helps to build a national culture and society. It also means fighting against *status-quoism* and elitism, enhances habits of frugality, punctuality and saving, doing away with binding traditions, and builds modern cultures.

Economic Development

Economic development is always invariably associated with social development. Thus, its entire fabric is defined in terms of *'inclusive'* human development and welfare under a framework of social, economic, political, and human rights in civil and justifiable and lawful conditions. These environments allow humans individually and collectively to be able to realize their potentials, enjoying a life of freedom, dignity and fulfilment. In brief, the process of economic development is a concrete form of positive progressive social action proceeding further and by the coordination of 'certain principal forces that reveal the inner structure of "logic" of an economy's development' (Meier, 1995). 'Economic growth is an essential means for enabling development', yet in and by itself, 'it is highly imperfect proxy for progress' (WDR, 1992). Economic development is a development process, e.g. a people centred economic growth plus structural positive qualitative incremental change in peoples' living conditions.

Standards and Consumption Baskets

Such all-inclusive intrinsic development is envisaged as the kind of a national carrier on which every resident is aboard. It is through the expanding broader active participation of more and more people in the ongoing development process whose qualitative dimensions 'extend beyond the growth or expansion of an economy's simple widening process' (Meier, 1995). Gunnar Myrdal termed this inclusive process of economic development as an accelerating instrument of *"upward movement of the entire social system"* (1968).

Nobel Laureate Amartya Sen brilliantly brings into focus some key elements of development. Accordingly, development progression incorporates deeper qualitative human life improvements enshrined in peoples' 'entitlement', 'freedom' and 'capabilities' to utilize fruitfully and digestively whatever the 'entitlements' generate and enlarge their "functionings" into people's life. He elaborates: "The entitlement of a person stands for the set of different alternative bundles that the person can acquire through the use of various legal channels of acquirement open to someone in his position" through the use of a person's own trades, employment and production (1990). These elements together enhance the social and personal welfare of peoples together with economic or material welfare (generated by increasing per capita income). Thus, "the capability to function" distinguishes the non-poor from the poor. As such, human well-being, that is being well is genuinely enshrined in a person's empowerment of making choices among expanding opportunities, aspects of freedom and resources whereby he realizes and enjoys such fulfilments as 'what he is, or can be, and does, or can do in terms of functioning' (see Box 1.2).

Sen has thus highlighted the real nature of poverty as it engrosses the life situation of the poor in the real world. Naturally thus, national gross per capita income (GNI) or per capita income or availability of goods and services do not suffice in the absence of freedom, 'entitlements' and 'capabilities', and opportunity of 'functioning' which emanate from 'entitlements'. In this analysis, 'entitlements' mean the person's choice to command certain variety of commodity (goods and services) bundles within the freedom, and opportunities and rights he possesses (Sen, 1981, 1985 and 1999).

Underdevelopment of the South

South's 'underdevelopment' is essentially viewed as an 'external' creation by the 'dependency' theorists but the 'colonial apologetics' (as addressed by Gunnar Myrdal) coming from the West blame the underdevelopment due to the region's own 'internal basic obstacles'. The West-propelled colonial and post-colonial and neo-colonialist structures that were imposed by them on South's continued exploitation is entirely ignored by these Eurocentrics. It is this mindset which constructed theories of race, slavery, exploitation of others,

Box 1.2: Stark reality of poverty, entitlements and capabilities: Amartya Sen

Poverty is not just a matter of being relatively poorer than others, but of not having some basic opportunities of material well-being—failure to have minimum "capabilities". The criteria of minimum capabilities are "absolute" in the sense that they must not vary from society to society, or over time, but peoples' deprivations are judged absolutely, and not simply in comparison with the deprivations of others in that society. If a person is poor because he is unable to satisfy his hunger then that diagnosis of poverty cannot be altered merely by the fact that others too also may be hungry (so that this person may not be, relatively speaking, any worse off than most others). A person's advantage is judged in this approach by his capabilities, viz. what can or cannot do, can or cannot be. The relative capabilities are of many different kinds, (e.g. being free from starvation, from hunger, from undernourishment, participatinig in communal life, being adequately sheltered, being free to travel to see friends, and so on. The ranking of capability vectors can be used to rank people's advantage vis-à-vis *others. But in the context of poverty analysis, it is a question of setting certain absolute standards of minimal material capabilities relevant for that society. Anyone failing to reach absolute level would then be classified as poor, no matter what his relative position is* vis-à-vis *others. Poverty in this view, is not ultimately a matter of incomes at all; it is one of a failure to achieve certain minimum 'capabilities'. The distinction is important since the conversion of real incomes into actual capabilities varies with social circumstances and personal failures" (1985).*

social Darwinism, colonialism and neo-colonialism, and dependent trade and dependent capitalism.

Denis Goulet (1971) has aptly well discussed this perpetuating misery: "Underdevelopment is shocking. (It is not revealed in a) mere statistic reflecting low income, poor housing, premature mortality or underdevelopment (but the entire web of chronic life-long inner sufferings) which prevail in the 'culture of poverty'. Chronic poverty is a cruel hell that is merely by gazing upon as an object."

Globally about three billion people are underdeveloped on the basis of poverty index. It is more frustrating because power of wealth, capital, knowledge and technologies are all existing globally to eliminate this condition. Human life situation under this condition is not truly indexed by GNP, GDP or the term 'underdevelopment' itself. The lives of the underdeveloped lack what Amartya Sen has so capably analyzed in terms of basic human rights as capabilities like freedom, entitlement, functionings and well-being.

Theotonio Santos (1989) submits how underdevelopment was created:

"Underdevelopment far from constituting a state of backwardness prior to capitalism is rather a consequence and a particular form of capitalist development known as dependent capitalism. Dependence is a conditioning situation in which the economies of one group of countries are conditioned by the development and expansion of the other. Dependence then is based on international division of labour which allows industrial development to take place in some countries while restricting it in others."

New Zealand's most perceptive geographer scholar, Keith Ruchanan (1968), affirmed Andre Frank's views and stressed straightway that:

> "...development of one centre—White North—resulted in the stagnation of actual retrogression of dependent areas. Capitalist development simultaneously generated development and underdevelopment, not as separate processes but as related facets of one single process."

Samir (1976) noted how the continued exploitation and stagnation of the periphery sapped away 'any internal dynamism' of the South. Doubly enough, the created forces of "internal colonialism", which created their internal agents, some selected local comprador 'elitist' *status-quoists*, in every dependent region to help continuing colonial structures. 'Internal colonialism' still continues Desi style by local politicians, bureaucrats and businessmen to corner away huge developments fund and putting in foreign banks.

It goes to the credit of Albert Hirschman (1965) who argued that there is simply no barrier to development. There is no ground to harp on the almost dead concept of inherent "obstacles" to development in the South as it is difficult to find a finite number of such 'reliable' hindrances like lack of certain resource, institution or attitude needed to be created or eradicated for "development to be possible". He outrightly rejects such so called obstacles as South's tropical climate, overpopulation, socio-cultural bottlenecks, intense traditionalism and lack of West-like 'modernization', fatalism, resistance to change, other worldly religions' (Max Weber's thesis), lack of 'achieving motivation', and so on, as advanced by other Western scholars, e.g. 'colonial apologetics' (as Gunnar Myrdal addressed such scholars in 1970).

Gerald Meier has stressed: "That the major weakness with 'the obstacles' approach to development is that it limits policy making. It forecloses the ability to perceive alternatives. It reduces the vision of different approaches to development. The very essence of policy making is to practice the art of the possible" (1995). In fact, it is well observed through concrete examples of the development scenario of four Asian Tigers, NICs, and over two dozen emerging economies led by BRIC countries today. Development in any or most countries is possible through developmentalist policy packages and acting seriously on them. As such, the South needs to forget the past exploitations, and finish off their own *status-quoist* 'national agents' and to rise above "some missing points" or the "Circle of vicious poverty" through new social-economic policy and enabling 'development environment' in their countries. They need to develop their strengths, which every country possesses in some measure or the other, and relax the strangleholds of the socio-cultural taboos, develop their human resources, through multidimensional approach to tackle slowly and steadily the level of absolute poverty, gender inequality, mass equalization of opportunity, increasing basic needs provisions and fair income distribution. All said, most developing countries including India grossly still suffer from their own newly evolved (as well as the descendants of the older elitists) large army of corrupt politicians, bureaucrats, civilians, corporation.

Population, Ecology, Development, Sustainable Growth and Sustainable Development

All the above aspects are intertwined in and fed by 'multiple relational synergies'. These ties, however, may become more seriously disbalanced during the long term human tenure on the earth, particularly if the humankind continues to behave in the current profligate use of ecological resources and health. The humanity's continued welfare depends on very positive programme of 'inclusive' development and economical unwasteful use of resources in both the developed and developing countries, but more seriously in the latter where most of the people suffer on account of underdevelopment. Participatory development is not possible unless the masses feel concretely in their daily biography that their living conditions, food, nutrition, sanitation, provision of safe drinking water, fuel, education, housing, opportunities for skill formation and employment, have been improving. IUCN: Caring for the earth, 1999, put up that people are urgently required to be made aware of the stark reality that continued development is not possible without inculcating a habit of resource 'conserving' behaviour. This makes people to adopt the ethic for living sustainably, and re-examining age-old values and altering their behaviour. Further, people *en masse* have to forego for ever to maintain an instrumental view of nature, ecology and earth and their resource processes. This requires reorienting our entrenched anthropocentric view and balance it with some geocentric centralities in our economies, environmental accounting and costing is necessary.

World Commission on Environment and Development (WCED)

The WCED report, Our Common Future (1987), is a very powerful document on sustainability of environment and development. The WCED Chairman, Gro Harlem Brundtland, the then Norwegian Prime Minister, asserted emphatically that environment and development are not either/or things: they are part of the same system. There is no alternative to either. She wrote that it was the increasing population, slackening development and the deteriorating environment of the world which led the UN to set up the WCED. The chairman wrote:

> "...the environment is where we all live, and development is what we all do attempting to improve our lot within that abode. Further, she put in paras 3 and 4: 'Yet the population issue is not solely about human numbers. And poverty and resource degradation can exist on thinly populated lands... *People are*

the ultimate resource; (but with the caveat that) improvement in education, health and nutrition allow them to better use the resources they command, to stretch them better. In addition, threats to the sustainable use of resources come as much from inequalities in peoples' differential access to resources and from the ways in which they use them as from the sheer numbers of people. This concern over the "population problem" also calls forth concern to human progress and human quality.

Nor are the growth rates the challenge solely of those nations with high rates of increase. An additional person in an industrial country consumes far more and places far greater pressure on natural resources than an additional person in the Third World. Consumption patterns and preferences are as important as numbers in the conservation of resources."

Some Critical Problems of WCED's Suggested Sustainable Development

The recommendations of the sustainability concepts in WCED (1987) champion the thrust of maintenance of the ecological stocks and quality of services, while they contrarily propose to accelerate the global economic growth rate up to 5 per cent a year to double the per capita income in just 14 years. This means that the economy will rise at a secular rate of 3 per cent or more yearly, in addition to about 2 per cent a year to neutralize or compensate for the annual population growth rate of the world. This is necessary to create and maintain the appropriate conditions for establishing a firm base for sustainable development to eventually occur and sustain in near future. These recommendations are based on certain anticipatory possibilities, some of which may hardly obtain:

- Technologies would develop meanwhile to consume much less energy and resources per unit of final product;
- The rich world redistribute their wealth towards alleviating the global mass poverty and themselves inculcate the culture of consuming less resources and give up their conspicuous consumerism and hedonistic life style;

- Population growth would set an equilibrium with the doubling of their per capita Income;
- All above factors would be leading to reduced rate of ecological resources and decelerate their depletion rate;
- High rate of resource conservation, substitution and recycling, and material use efficiency in both renewable (in time with their recovery cycle) and non-renewable resources to ensure the sum of the economy and extend them further smoothly;
- High level of economic decentralization would ensue to reduce the diseconomies of large scale operations; so also political decentralization and devolution of power will lead to peoples' participatory stakes in the development process; and
- Narrowing disparities and inequalities among population groups at all spatial scales, internationally and intra-nationally.

These recommendations are highly laudable, but even if they do succeed the high economic growth rate of 5 per cent per annum may be highly detrimental to the ecological resources before sustainable development would occur. World Development Report in 1992 elaborated on the theme of such anticipations:

'Efficiency gains from economic policies will often reduce the demand for natural resource inputs. Environmental policies can reinforce efficiency in resource use and provide incentives for adopting less damaging technologies and practices. The investments that are induced by environmental policies will change the way in which goods and services are produced, (and distributed and consumed) and may result in lower output but will also generate benefits that can increase human welfare'.

This analysis also invites certain critical comments. Certainly, positive environmental policies would induce efficiency gains but whether the lower output with the given distributive system providing in the world (as clearly seen currently what a paltry per cent share of the international and national income and purchasing power is delivered to the lowest quintile of population) will emerge as acceptable proposition to lead to the poor people's welfare is too heavily loaded to be acceptable to the champions of growth!

Sustainable 'Economic Growth' and 'Sustainable Development' and Population

Whatever the connotation, sustainability in practical and result providing concept so far generally remains at best a myth, in fact, repeated by all organizations, or individuals as a rhetoric. Since it is a mere recommendation, without even any barking, leave out biting teeth, at any level, it remains merely a paper tiger.

The famous economist, Herbert Daly (1991) suggested that 'Sustainable growth' and 'Sustainable development' are mistakenly often used interchangeably, more or less synonyms, carrying identical meaning or objective. In reality, 'growth' refers to quantitative expansion in the scale of the physical dimensions of the economic system, while 'development' refers to qualitative change of a physically non-growing economic system in a state of dynamic equilibrium maintained by its environment (1991). For instance, 'the earth is not growing, but has ever been developing in human economy and increasing resource providing system. 'Any physical subsystem of a finite and non-growing Earth must itself also eventually become non-growing' (1991). Therefore, the term 'sustainable growth' indicates an 'eventual impossibility'. In contrast, the term 'sustainable development' indicates a continued qualitative development on this physically finite earth system or its major subsystems or spherical levels. In reality, the earth in terms of developmental and resource-churning, serves the Einsteinian meaning of the phrase "finite but unbounded". 'Development' can be expandingly sustainable, but not growth. This is because, as Daly explains:

> 'What is being sustained is a level, not a rate of growth of physical resource use. What is being developed is the qualitative capacity to convert that constant level of physical resource use throughput into improved services of satisfying human wants' (1991).

Thus 'development' qualitatively enhances the productivity of a factor or input, e.g. of labour per hour of work, an input in production, or such like. Production can be increased to a larger quantity by cultivating more land, but increases in yields per unit of land can give higher total output from less land. Human population has grown to approach 7 billion, by raising productivity of resources, not at all the physical size of the earth.

Land productivity in yield per unit size has gone up several fold because of development process energized through rise in technology of use and input of knowledge and skill and experience. This is what is termed as the rise in human capital than the physical capital in size or quantum of population. Poverty can be dissolved by this qualitative development of every child, man or woman through balanced food, provision of nutrition, maintenance of health, raising motivation, building character and providing quality education and skill at different tiers. Equality of opportunity to all is a *sine qua non* for human capital rise in a country.

Human population size was limited and slowly growing for lack of many mechanisms in the past centuries. Physical natural resources were more extensive per capita, with smaller population. Today, natural resources are relatively setting the limits currently in the developed countries. In large developing countries with rapid population growth, both types of resources, e.g. human capital as well as land resources per capita is limited. While the natural resources are by nature of a diminishing type resource, human capital is ever-growing. However, some neo-classical economic enthusiasts put that manmade capital can be all-through a near-substitute of natural resources. It is a misnomer. Daly puts that substitution in economic theory is reversible, but substitution of man-made capital for natural resources is frequently irreversible. Both natural capital and man-made capital are complements to each other more than substitutes.

Sustainable development ultimately implies a statically growing or more or less stationary population. If large population is involved, it will be necessarily at low level of per capita resource use and wealth creation and *vice versa*. Moreover, sustainable development is also incompatible with constantly growing technology. Technologies growing at a faster productivity-raising rate, even if leaving some negatives, definitely increase resource productivity, thus reducing the pressure on natural resources stocks to yield increased flows and per unit their resource productivity of the stocks. To Daly (1991),

> "Sustainable development implies a different direction of technical progress: One that squeezes more service per unit

> of resource, one that is efficiency-increasing rather than throughput-increasing, one that does not sacrifice natural resource productivity, and if necessary will sacrifice labour and capital productivity."

Harvard Professor Payotou (1991) also meant the same view that by increased technology in quality, productivity will be rising incessantly. Both the terms 'sustainable development' and 'socially and environmentally responsible development' mean almost the same, 'raising the human welfare all the more than ever.'

Conclusion

Mankind has been confronted in the current age with an unprecedented predicament, largely of its own making. Population and human economy together have reached a critical stage with an increasing negative impact upon the environment and ecological set-up of their earthly home. It is the environment which offered the earthly space for human ecumenal global habitat, and its entire edifice of basic resources which has been used for the growth of population and economy's rise. Mother Nature also gave to its prodigal human child a unique equipment, e.g. the living brain with the powerfully functional 'mind' pocketed into it. It is this equipment which has over time under great evolutionary pressure given birth to the *Homo sapiens* and further to the great 'thinking' species of the *Homo sapiens sapiens*, the 'wise' modern human race. Mankind can think, imagine, calculate and plan and do all that it has performed so far as a master of the earth to go up from the early primal bare survival stage to that of the space-probing age as the Homo cosmos. But while arriving at this stage, humans have since very early stage of evolution, began to carve out a 'generalist' way of life, with a free-breeding strategy, unlike the other non-human species' 'speciated' confinement to the Natural economy, and over time producing own food and setting up an ever-upward moving Eltonian human economic, cultural, and technological growth set-up. While doing this, humans have continually been modifying, changing and transforming the First nature into their own desired Secondary Nature with a reckless and inefficient profligate exploitation of the environmental and ecological resource edifice and processes. This profligacy has carried the human world towards the edge of endangerment to its own continued survival and tenure on the earthly home.

The International Union for Conservation of Nature (IUCN) since 1980 has been warning mankind to live a resource-conservative sustainable way of life to save and increase the life-supportive structure and stocks of ecology and learn that the economy has to be brought into being a subset of the larger ecology. The United Nations established the World Commission on Environment and Development (WCED) in 1987 with the same objective. Since mankind is obsessed with an impulsive urge towards further economic growth and development, the Commission's recommendation has strongly recommended to impress upon the humankind that environment and development are not either/or things: they are part of the same system; there is no alternative to either. While discussing the issue of sustainable growth and sustainable development, the Harvard economist Herman Daly emphasized that in reality, 'growth' refers to quantitative expansion in the scale of physical dimensions of the economic system, in contrast, development refers to qualitative change in a physically non-growing economic system. As such, while continuing 'sustainable growth' indicates an 'eventual impossibility', or has essential limitations, 'sustainable development' indicates a continued qualitative development in a physically non-growing system. 'Development' provides improved services on and on in a territory's physical resource use, subject to two caveats: one is that the territory must control its population to a statically growing or stationary size, and the second is that development must be equipped with not a 'constantly growing technology', but ever-faster growing productivity-raising technology, one that squeezes more services per unit of resource in an economy. The humanity must tackle the challenges of elimination of absolute poverty, gross inequality, population explosion and the environmental/ecological damage on the way to realistic sustainable development.

REFERENCES

Commoner, B. (1972). *The Closing Circle*, Jonathan Cape.

Daly, H. (1991). Sustainable development: from concept and theory to operational principles, in (eds.) K. Davis and M.S. Bernstam, *Resources, Environment and Population: Present Knowledge and Future Options*, New York: Oxford University Press, 25-43.

Ehrlich, P. (1994). Too many rich people. *Our Planet*, 6(3), 12-13.

Frankl, V.E. (1984). *Man's Search for Meaning*. Revised and enlarged ed. Washington, Square Press.

Goulet, D. (1971). *The Crucial Choices: A New Concept in the Theory of Development*. New York: Athenium.

Hardin, G. (1968). The tragedy of the commons. *Science*, 162, 1243-48.

Harrison, P. (1994a). Not just a numbers game. *Our Planet*, 6(3), 5-8.

——(1994b). Population facts that can spur women's well-being. *People and the Planet*, 3(3), 6-9.

Heilbroner, R.L. (1972). *The Economic Problem*. Englewood Cliffs, N.J., USA: Printice-Hall. 3rd ed.

Heller, W. (1973). Ecology and economic growth. Reprinted from Monthly Labour Review—*Economic Impact,* No. 1, 34-39. (Part of the debate on "Economics *versus* Ecology: Two Views").

Henderson, H. (1973). Challenging traditional values. (Part of the debate on "Economics *versus* Ecology: two views") *Economic Impact*, No. 1, 27-33

Hirschman, A.O. (1965). *Obstacles to development: a classification in a quasi vanishing act. Economic Development and Cultural Change*. Chicago: University of Chicago, 385-393.

Keyfitz, N. (1991). Toward a theory of population-development interaction. in (eds.) Kingsley Davis and Mikhail S. Bernstam, *Resources, Environment and Population: Present Knowledge, Future Options*. London: Oxford University Press, 295-314.

Kyte, R. (1994). Forum: two views on population and the environment. in Joe Kirwin (ed.), *Our Planet*, 6(4), 21.

Leaky, R.E. and Lewin, R. (1982). *Origins: The Emergence of and Evolution of Our Species and its Possible Future*. A Futura edition: First Futura ed. 1982. (First pub. 1977). Bury St. Edmunds, Suffolk. England.

Lewis, A. (1970). Is economic growth desirable? Reprinted from 'The theory of economic growth' (1955). S. Spiegelglas and C.J. Welsh (eds.) *Economic Development: Challenge and Promise*. Englewood Cliff: N.J., USA: Prentice-Hall, 17-29.

Meier, G. (1995). *Leading Issues in Economic Development* (6th ed.), Oxford: Oxford University Press.

Midgley, M. (1995). *Beast and Man: The Roots of Human Behaviour*. London: Routledge.

Myrdal, Gunnar (1968). *Asian Drama—An Inquiry into the Poverty of Nations.* Harmondsworth: Penguin.

Myrdal, Gunnar (1970). *The Challenge of World Poverty.* (3 Vols.), New York: Vintage Books.

Nilkeni, N. (2008). *Imagining India: Ideas for the New Century*. India: Penguin.

O' Sullivan, P. (2004). Global change and the global economy. *Global Change News Letter* (IGBP), No. 58. June 20, 16-19. IGBP: International Geosphere-Biosphere Programme.

Odum, H.T. (1971). *Environment. Power and Society*: Wiley.

Our Common Future (1987). Report of World Commission on Environment and Development (WCED) by Gro Harlem, New York: Oxford University Press.

Polanyi, K. (1944). *The Great Transformation*. New York: Farrarand Rhinehart.

Prakash, S. Sustainable development and resource management. *National Geographer.* 26 (2), 121-139.

Pretes, M. (1997). Development and infinity. *World Development* 25 (9), 1421-30.

Radhakrishnan, S. (1966). An idealist view of life. (Hibbert lectures) in 1929, London: George Allen and Unwin Ltd.

Rifkin, J. (1992). Bovine burden. *Geographical Magazine*. 64 (7), 10-15.

Samir, A. (1976). *Unequal Development*. New York: Monthly Review Press.

Santos, T.D. (1989). *The crisis of development theory and the problems of dependence in Latin America*. Siglo 21.

Schumeli, E. (1980). The right to self-re-dization and its predicaments, in A.S. Rosenbaum (eds.) *The Philosophy of Human Rights: International Perspectives*, West Port, Conn, USA: Greenwood Press, 151-166.

Sen, A.K. (1981). *Poverty and Famines: An Essay on Entitlement and Deprivation*. Oxford: Oxford University Press.

Sen, A.K. (1985). *Commodities and Capabilities*. Amsterdam: North Holland: Elsevier Science Pub. Co.

Sen, A.K. (1999). Development as Capability Extension, in (eds.) K. Griffin and K. Knight, *Human Development and International Strategy for the 1990s.* London: Macmillan.

Simon, J. (1981). *The Ultimate Resource*. N.J., USA: Princeton University Press.

Waddington, C.H. (1978). *The Man-Made Future*. Jonathan Cape. Paladin: Basic Books.

Wilson, E.O. (1978). *On Human Nature*. Harvard.

World Development Report (1992). *Development and the Environment*, New York: Oxford University Press.

2
Population, Development and Environment: Issues and Challenges

M.H. Qureshi

Introduction

In the context of contemporary debate on population, development and environment, it is the population which plays pivotal role because as an active agent of change it tries to maintain balance between the remaining two which is the core issue of any discourse. The human beings open their eyes as new born babies in the natural environment and as they grow they become not only its part and parcel but also contribute in the making and sustaining of the social environment. Geographers are interested to study both, the environments, i.e. natural/physical as well as social. The natural environment is given to the human beings with all its gifts of resources. The social environment is the creation of the institutions organized by the society for the accelerated pace of their economic development. Natural environment provides the base for economic development and the societies interact with it with the tools provided by science and technology. As the economic development proceeds, the tools are improved, the capacity of the society to utilize resources increases, the natural environment gets affected. Generally this effect is negative but it is also positive if the utilization of the resources is done in a planned manner. The environment is very basic to the existence of population. It provides the resources and indicates the direction of development. Human beings understood this indication and proceeded in that direction on the path of their economic well-being. The more primitive societies depended directly on their immediate environment for their sustenance. They depended on the edible plants and animals with collection, gathering and hunting as their major economic activities. Their tools were primitive and therefore, they could not inflict large scale damage on the environment. The population increase threatened the sustainability which led to the process of sedentarisation as well as domestication of plants and animals. It was not by chance that the people in the Nile valley, Mesopotamia, Indus and Ganga valley became farmers; people in the coastal villages became fishermen, lumber persons in the forested villages and miners in mineral rich areas. It was the rich soil and availability of water which made the people farmers in the river valleys. It was not simple environmental determinism in choosing the activity but it was also due to the limited choice which was available to the people under the low level of technological capability. It was a situation of necessity rather than of freedom. Therefore, the human beings constantly strove to improve the level of technology in order to loosen the shackles of environmental forces. The technology, as it was invented, reduced the drudgery of labour, increased its efficiency and provided leisure to attend to the higher needs as human beings cannot live by bread alone. Thus, the invention and constant improvement in technology became a major concern of the societies. But the technology is not an innocent homogenizing factor, helping everyone in the same measure. Three statements can be made about technology. Primitive technology is income neutral. It is available to all with a little investment. If it is a simple technology like a sickle, axe and spade, a wooden plough, a dug well, even a poor person can afford it. Technology of an intermediary nature is so income biased that it cannot be obtained unless one has investible surplus. It is not available to all and can be obtained by people who have capital for investment. While it is possible to obtain a dug well, a tube well is out of reach of many farmers. A highly sophisticated complex and large technology becomes out of reach of individuals. Such technology has to be brought under social control. Once such technology comes under social control, it is available to all after a nominal payment. An individual, for example, can neither own nor operate a canal but she/he can irrigate her/his agricultural land by paying the water cess because canals are dug and maintained by the state.

Human beings interact with the environment with the help of technology and simultaneously create institutions for accelerating the pace of their economic development. These institutions become rigid with the passage of time and start decelerating the pace of development. Such institutions, then, require amendment and reform. Best example is our constitution which can be amended by the Parliament as and when the need arises. The example of a rigid institution is our caste system. One can change her/his religion but cannot change her/his caste. Moreover one's caste is a matter of accident as one has no option in choosing her/his parents. The caste gets strengthened with socio-political polarization. There seems to be a strong equation between *roti, beti* and vote. The idea that the vote will go to those individuals with whom one inter-dines and with whom one enters into matrimonial relations. The distortions in the political processes have been strengthening the caste polarization.

Thus, the human beings occupying a central place in the triangle of environment, technology and institution interact with the environment with the help of technology and create institutions to regulate the development. The population growth has to be sustained and in order to accommodate the increase in population, technology has to be constantly innovated and improved. This constant improvement in technology routed through scientific research ensures constant development. Development is qualitative change, always having a positive value attached to it and it is, of course, unidirectional. There may be decay but there is no reversal of direction of development. A tree can never be a plant again and a grown woman/man can never become a child. Economic development is not class neutral. It helps certain classes of people more than the others. Primitive societies are, generally, more integrated. The societies get differentiated as the economic development takes place. At the initial stage of economic development the society gets differentiated but it is a situation of integrated differentiation. The differentiation gets accentuated with further development ushering the era of mass production as well as mass consumption. The attempt at that stage is made to obtain a differentiated integration with intervention of the public policy. The differentiation leads to the appearance of disparity which is not desirable for the establishment of a cohesive society. While diversity is to be celebrated, disparity has to be avoided as it leads to discrimination and exclusion.

Economic development is initiated by utilizing the resources. It is generally said that resources are not, they are made. The mineral oil did exist for a long time before the invention of the combustion engines. It use to ooze out of the rocks in Pennsylvania and was sold as Kirk's oil as medicine to be massaged to relieve the muscular pain. Its utility later made it possible to extract it for the first time in 1857 at a place in Romania known as *Ploesti* after the development of drilling technology. The extent of development depends upon the resource endowment and capacity to utilize these resources with the help of the available technology. This capacity to utilize resources increases with further inventions, innovations and adoption of technology. In the initial stages of economic development, societies are characterized by what Gandhiji termed as, "production by masses". At this stage of development, the damage to environment is the least. As the technology becomes more sophisticated, damage to environment also increases. The industrial revolution created the need and urge for searching tropical raw materials as well as markets by the industrial powers which ushered the era of colonialism. The industrially developed countries started "mass production" with market orientation. The over-utilization of resources and the industrial production process brought in the signs of environmental degradation. The industrial societies also became the societies of mass consumption. The equation has now clearly emerged in the contemporary world. The mass production leading to mass consumption which in turn leads to intense competition which has become the cause of social conflicts and violence. The unplanned over-utilization of resources by people under political patronage is violence against environment. We must realize that violence against environment is also violence against humanity. Violence against environment will increase the vulnerability of the society to disasters. Let us examine the interfaces between environment, population and development.

Environment and Population

We have to imagine a primitive society in its early history of existence with very limited population and limited resources subsisting on its immediate environment. The people largely depended on two resource endowments, namely (a) edible plants with the quality of providing energy and being palatable and non-poisonous and (b) animals hunted using primitive tools. The immediate niche was able to support limited population. As the population increased the carrying capacity of the immediate environmental niche was not able to support the expanding population. The area of operation, thus, expanded and two societies existing in different niches came into conflict and were forced to define their areas of operation. It was also realized that the destructive economic activity of uprooting of plants, collection, gathering and hunting have to be replaced by domestication of plants and animals. The domestication led to sedantrisation of human settlements. Though the birth rate at this stage of development was high, the growth of population remained low due to high death rate.This fact clearly indicates that the uncertainty of survival of the children at the low level of economic development leads to higher birth rate. As the resources are important for economic development, population is also an important factor as human resource in ensuring economic development because the population acts as producer as well as consumer. People create and use technology to increase their productivity and the productivity of land. The primary production helps in the development of secondary sector as well as the tertiary sector. The economic and social relations go on becoming complex and the effect on the environment is obvious. The scientific achievements ensure better quality of life and development becomes responsible for bringing in the demographic transition and restricts the population growth through its different stages. Unfortunately the development process is not uniform across the regions and societies. Hence, the contemporary world is highly differentiated with regions of high level of development and regions of low level of development coexisting together, designated as the first and the third world. The development of underdevelopment is not because the underdeveloped countries lack resources but they lack in technology and the transfer of technology is not the priority of the developed world. The technology available in the market is very costly. Then the underdeveloped parts of the world are told to opt for appropriate technology. One of the examples of appropriate technology is "*buggy*" adopted by the farmers in the northern plains of India which means modification of the wheels of a bullock cart into wheels with rubber tyres and ball bearings but a bullock or a buffalo providing the traction power. According to the "First World," Third World should not use engine for traction as it consumes petroleum and petroleum pollutes the environment. The question thus arises, what is appropriate to the Third World is not appropriate for the First World and *vice versa*. The Third World should not use petroleum as it is required in the developed world for further development. The Third World should not develop and put a carbon cap and limit emission. Consequently, so called highly developed and highly primitive societies coexist even in the present twenty-first century. The developing countries are sermonized by the developed world to control their population. The developed societies could control their population by using the development as the best contraceptive. The underdeveloped or developing societies are passing through different stages of demographic transition according to the different levels of their economic development. While the developed countries have to depend on expatriate labour force, the developing countries have surplus labour force, both skilled and unskilled. This population should be taken as asset in terms of human resource. The proper planning of education and capacity building may make our large population as asset rather than a serious liability. Had the population been a liability, productivity would have never been worshipped in different societies across the world. Many rituals are performed to get a child and many more rituals are performed after the birth of the child. The Children both, boys and girls, are sacred how can they be treated as unwanted and undesirable? The basic question is how to limit their number and how to build them as the most productive human resource. Can we not apply the same dictum that development is the best contraceptive in case of the poor societies of the Third World? A high population growth in developing countries is the function of their economic under-development. India also has experienced high population growth rate due to the poverty of a large segment of her population. Poverty is an important cause for large family sizes. As the social and economic development takes place, the growth rate falls. If we examine our own families we generally do not have the same number of children which our grandfather or the father had? There may be some exceptions but, by and large, we have reduced the number of our children as we are influenced by the Principle of Social Capillarity. It is a fact that when one attains certain status in the society, she/he would never like that our children slide down in the social ladder from the place which we have already achieved. Then, that is the incentive for deliberate control of our population. Poverty will always act as breeding engine. The poverty stricken marginalized groups will continue to have larger family sizes till they also become the participants in the process of development and share its benefits. The growth of GDP is essential but at the same time the concentration of wealth in few hands has to be avoided. The modern society is standing on a tripod supported by three legs of the *State*, *Market* and the *Civil Society*. If the State goes into a withdrawal mode, the Market starts dominating. In such a situation the Civil Society has to play an effective role to maintain the balance.

Indian Experience of the Interface among Environment, Population and Development

India has a diverse Physical Environment. The country has a specific structure which has provided it a unique personality. The Indian sub-continent has been perceived differently by people of various regions of the world. Susan Gole has tried to unravel the perception of others about India in her book, *India within Ganges India Intra Gangem,* the Latin name used by Europeans for many centuries when they were referring to what we now call the Indian subcontinent or South Asia. *"There were many Indias, since the term was used for all known or unknown land beyond the river Indus, and even for some areas on the African continent. Hence, our India was known as India 'on this side of Ganges' to distinguish it from 'India beyond the Ganges' which meant mainland South-east Asia, 'India superior' which might extend to northern China or in more general term "India Orientalis" which includes all the islands, from Australia to Japan and beyond."* Thus, has been the fascination for this country in the minds of the outsiders. It will not be worthwhile in the address to acquaint the audience with the diversity of the physical environment of India as majority of the delegates come from Geography background. But it will not be out of place to mention about the occupancy of the land by human population in the hoary past making India one of the oldest settled countries of the world. India has been able to absorb waves after waves of migration of people due to its rich physical resource base in the form of fertile soil, plenty of water flowing in her rivers, diverse natural vegetation and rich biodiversity. In the process of her peopling the human settlements sprang up in the river valleys, forested areas, hill slopes and coastal plains. The size of the country provided isolated niches in which the economic and social development started taking place. The regional structure matured over long historical period. The people with superior technology spread fast and the people with low level of technology shrunk in limited hilly and forested terrains. The diversity of terrain and the distances between the different ecological niches translated itself in the development of so many dialects, languages, art forms, fairs, festivals and different economic activities. Human achievements can be gauged by the skills developed in the production of crops, livestock and specialization in producing fine cloths and artifacts learnt through the parental skills.

The colonization of the country changed the pattern of development. Two basic elements were introduced which set the process of pauperization of the artisan classes. One was the introduction of port-enclave model of development in which the hinterland was connected with the ports. The other was the introduction of railway network to link the local markets to regional markets, the regional markets to national markets and the national markets to the international markets. The local manufacturing in the form of handicrafts and household industry suffered. The artisans could not stand the competition with the machine made goods which was not only produced in large quantity but also at cheaper costs. With the demand of raw materials, the cropping pattern changed. The introduction of ownership systems of land through different settlements, the agrarian relations also underwent drastic change. A few enclaves of industrial development were created to process the raw materials to be exported for further value addition or for trading with other colonial powers. The ports were developed to facilitate the suction of resources through the road and railway networks. The cropping pattern in the Kolkata hinterland was adopted not to feed the people but according to the requirements of the colonial powers. Indigo cultivation was introduced to be used in the cloth mills of Manchester and Liverpool for bleaching and colouring purposes. The farmers were forced to grow indigo. The development of the chemical industry made the indigo redundant. The area under indigo was replaced by opium again for the purpose of trade. The *Neel- ki- kothi* and *Afeem- ki- kothi* are still the part of our memory of the people as well as part of the landscape in Eastern India. The cultivation of jute replaced the opium as the demand of the product went down in the international market. Thus, attempt to force the farmers to grow industrial crops lead to the shortage of food grain production. The industrial activity was confined in the immediate hinterland of the three ports. The islands of development existed in the vast ocean of underdevelopment.

The independence of India ushered a new era of opportunities and challenges. The partition of the country was the worst tragedy which brought immense misery and tragedy to the population of the Indian subcontinent. Large scale population displacement took place which was not a voluntary migration but coerced displacement. The settlement and rehabilitation of the displaced population was the foremost priority. The initiative of development was taken by introducing drastic land reforms, abolition of *zamindari, consolidation of land holdings* and correction of agrarian relations. The beginning was done but a lot was there to be done. The country embarked upon the path of economic development based on development of infrastructure ensuring agricultural and industrial growth under the centralized planning strategy. The large dams in the form of multipurpose projects were constructed but such schemes had long gestation period and ameliorating the food shortage became the main concern in the fifties and early sixties. The development of irrigational facilities ensured quick adoption of Green Revolution technology. Unfortunately, the effects of Green Revolution were confined to only those areas where assured water supply for irrigation was available. Vast tract of the country was drought prone and has the potential for the production of pulses and oilseeds along with the scope for livestock farming. Since Green Revolution was based on the package of assured irrigation, HYV seeds and fertilizer, its diffusion was difficult in the dry areas consequently different strategies of development were adopted. The country was able to come out of the food shortage and to some extent the peasant farms also witnessed the increase in land productivity. The intensive irrigation based on pumping of water by tube wells has now created the problem of depletion of ground water aquifers as withdrawal of water far exceeds the recharge. The introduction of submersible pumps has further aggravated the water problem. The shortage of water for irrigation has also created water markets in the villages of the Green Revolution areas. The installation of a pumping set in one's field provides the large farmers the opportunity to draw water from much larger areas for sale as the field boundaries on the surface are not extended to separate the ground water aquifers down below. This is a case of selling a resource as a commodity which does not belong to the seller. This situation is bound to create haves and have-nots in the same society leading to social tension. The bad effects of over-irrigation, due to the adoption of water intensive crops, have already appeared in the form of salinisation, alkalization and water logging in Punjab Haryana and in Indira Gandhi Canal command area.

The development as a concept, no doubt, is value positive but it also has some negative impacts, if not managed properly. Since there is no zero environmental impact of any developmental effort, there will be some impact on environment with the initiation of developmental work. Even simple tillage may initiate the process of erosion. Development of the highways in the hilly areas weakens the slopes resulting in slope slumping and landslides. Industrial activity based on the energy derived from the fossil fuels is bound to pollute the air. Untreated industrial effluents discharged in the water bodies render the water unfit for domestic use. Can we think of any economic activity which does not impact the environment? However, tillage of soil cannot be avoided for agricultural development. Highways have to be built for efficient transportation to connect the places of production and consumption. The processing of raw materials in industries using coal and mineral oil, the movement of vehicles using petroleum will add carbon dioxide and carbon monoxide in the atmosphere and create greenhouse gases. There is no denying the fact that development is essential for fulfilling the consumption needs of different segments of the population. These needs may be necessities, comforts and luxuries depending on the income differentials. The basic question, then, is how to minimize the impact on the environment and make the developmental process less painful to the life-forms.

The development through the route of industrialization and tertiarisation will create nodes in the form of urban centres with concentration of population on the nodes *which* are intersection points of transportation network. Expanding urbanization has its own environmental problems. Urbanization, unfortunately, does not mean urbanism of urban people. The urban spread has created intense competition for land on the urban fringe between the urban and rural land uses. The urban society has to deal with congestion, shortage of housing, provision of essential services, waste disposal and slum and squatter population. India's urbanization is expanding along with the challenges of management of the environment.

Conclusion

The following challenges are there which the social scientists have to examine with serious research inputs at the micro levels:

(a) Degradation of land and water as basic natural resources.
(b) Depletion of soil fertility and fluctuations in land productivity.
(c) Depletion of bio-diversity. Large and diverse areas and have specific issues in each region. The generalizations of the studies are going to be related to specific regions but these studies may be used as empirical examples for generalization on a wider scale. The present generation is technologically empowered as compared to my generation.
(d) Pollutions: air, water and sonic, reducing the quality of life.
(e) Climatic change and dangers ahead.
(f) Political economy of development and political ecology.
(g) Development Projects and displacement of Project Affected Families (PAFs) and their rehabilitation.
(h) Housing: Demand and supply gaps in rural and urban areas.
(i) Competition for land on the rural urban fringe and land market. The social conflicts created by speculative building activity have to be resolved.

These are suggestive but do not provide a comprehensive list. The older generations of Geographers were handicapped due to the lack of facilities available to the present generation and were technologically challenged. The present generation has access to very powerful tools for analyzing huge body of data. Integrity of the researcher is an essential ingredient of any research, big or small, in order to avoid the fiasco which the world has witnessed recently regarding the findings of some glaciologists about the glacier melt in the Himalayan region.

We should also examine the Indian's conservationist ethos. There are communities which have a tradition of conservation of environment. We have large store of traditional knowledge. Can we retrieve some of it for our micro regional benefits? Are the water conservation methods new for Indians? Why were the trees sanctified and avoiding the damage to vegetation was a matter of faith? Is there any other example of saving the trees from felling by embracing them and getting killed? What has then happened to us since then? It seems the greed of some people has pushed the ethos in the back waters. The country still has enough to satisfy the needs of her people but, as Gandhiji put it, it is not enough to satisfy the greed. Population, Development and Environment have their own dynamic relationship. This dynamic relation has to be balanced. The balance has to be ensured by accommodating the aspirations of the population, by reducing the regional imbalances and by protecting the environment.

3

Population and Environment Relationship: The Context of Climate Change

R.B. Bhagat

Introduction

There is general agreement that human beings have already had an overwhelming impact on environment. The climate change is one such impact which is the most devastating to the mankind. There has been climate change in the geological history of the earth, but the climate change that the mankind is confronting today is anthropogenic which is destroying the delicate balance between the man and nature over the earth. The fundamental cause of climate change is the nature of development and the philosophy of development that we follow so arduously to promote ourselves in the comity of nations. Development is measured in terms of national products, which is defined as production of goods and services during a financial year. On the other hand, it does not discount the loss to the environment and biodiversity and also loss of human health due to air and water contaminations. There is a need to incorporate methodology of green accounting in our national accounts statistics to change the calculation of national income from Gross National Product to Gross Nature's Product. Also much of our development process is highly energy intensive and consume increasing amount of fossil fuels leading to the release of greenhouse gases and contributing to the global warming and climate change. This paper attempts to presents a synthesis of the nature of relationship between population, development and environment in the context of climate change. It also recapitulates some of the policy issues regulating environment and development at the national and international levels.

Greenhouse Gases, Global Warming and Climate Change

Climate change is the result global warming in the wake of rising greenhouse gases in the atmosphere. The greenhouse gases play a crucial role in regulating the heat balance of the earth. It allows the incoming short-wave solar radiation to pass through the atmosphere relatively unimpeded, but the long wave terrestrial radiation emitted by the earth's surface is partially absorbed and then re-emitted by a number of trace gases in the atmosphere. These gases known as GHGs (greenhouse gases) are—water vapour, Carbon dioxide, Methane, Nitrous oxide and Ozone in the troposphere and in the stratosphere. The greenhouse gases (carbon dioxide, methane etc.) prevent some of the heat radiation from the earth from escaping into space. The concentration of greenhouse gases is increasing which is presumed to raise the temperature of the earth in a long perspective. Eighty per cent of the effect of the greenhouse gases is caused by carbon dioxide.

As a result of the greenhouse gases, the temperature of the earth surface is rising. This rise in the earth's temperature is known as global warming. Global warming can have major physical, environmental and socio-economic consequences. This is also referred as climatic change. The global warming could cause changes in precipitation patterns, ocean circulation and marine systems, soil moisture, water availability and sea level rise. This would make an impact on agriculture, forestry and natural eco-systems. A rise in sea level threatens sections of all the coastal states of the country. A rise in sea level would inundate wetlands, erode shoreline, exacerbate coastal flooding, increase the salinity of estuaries and aquifers and impair water quality.

Along with industrial pollution, vehicular pollution is the major source of air pollution and increasing greenhouse gases in the atmosphere. Total number of vehicles was 754 million in 2000 at the global level which would increase to 1116 million by 2020. A large number of them are concentrated in the developed countries (De Souza, William and Meyerson, 2003). The important aspects of vehicular pollution that are affecting the air quality are types of engine used, age of the vehicles, poor road conditions and congested traffic. The principal vehicular pollutants are carbon monoxide, oxides of nitrogen, hydrocarbons, suspended and particulate matters, a varying amount of sulphur dioxide depending on the sulphur content of the fuel and lead compounds.

The rising greenhouse gases in the atmosphere are attributed to the rising affluence, increasing consumption, technology and rapid population growth. The relationship between all these factors is not simple and straightforward. The section below describes the nature of relationship between these factors.

Climate Change, Population and Development

There are people who are skeptical about the global warming and climate change. But there are enough evidence and research to show that the temperature of the earth surface has risen close to 1 degree Celsius in the past 100 years. More importantly, if recent trend continues or accelerates as many scientists predict, the earth's temperature may rise another four to six degrees by 2100. There is no doubt that this has catastrophic effects globally, but will more affect the people living in tropical areas (IPCC, 2007). By coincidence most of the poor people of the world live in the tropical countries. India will be severely affected by global warming as food production is likely to be affected, on the other hand population will continue to grow up to 1.5 billion by the middle of this century.

When you look at the role of population in global warming, population is certainly an important but not the overriding factor. It is the rising market driven consumerism which has been compelling people to use increasing quantity of fossil fuels globally particularly by the rich people. In this context, slowing down of population growth may not be a very effective strategy of controlling global warming.

The world population was 2.52 billion in the year 1950 which increased to 6.8 billion in 2009 and is likely to reach 9.1 billion by the year 2050. While the population size will remain almost stationary in the more developed world, the most of the population growth would take place in the developing world with Asia alone harbouring more than half of the world's population (5.2 billion) by the year 2050 (UN, 2009). As most of the growth in population size is going to occur in less developed countries, this is likely to pose challenge for the economic growth and pressure on environmental resources in the developing countries. Furthermore, most of the population growth in the developing countries also is likely to be concentrated in the urban areas. This has implication for increased demand for energy and water resources in the urban areas. It will also pose challenges for the management of increased waste generated and air pollution.

The population growth in India is much higher than the world standard. The Indian population has grown nearly four times during the last century from 238.4 million in 1901 to 1027 million in 2001. The latest figures from 2011 Census shows India's population enumerated at 1210 million in 2011. India thinks to achieve a stable population by 2045 as envisaged in the National Population Policy 2000. As such, by the middle of this century India is likely to add approximately 400 million more population and the total population will be nearly 1.6 billion.

Earlier efforts by Malthus (1798) and latter by Boserup (1965) elucidated the relationship between population growth and development. While Malthus argued that population growth is the root cause of poverty and human sufferings, Boserup explained how technological advances and increased innovation in the agriculture was the result of increased density of population. However, both the views throw an alternative way of explaining the relationship between population growth and development. Recently environmental issues are emerging important in the relationship between population growth and development.

PAT model is generally used to assess the impact of population growth and economic development on environment. The model proposes a simple relationship between the three as follows (Ehrlich & Holdren, 1971):

$$I = (P) \times (A) \times (T)$$

Where I = a measure of environment impact

P = population
T = technological efficiency (use of energy in the production of per unit GNP)

The PAT model identifies consumption and production patterns which are proximate factors of environment degradation. It is simple to apply also. Several researchers have applied the PAT model to estimate the relative contribution of the factors in the model. According to Bongaarts (1992), the contribution of population growth will be 50 per cent in the increase of CO_2 during 1985-2025 and will decline to 27 per cent during 2025-2100. These figures are based on projected population and CO_2 growth rates. However, the PAT model is criticized for not taking into account several parameters, as such poverty and governance in influencing global warming and climate change (Bhagat, 1997).

There is a tremendous pressure on the environmental resources to produce more food in view of growing population. While this can be done, it all depends upon whether the increased food production is achieved without degrading the environment. The history of agricultural development in India is not encouraging as much of the agricultural production in the past has been achieved with heavy doses of chemical fertilizers and depleting the ground water resources. Wherever, surface water was available through canal irrigation, the water was used in excess leading to expansion of wastelands as a consequence of the growth of salinity and alkalinity in the soil.

Although the industrial production in the country has been well above 6 per cent per annum during the 1990 and close 8 per cent during the 2000, the quality of urban environment deteriorated rapidly during this period. It is evident in increasing air pollution and declining quality of water and in the sanitation condition in the urban areas. A study on the ambient air quality recorded for 23 cities in the country reveals that CO_2, Suspended Particulate Matters (SPM) found critical in many cities. Although Sulphur dioxide and Oxides of Nitrogen levels have registered an upward trend; they remain well within the National Ambient Air Quality Standards in all the cities. In addition to these common air pollutants, some of the toxic and carcinogenic chemicals are being detected in urban air. On the other hand, very little monitoring of these pollutants is being done until recently

(Planning Commission, 2002). The level of waste in the event of rising level of development particularly in the large cities is enormous. Development process generates waste both at place of production as well as at the place of consumption. The safe disposal of industrial by products and household wastes are serious challenges for the environment. Open dumping and uncontrolled land filling are in most cases the main disposal methods. In such places, the uncovered garbage is the fertile ground for bacteria and viruses to develop which cause the occurrence of diseases.

Development based on the mindless exploitation of nature is unfolding disaster in several forms, e.g. deforestation, desertification, land degradation, erosion, salinization, siltation, climate change, while often more gradual, has been growing rapidly in recent decades due to a combination of increased demand, improved technological means of exploitation, and the lagging pace of conservation and control. Development projects—generally dams and irrigation projects—often involve forced resettlement and affect many aspects of human security. In India, for example, it has been estimated that over 20 million persons have been uprooted by development projects in the past three decades.

The foregoing discussion clearly shows that environmental changes are taking place both at the local, regional and the global levels (see also Table 3.1). Although environmental changes at various levels are interlinked, it is the global environmental changes that are of prime importance for climate change. The global environmental changes having implications for climate change is crucially linked to energy consumption based on fossil fuels. The global concerns in the environment and climate change took place in the late 1980s onwards raising the issue of equity among the nations (Pebley, 1998).

Table 3.1: Environmental Changes: Local, Regional and Global Levels

Local	*Regional*	*Global*
Heavy metals in air, soil and plants, e.g. From industrial emissions and Discharges—Noise, Smell, Air pollution.	Eutrophication, Contaminants in the soil, Landscape changes due to mining or agriculture.	Changes of the climate due to ozone depletion and the greenhouse effect.

Source: Central Statistical Organisation (1999).

Development, Energy Consumption and Climate Change

The present development process heavily relies on non-renewable sources of energy, i.e. fossil fuels like coal, petroleum and natural gases. The exploitation of these sources of energy has also increased manifold in the recent years. Energy demand in future will also likely to grow. The share of the developing countries in energy use is likely to increase from 26 per cent in 1991 to 40 per cent in 2020 (WRI, 1994-95). If this demand were to be met with the use of fossil fuels, the environmental consequences will be dire at both local and global levels. On the other hand, in spite of increased use of energy in developing countries, the per capita consumption is likely to be still much lower. Presently, the per capita consumption in developing countries is less than one-sixth that in the industrialized countries and less than one-tenth that in US. A comparison of the intensity of energy use across selected countries is shown in Table 3.2. It may be noted that in comparison to India, the intensity of energy use is 17 times higher in US and emission of CO_2 is also 17 times higher than India on per capita basis. This shows the differential responsibility of different nations in causing global warming and climate change. Although greenhouse gases have mainly contributed by the industrialized nations, the impact of climate change would be global and population belonging every section of society will be affected.

Table 3.2: Income, energy consumption and CO_2 in selected countries compared to India (if India =1 unit)

Country	*Income per capita*	*CO_2 per capita*	*Commercial energy consumption per capita*
US	15	17	17
UK	10	8	8
Japan	12	8	8
South Africa	4	8	5
Brazil	3	2	2
China	2	2	2
India	1	1	1
Pakistan	0.7	0.6	0.9
Bangladesh	0.7	0.2	0.2
WORLD	3	4	3

Source: Population Reference Bureau (2002).

The poor people living in developing countries would be hit hardest. This raises an important question of differential responsibility and differential vulnerability across the nations at the global level.

An environmental friendly, sustainable and affordable source of energy is crucial for the sustainable development. Hydroelectricity, thermal power, nuclear, solar and wind sources of energy will play a vital role in safeguarding the global environment.

Implications of Climate Change

The consequences of climate change due to global warming are enormous on various sections of population. The poor, women and children living in developing countries would be the worst victims due to extreme weather events, increased food insecurity, loss of habitat and livelihood. The health impact of environmental degradation falls into two broad categories. The first is a lack of access to essential environmental resources such as clean water, food, shelter, fuel and air. The second broad category is exposure to hazards in the environment. These hazards include biological agents—microorganisms such as bacteria and viruses and parasites that contribute to the huge global burden of infectious diseases. Biological agents are implicated in diseases from diarrhoea to acute respiratory infections, malaria, ulcers, and some cancers. Apart from health implications of climate change, it is also likely to affect the agriculture and food security of the different countries. Many poor people are likely to lose their livelihood and will become environmental refuges. The impact of climate change on human population is depicted in Fig. 3.1.

The climate change is likely to exacerbate the existing inequality at the global, national and local levels. Therefore, the mitigation of climate change will be helpful to the poor in the long run. However, it may be important to bear in mind that mitigation of climate change should not be seen as hindering economic and human development. In fact economic and human development based on renewable sources of energy is the need of the hour which should be incorporated in our mitigation strategy of stalling climate change. Along with mitigation, human beings also have the ability to adapt with the aid of new technology and resorting to migration to the relatively colder areas.

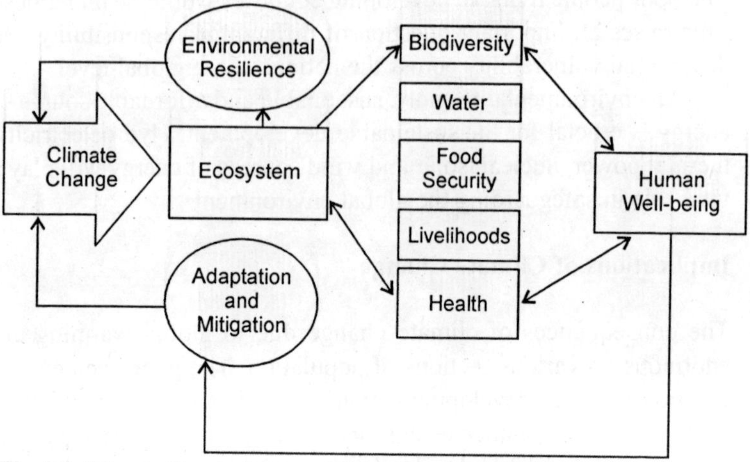

Fig. 3.1: Climate Change Impacts: Mitigation and Adaptation

Source: ICIMOD (2010).

Adaptation strategies would be more apt for the tropical areas compared to temperate and colder areas which may even benefit with global warming in the short run. Many part of the temperate and cold area are now not fit for agriculture and it is possible that with rising temperature these areas may benefit in agriculture and food production.

Environmental Policies and Regulations

There have been many global efforts to protect environment and mitigate climate change since UN Conference on Human Environment at Stockholm held in 1972. Subsequently, The UN General Assembly established the World Commission on Environment and Development under the headship of Brundtland in 1983. The Commission submitted the report entitled "Our Common Future" in 1987. In this report the concept of sustainable development was defined as "ability to satisfy the needs of the present generation without compromising the needs of the future generation" (UNCED, 1987). As a sequel to this report United Nations Conference on Environment and Development (UNCED-Earth Summit) was held in Rio de Janeiro in 1992. The Rio Conference represented a turning point in the way we look at environment and development. At the Earth Summit, world leaders adopted Agenda 21, a blue print to attain sustainable development in the Twenty-first century. The Rio summit was followed by Earth Summit held in Johannesburg in 2002. It aimed to build on the Rio Earth Summit of 1992. The UN has planned to hold third Earth Summit in Rio-de Janeiro in 2012.

Apart from Earth Summits, there were several protocols, treaties and conventions that have been adopted at the international level. The Montreal Protocol was adopted in 1987 which envisioned ceasing the CFCs (the ozone depleting substances) production by the year 1996. The Protocol turned to be very successful effort in curbing the consumption of CFCs by more than 70 per cent in a decade's time (World Resources Institute, UNEP, UNDP and World Bank, 1998). Another protocol very popularly known as Kyoto Protocol fixed the targets of greenhouse gases reduction by the developed countries by 2012. The Kyoto Protocol adopted in December 1997 was considered to be the first international attempts to place legally binding limits on greenhouse gas emissions from developed countries. It considered the six gases, namely carbon dioxide, methane, nitrous oxide, hydrofluorocarbons (HFCs), perfluorocarbons (PFCs), and sulphur hexafluoride. The Protocol specifically aimed to cut the combined emissions of greenhouse gases from developed countries by roughly 5 per cent from their 1990 levels by the year 2012 (see Table 3.3).

The Kyoto Protocol provided that if a country's emission level falls below the treaty limit, then it can sell the credit to another nation which in turn can use the credit to meet its own obligations. However,

Table 3.3: Target of greenhouse gas emissions reduction by 2012, Kyoto Protocol

Country	*Target (% change from 1990 emissions)*	*Country*	*Target (% change from 1990 emissions)*
Australia	+8	Iceland	+10
Bulgaria	–8	Japan	–6
Canada	–6	Norway	+1
Croatia	–5	Poland	–6
Estonia	–8	Russian Federation	0
European Union	–8	Switzerland	–8
Hungary	–6	US	–7

Source: The World Resources Institute, UNEP, UNDP and World Bank (1998).

this Protocol has not been implemented due to lack of consensus among nations. The US, which is the biggest contributor of greenhouse gases and where this is likely to increase by 34 per cent in 2010 from the level of 1990, does not find the Protocol amenable to its interests. The recent international effort on climate change negotiations has virtually dumped the Kyoto Protocol which is evident in the failure of Copenhagen Summit (held in December 2009) to arrive at a legal binding on the emission of greenhouse gases by the rich countries. On the other hand, India voluntarily declared to cut the carbon emissions intensity of GDP by 20-25 per cent by the year 2020 from the level of 2005. As a result the Planning Commission of India has now accepted that the 12th Five Year Plan will be high, inclusive and low-carbon growth (Ramesh, 2011).

In India concern for environment started when Government of India appointed a National Committee for Environmental Planning and Co-ordination in 1972. In 1980, Department of Environment was created within the Ministry of Science and Technology. In the meantime the constitutional provisions in terms of the 42nd Amendment to the Constitution effected in 1976 imposed an obligation on the part of state (Article 48A) and the citizens (Article 51A.g) to endeavour to protect and improve the environment and to safeguard the forests and wildlife of the country.

As we know that a major accident in an industrial plant of Union Carbide took place in Bhopal in December 1984. The release of the deadly methyl isocyanate gas from the plant killed more than 2000 people. This incidence was a turning point to apply the stringent government measures to protect people and the environment by enacting the comprehensive law on environmental protection in the country. Following this, Ministry of Environment and Forests was created in the Central Government in 1985. The Ministry of Environment and Forests reviewed the earlier laws on ***Water Prevention and Pollution Act, 1974 and Air Prevention and Control of Pollution Act, 1981*** and framed a comprehensive Environmental Protection Act in 1986. The Act authorized the Ministry of Environment and Forests to initiate action to close down any firm that violated the Act's restrictions regarding effluent discharge, and permitted individuals to initiate legal action against any one violating the Act. Ministry of Environment and Forests acts through Central

Pollution Control Board (CPCB) and State Pollution Control Boards (SPCBs). Since passage of the Act in 1986, the Ministry of Environment and Forests issued numerous closure notices. However, firms routinely obtain stays against closure, shifting decisions to the judiciary and stalling action for years. The Central Government has now formulated a National Environmental Policy 2006 for providing an overall direction to the sustainable development of the country.

It is worthwhile to mention that the Honourable Supreme Court and High Courts have played an active role in the enforcement of constitutional provisions and legislations relating to environmental protection. The fundamental right to life and personal liberty enshrined in Article 21 of the Constitution has been interpreted by the courts to include the right to pollution-free air and water (Curmally, 2002). There are many public litigation petitions filed by the individuals and NGOs before the courts. The courts have pronounced, invoking the provisions in the Constitution, that it is the fundamental duty of every Indian citizen to protect and improve the natural environment and to have compassion for living creatures. But, all these efforts have not yielded very satisfactory results. The reasons are many. Among them the technical capabilities of pollution control boards and their understaffed position, poor commitment at the corporate level, lack of comprehensive industrial zoning policy and dearth of relevant cost effective technologies are some of the important factors hindering environmental protection in the country. Also the research and development of the alternative sources of energy like solar, wind and tidal is slow and not cost effective so that they could be alternative to the fossil fuels. This would be the ultimate challenge in mitigating environmental degradation and climate change in the coming decades.

REFERENCES

Bhagat, R.B. (1997). Interrelationship between population and environment with special reference to greenhouse gases: the result from cross country study, *Indian Journal of Regional Science*, 29(2), 89-101.

Bongaarts, J. (1992). Population Growth and Global Warming. *Population and Development Review* 18, 299-319.

Boserup, E. (1965). *The Conditions of Agricultural Growth.* Chicago: Aldine.

Central Statistical Organisation (1999). *Compendium of Environment Statistics*. Ministry of Statistics and Programme Implementation, Govt. of India.

Curmally, A. (2002). Environment and rehabilitation. in M. Sebastian (ed.) *India Infrastructure Report 2002: Governance Issues for Commercialization*, New Delhi: Oxford University Press.

De Souza, Roger-Mark, William, J.S. and Meyerson, F.A.B. (2003). Critical links: population, health and the environment". *Population Bulletin*, 58 (3), Washington, DC: PRB.

Ehrlich, P. and Holdren, J. (1971). Impact of population growth. *Science*, 171, 1212-1217.

ICIMOD (2010). *Climate Change Vulnerability of Mountain Ecosystems in the Eastern Himalayas*, Kathmandu: International Centre for Integrated Mountain Development.

IPCC (2007). *Climate Change 2007: Summary for Policy Makers*, Fourth Assessment Report of the Intergovernmental Panel on Climate Change, Cambridge: Cambridge University Press.

Malthus, T.R. (1798). *First Essay on Population*. Reprinted, London: MacMillan.

Pebley, A.R. (1998). Demography and environment, *Demography*, 35 (4), 377-389.

Planning Commission (2002). *National Human Development Report-2001*. New Delhi: Government of India.

Population Reference Bureau (2002). *Making the Link: Population, Health and Environment*, Washington DC: PRB.

Ramesh, J. (2011). The Limits to Growth Revisited. Convocation Address, Mumbai: Tata Institute of Social Sciences.

The World Resources Institute, UNEP, UNDP and World Bank (1998). *World Resources, 1998-99*. Oxford University Press: New York.

UN (2009). *World Urbanization Prospects: The 2009 Revision*. New York: Population Division, Department of Economic and Social and Social Affairs, UN, 2002.

UNCED (1987). *Our Common Future*. United Nations Commission on Environment and Development, New York: Oxford University Press.

WHO (1997). *Health and Environment in Sustainable Development: Five Years after the Earth Summit*, Geneva: WHO.

4
Contextualizing Population in Developing Societies: Third World Perspective

Anand Prasad Mishra

Introduction

The debates on population and development have many dimensions which needs fresh orientation towards the explanation of nature in development studies. Most of the formulation on the question of population gets derivation through colonial mind set. Malthus' view on population is still dominating in most of our policies and academic discourses. However, the ground reality always contradicts the imposing Western philosophy on population and reminds us our tradition to consider people as assets for resource enhancement process and future growth. The geographical background is another factor which differentiates the tropical nations in a region for demanding more people to manage their resources. The vast chunk of arable land (more than 50 per cent arable land resources of World) in tropical countries developed agrarian economy through efforts of its large population size. Most of the developmental fruits became possible only through its people and livestocks energy. In India, milk and son (*doodh* and *poot*) were always considered as a symbol of pride. There is common blessing for younger people from older generation to increase their family size and livestocks, i.e. '*doodho nahao, pooto fallo.*' This belief explains the policies on population, especially for tropical economy. But the colonial impact deviate our policies and paradigms in development studies and gave space for colonial design in contemporary socio-economic structure. The Indian subcontinent has different background which offers its people to enjoy resources and also enhance its base for future. Susan Gole has tried to unravel the perception of others about India in her book, "India Within Ganges": *"India Intra Gangem"*—the Latin name used by Europeans for many centuries when they were referring to what we now call the "Indian subcontinent or South Asia." "There were many Indians, since the term was used for all known or unknown land beyond the river Indus, and even for some areas on the African continent. Hence, our India was known as India 'on this side of Ganges' to distinguish it from 'India beyond the Ganges' which meant mainland 'South-east Asia.' 'India superior' which might extend to northern China or in more general term *India Orientalis* which includes all the islands, from Australia to Japan and beyond." In India about the occupancy of the land by human population is quite evident and it considered one of the oldest settled countries of the world. Through its resources and rich heritages, India has been able to accommodate migration of people and never considered it problems for future.

In the process of her peopling the human settlements sprang up in the river valleys, forested areas, hill slopes and coastal plains. The size of the country provided isolated niches in which the economic and social development started taking place. The regional structure matured over long historical period. The people with superior technology spread fast and the people with low level of technology shrunk in limited hilly and forested terrains. The diversity of terrain and the distances between the different ecological niches translated itself in the development of so many dialects, languages, art forms, fairs, festivals and different economic activities. Human achievements can be gauged by the skills developed in the production of crops, livestock and specialization in producing fine cloths and artifacts learnt through the parental skills (Qureshi, 2010). Contrary to this European thinkers identified people as a disturbing element to nature and environment. Marsh (1965) observed, 'man is everywhere the disturbing agent. Wherever he plants his foots, the harmonies of nature are turned to discords.' Richards Grove (1990) found, 'the way in which European ideas of nature were transformed from the fifteenth century onwards by global trade and travel. These changes developed further with the rise of colonial trade and conquest.' The emerging global developmental philosophy, 'created a new kind of Eurocentric planetary consciousness' (Pratt, 1992). This Euro centred ideology has vital impact on our population policies and concern. Most of our policies on population derived its conceptual orientation from colonial routes, especially from the teaching of Malthus. But Malthus was never relevant for Europeans and only became champion for population in most of the 'Third World' countries, where colonial legacy is predominant. Moreover, the tropical reality contradicts colonial design of population debate and also put a fresh multiple ideas on the issues of population and development. The present paper is an attempt by author to incorporate genuine debate on population policy.

Colonial Legacy on Population Policies

Robert Malthus was man who became champion for colonial interest on the issues of population, human condition and development, when he published: *An Essay on the Principle of Population* in 1798. The background of Malthus is very much pertinent to mention here in process of any formulation on population policies. Thomas Robert Malthus life period was 1766-1834.This period also experienced the impact of French Revolution (1780-1789) and European Renaissance of eighteenth century. The European social dynamics and the process of historical development led their emphasis at lower level on Malthus' views and many prevailing aspects have contradicted the issues of population and development. In 1804, Robert Malthus accepted a post at East India Company's College at Haileybury, which prepared young man to rule over India, where he remained until his death. He had eight children. He occupied a position of comfort in an intellectual atmosphere of optimism, but was compelled by the rigour of his intellect to argue that nature condemned the bulk of humanity to live in the margin between barely enough and too little. Finally, Malthus' message as a teacher fell on the ears of future colonial bureaucrats who would guide or preside over the destinies of India (Ehrlich,

Ehrlich & Holdren, 1977). After independence, the colonial mind set has bigger impact on our policies for development and it neutralized our real thinking process.

The subaltern views put entirely different orientation on population which makes an alternative model in order to contextualize number of human beings to its development process. It also contradicts the Western concept of world development which always claim—the ultimate end of modernization and development. However, Western hegemony legitimize most of the Third World ideological position towards its liking end, i.e. industrialization and Malthus' view on population. This happened only due to colonial legacy on our ideological orientation. Though most of the Third World countries followed the philosophical orientation of Malthus by ignoring its historical and geographical legacy. While on other hand, most of the developed nation never considered population issues as per Malthus' view and in most time of historical development ruling class tried to oppose it. Developed nations are characterized by industrial economy and its social-cultural environment has different philosophical orientation towards life and development. Francis Place, the labour leader, for the first time tried to highlight the issues of population control in most of industrial complexes of Europe and to promote his thought among masses he wrote a treatise on population policies entitled *Illustrations and Proofs of the Principle of Population* in 1822. Francis' effort got recognition in Europe and United States of America and consequently various literature on issues of birth control was appeared during 1830-1870.The ongoing debate and civil societies activities tried to explore all possible methods for population control, especially in industrial workers, in developed nations. However, the religious leader and policy-makers used state machinery to neutralized the efforts for population control measures rather to support it. To suppress the movement for population control the then government passed Comstock Law in 1873 in Britain and United States of America (U.S.A.). This showed the dichotomy of European as well as American societies on the issues of population policy, which opposed Malthus view on population. The historical evidences proved that Malthus' teaching was only relevant in nations where East India Company had established its empire in tropical south. The relevancy of Malthus was only possible through colonial interest and nexus. To start any serious discussion on population policy, colonial linkages should be analyzed and its suitability to concern society and nation also examined in a scientific way.

Geography and Population

Geographical background of a space has association with prevailing condition of physical, cultural, economic and political dimension of landscape. James (1952) observed, "the phenomenon associated in a particular place is unsystematically related because they are produced by different processes. Geographer's particular mission is to study each process as it operates in particular places, and as it is actually modified in its action by the presence of other unsystematically related phenomenon grouped naturally together on the face of earth." This conceptual orientation gives insight that population may follow many aspects of geographical background and have strong linkages towards land, water and other resources of concerned regions. On this account, the perspective on population, especially in agrarian societies, derives its shape through prevailing conditions of diverse geographical setting. Geographical variation in availability of land, forest, water, energy and other resources have contributed to the issues of population and development in many ways. The world has 3.19 billion hectares of arable land, of which 1.67 billion hectares of land lies in tropical zone (more than 50 per cent). The warm temperate zone and temperate zone have arable land 0.50 and 0.90 billion hectares respectively. The distribution patterns of land resources are a significant component which put a condition before societies and human being to develop its perceptions on the issues of population ideology.

Hudson (1969) extended his study on settlement and observed as morphology seems to concentrate solely, on spatial properties almost the exclusion of process. In this context, the evolution and development of Indian villages and its people have revealed the ongoing process of economic development and dynamics of social organization that prevailed in a particular geographical background. In spite of having a good and substantial resources base most of the village population of Asia and Africa suffered from poverty and underdevelopment, especially in modern industrial era. These poor countries said to be poor because their peoples were the prisoners of tropical climates that discouraged effort and innovation, or because they lacked the mental faculties that could be taken for granted by Caucasians (Halter, 1975; Pratt, 1992). The colonial design of development studies considered this prevailing condition of tropical climate as a factor for development in village's economy and society. On this account, colonialism was a necessary consequence of this division between rich and poor, temperate and tropical, able and less able. It was the white man's burden to look after—and slowly improve—the native population of Africa, Asia, Latin America and Pacific. This was the West's duty towards the rest of the world, or to what said has called the West's others (Said, 1999).

The Western colonial ideology always seeks to logic, which indorsed the legitimacy of their design of exploitation and expansion. Contrary to colonial logic Indian people and its socio-economic systems have many glorious chapters in its path of development and command over the resource base. The Indo-Gangetic plains have long heritage of village economy and its prosperity through development process. It is due to alluvial filling, which is one continues and conformable series of fluviatile and sub-aerial deposits, mainly composed of unconsolidated beds of clay, sand, gravel and their mixture in varying proportion. Oldham estimated the depth of the alluvium between 4000-6000 metre. Alluvial filling on an average is 1300-1400 metre, decreasing gradually southward, a zone of over 8000 metre depth runs all along Himalaya (Singh, 1993). Actually, the geographical background and its resource base have given the shape of Indian villages and its people as an alternative to Western models of development. Under this prevailing situation colonial counterpart of Western thinkers always tried to blame the Indian village system as backward and uncivilized one. But it is not true. Now there is a need to identify and study the Indian village productive system and the role of residing people to enhance resource base and development level.

Resources and Population

Natural resources are "factor of production'. The production of resources evolved through the interaction of labour and other forms of capital, technology to nature. Zimmerman (1944) observed, resources are not pregiven natural factor around which society fashions its production processes but rather their usefulness is conditioned by cultural, historical, technological (and other knowledge) and geographical circumstances. In fact resources are produced by human efforts through their creative mind. Human beings and their cumulative knowledge in a integrated way frame the organizational network of population which have shape and reshape of social structure and its ideological orientation. The level and storage capacity of cumulative knowledge play a major role in the process of resource evolution and development. Population is a factor for all types of cumulative knowledge and it also a carrier which transfers this creative knowledge from generation to generation.

Considering this aspect on natural resources people have value only when employed as factors of production. Moreover, resource depletion caused by human being led to as contribution to national health; and also contributes to gross domestic product in national wealth. While conservation does not (Steers & Litz, 1993). At local level resources are viewed in the same way. Job and income are generated when natural resources are exploited as a factor of production.Unempolyment and poverty result from their preservation (Powar, 1996). Considering the nature of resources and its relation to population one may draw conclusion that more population ensured the maximum utilization of resources and multiple chances of involvement of factor of production of nature. Thinner population size restrict this production process. The nature has potentiality of infinite resources and it can only be endeavoured by human labour and skilful knowledge. In this process human labour, both physical and mental, can pay a vital role towards the expansion of resource base in a specific geographical region.

Resources are the engine of all modern development and inadequacy of this may minimize the scope of further development. The role of resources in development process is very much important and this were motivated the geographer and other social scientist to start their study for various aspects of resources and its relation to population. After Second World War, especially since 1970-80 a growing concern about population and space became centre for debate among geographers and social scientist to inquire the dynamics of population *versus* development. Simmons (1996) argued that each human civilization was sustained by a particular set of resources and technology for its exploitation. We may find it in case of emergence of industrialization over a period of time in a specific space with social formation. The industrialization based on fossil fuel energy put a turning point in the history of human relation and its role in global geopolitics. In nineteenth century industrialization took place through the exploitation of coal, then after the twentieth century had transformed into its resource base on oil and gas. Thus, the changing pattern and its dynamics of resources production by human efforts are a significant factor in shaping the relationship between human societies and natural environments (Bradshaw, 2005). On the issues of resources and population development, Rees (1985) observed, "resources are defined by society not by nature". The high income countries accounted for 16 per cent of world population, but it produced 80 per cent of global Gross National Product (GDP) and shown share of consumption for global energy about 51.8 per cent (World Bank, 1998, 1999). These emerging view on resources revealed the effectiveness of human being in process of production of nature as useful to society.

In contemporary development studies the debate on population and resources are quite interesting and its emerging dynamics attract researchers and policy makers to reshape future policy on population. On this account the conceptualization of resources in a scientific way is quite significant which need a fresh understanding on the issues. This debate can be categorically explained in following ways:

1. The concept of resources are purely functional, inseparable from human wants and human capabilities. This functional concept of resources has deep association to the dynamics of social progress. The physicist claims that quantity of matter and energy in the universe is constant, the social scientist replies that nothing is constant that everything is in flux (Peach and Constantin, 1972).
2. Knowledge is truly the mother of all other resources (Zimmerman, 1944). Through knowledge man has acquired control level and also expands their resource base from primitive man to relatively comfort and secure modern man. The evolution of knowledge—upgradation of creativity of

human being—has constructed many cultural and scientific changes in societies. This aspect may be useful in policy formulation on population.

3. Technology, which is most advance part of human creativity added value to the natural stuffs may consider as a factor for population dynamics. The progress in science and technology has made population as a incentive for furthering the growth and expansion of resource base in a specific productive region. In this context the quality and quantity of population obviously became the core agenda for any rational policy on population.
4. The evolution of resources, which exist in nature as neutral stuff became possible as a contribution of modern sciences. However, the impacts of sciences on human productiveness are cumulative—one invention leads to another. A new discovery enhances the value of an old. The inventor of today stands on the shoulders of his predecessors and they, in turn, reaped the benefits of past performances. These historical facts about the evolution of science and technology have put many new things before us on the issues of population policy.

Resource Potentialities and Human Capital

The nature has tremendous resource potential and on this account since early 1980s a growing number of economic geographers have put nature at the centre of their inquiry. In contemporary period number of debates is going on around the multiple dimension and dynamics of population, which explained the role of population in developing economy both in positive and negative way. Smith (1998) studied that nature and societies are inner-related from very start. In continuation to this debate Harvey (1974) wrote his famous thesis entitled, "Population Resources and Ideology of Sciences" and he criticized the concept of resource scarcity hypothesis and writing of Neo-Malthusianism of early 1970s. Harvey rejected the argument of over population. He said, "there are too many people in the world because the particular ends we have in view (together with the form of social organization which we have) and the materials available in nature, that we have the will and the way to use, are not sufficient to provide use with those things to which we are accustomed" (Harvey, 1974). The production of resources through interactive processes of human being depends on the level of accumulative knowledge of people. The historical evidences of population growth may open some hidden facts in contextualizing the agenda of population in developing societies. Prior to 1650 there are lack of historical evidences and reliable data of population size and its changing pattern. Agriculture was unknown before about 8000 B.C.; prior to that all human groups made their living by hunting gathering. No more than 52 million square kilometres (20 million square miles) of earth's total land area of some 150 million square kilometres (58 million square miles) could have been successfully utilized in this way by our early ancestors. from this and the population densities of hunting and gathering tribes of today, it has been estimated the total human population of 8000 B.C. was about 5 million people...The size of population at various times from the agricultural revolution to modern time have different pattern (Ehrlich, Ehrilch and Holdren, 1977).

The historical facts about population, development and availability of resources revealed that the expansion of resources in a specific mode of production mostly confined with level of human progress. Resources like civilization is a dynamic concept which with the upgradation of creativity and knowledge have explored new opportunity to human being. In this process the quantity and quality of population itself played a major role. Population as most of natural resources prevailed in geographical space in form of neutral stuff. By implication of creativity and labour it turned into resources and presently classified as 'human resources'. The condition of social organization, freedom to individual, and opportunity to share in development processes are the some basic prerequisite for proper expansion of resource base in geographical space. To understand the debate on resources and population, Table 4.1 may through some light.

Table 4.1 revealed the potential availability of various energy resources on our earth system. However, the contemporary energy crisis is serious which on many account warned us that all these have come to surface by increasing population pressure. After 1970, many hypotheses came to academic debates and blamed most of the Third World population *versus* development for present day crisis. The neo-colonial perspectives even today tried to dictate developing nation on the issues of independent development strategy. The geo-spatial position of tropical nation is another factor which never be considered in formulating policy for R&D in energy sector. Most of the R&D on energy sector devoted for coal, gas and petrol and it accordingly to geo-spatial location of Western world fulfilled their interest.The case of solar and other non-conventional energy resources for R & D have always cornered by Western scientific world. The tropical nations of world have more than 300 days of solar ray's exposer (Table 4.2) which may be exploited for our energy requirement but it never happen so due to lack of human capital in concern region. In same ways other resources are found most oftenly in tropical areas but its

Table 4.1: Energy Flow and Storage in the Physical World

Sl. No.	*Storage*	*Energy of Power Trillion MJ*
1.	Energy released in a large volcanic eruption	100
2.	Chemical energy stored in all living organisms	30,000
3.	Energy released in a large earthquake	100,000
4.	Chemical energy stored in dead organic matter	100,000
5.	Heat stored in atmosphere	1,300,000
6.	Kinetic energy of Earth's rotation on its axis	250,000,000,000
	Flow	*Million MW*
1.	Tide	3
2.	Heat flow from Earth's interior	32
3.	Conversion of sunlight to chemical energy in photosynthesis	100
4.	Conversion of sunlight to energy of motion of atmosphere	1,000
5.	Sunlight striking top of atmosphere	172,000

Source: Ehrlich, Ehrilch and Holdren, 1977.

Table 4.2: Total Annual Insolation at Various Latitudes

Latitude	*Thermal*	*Latitude*	*Thermal day*
0°	365.0	50°	249.7
10°	360.2	60°	207.8
20°	345.2	70°	173.0
30°	321.0	80°	156.3
40°	288.5	90°	151.6

Source: Finch and Trewartha, 1949.

utilization confined to minimum level. Actually, the presence of resources in any specific region is an outcome of individual capacity building processes through social, cultural and political freedom and evolution of knowledge activities in space and time.

People as Useful Stuff

Resources are produced by people through interaction with nature. All mechanical revolution emerged in evolutionary way with the interaction of cumulative knowledge of human being. James Watt discovered steam engine in 1776 which revolutionaries the capacity building level in contemporary world. This invention remade multiple resource opportunity to society, i.e. coal, petroleum and gas. Peach and Constantin (1972) explained this in following words:

> "The old machinery, not only of material civilization but of most refined cultural patterns as well, was designed to meet the great emergency of an ever-growing population pushing with increased vigour and rising impatience against the subsistence levels of the earth. Man faced a chronic condition of scarcity of food, of land, of everything. The inevitableness of scarcity has yielded to the possibility of plenty; at times even overwhelming superabundance threatens. Therefore, the machinery of social and economic institutions had to be adjusted to new dangers. Philosophy, which had long served as a mean of escape from the limitations of human abilities, has had to make an about face to meet the needs of the Prometheus who feel his powers as clearly as the old adams realized his limitations. The positive attitude towards population increase, which was demanded in the agriculture world because of the need of labour and markets, had to be reversed to meet the emergence of technological unemployment and the man-replacing power of machines and engines. Agriculture, for millennium the backbone of civilized life lost its primacy and was forced to take its place behind mining, manufacturing, commerce and finance. Thus the mechanical revolution necessarily involved a social revolution in every field of human thought and action."

The above observation on issues of population, resources and development highlight about constraints and possibilities of human population in consuming and exploring the horizon of resources. Therefore, human population itself presents its faces both in a positive and negative way. The positive side of human number depends on the level of development opportunities for enhancement of people creativity which mostly based on freedom, social harmony, and status of cumulative knowledge inherited from generation to generation. In this way, human being like other natural resources prevailed on earth surface in form of neutral stuff and its usefulness became possible through the expansion of creativity and R&D in specific social formation and organizations.

The evolution of science and technology in Europe occurred through social and cultural revolution which established the rational socio-political order in then society. These historical phenomenons have contributed towards the process of evolution for knowledge and resource expansion. Prior to that European nations and its people have suffered from resource scarcity and problems of over population; it also produced many negative consequences among people and society. In continuation to its development and scientific progress human being with their labour (both physical and mental) had contributed many new things to their nation and society. In this background the European population took active part in their nation-state building process and also proved its utility for generation of resources and other upgradation in social space. On this account population itself has all grounds to be classified as resources and its level of development depend on the level of expansion and production through various means. The another aspects of population as resource were discussed in International Conference on Population and Development (United Nations, 1994) in Cairo, and known as Cairo declaration which considered population as depletable and non-renewable natural resources, which need a conservation policy for its proper use. By definition and others parameter for resources population has fulfilled all requirements to become natural resources. This scientific interpretation put logic before us to rethink about Malthusian views on population. The concept of overpopulation is a product of underdevelopment.

Third World Scenario

Most of Third world countries suffered from the problems of stagnancy in economy, backwardness in knowledge and social structure, territorial disputes and other unscientific cultural elements in society. All these prevailing conditions create multiple problems in various layers of social configuration. On the other hand geographically Third World countries are in tropical zone which has substantial resource base in its natural environment. In spite of having rich resource base most of Third World countries classified as a region of poverty, unemployment, underdeveloped, environmently degraded and most deprived area of globe. Colonial mind set club with Malthusian Philosophy of population blamed that all kinds of deprivation have occurred through high population pressure. It also ignored the legacy of colonial exploitation. The most important feature of the unevenness of population growth across the world is the disproportionate concentration of growth in Less Developed Regions (LDRs). Over the half century, these regions have accounted for fully 89.00 per cent of the World's 3.4 billion extra people (Champion, 2005). Moreover, as per United Nations Development Programmes (UNDP), Human Development Report 2009 (United Nations, 2009), the Human Development Index (HDI) for India for 2007 was 0.612 on the basis of which India is ranked 134 out of 182 countries of the World. The HDI for different Third World countries revealed the level of development in context to its population and economic growth. However, the growth of a country influenced by its physical resources (land, mineral and other raw material) and human resources (number of people and their level of skills).

Table 4.3 reveals the patterns of economic development in most populous countries of world which highlights many new things about population *versus* development. In most of the underdeveloped countries of Africa, Latin America and Asia number of people are less and also they have good resource base but their level of development is quite unsatisfactory. The impact of colonization may be considered as major consequences towards structural problem in its socio-cultural networking, which minimize the capacity building process of human skill. Human labour as tools for production of nature requires favourable condition for its maximum output. The needed

Table 4.3: The Ten Most and Selected Least Populated Countries and Their Per Capita Income, 2000

Sl. No.	*Most Populous*	*Population (million)*	*GNP per capita (U.S. $)*	*Least Populous*	*Population (thousands)*	*GNP per capita (U.S. $)*
1.	China	1261	840	Saint Kitts-Nevis	41	6,660
2.	India	1016	460	Antigua & Barbuda	68	9,190
3.	United States	282	34,260	Dominica	73	3,260
4.	Indonesia	210	570	Seychelles	81	7,310
5.	Brazil	170	3,570	Kiribati	91	950
6.	Russia	146	1,660	Grenada	98	3,520
7.	Pakistan	138	470	Tonga	100	1,660
8.	Bangladesh	130	380	Saint Vincent	115	2,690
9.	Nigeria	127	260	Micronesia	118	2,110
10.	Japan	127	34,210	Sao Tome	149	290

Source: World Development Report, 2011 (New York: Oxford University Press), World Development Indicators (Washington, D.C.).

favourable environment could be made through freedom to human being. Marx explained it as, "the property therefore which labour-power in action, living labour possesses of preserving value, at the same time that it adds it, is a gift of nature which cost the labourer nothing, but which is very advantageous to the capitalist" (Marx, 1984). In this context, man and their skill and labour are the primary factor for resource expansion, while money stand as secondary.The Third World population deprived in many ways by colonial exploitative nexus, and their human resource development processes got less chance to achieve its optimum level. The poverty of Third World is not due to its population but colonial exploitation and the presence of independent initiatives are the some major factors, which may be considered in any policy discussion for population.

Towards Rational Approach

Malthus spectre of population has been haunting world even today since its origin of more than two hundred years, and population growth became many times. The science and technology are the parallel aspects for population policy and any debate on the issues of population this contemporary achievement in knowledge should be taken into account by policy makers. To start discussion on population policy we should more careful because it reflects the whole range of our need for socio-economic development. Jain (1972) says that one may be inclined to treat policies of socio-economic development within the ambit of population policy, as population is all pervading element in all spheres of developmental activities. This approach gives us an independent thinking about Third World development strategy and also has made ground for anti-colonial policy framework. Many Malthusian aspects of population are not fit to present day realities which need a fresh understanding in formulating the policy of Third World development strategy. The different parts of the world have varying nature of geographical phenomenon and natural resource base in its surroundings. These geographical backgrounds get reason to be incorporated in any genuine policy for development so that the benefit should reach to the people. Population policy in context to above-mentioned debate put many questions on the relevancy of colonial mindset about our human population. It required rediscovery of our routes and thinking prior to colonial penetration in most of tropical nation, including Indian subcontinent. On the basis of previous discussion, following facts should be considered by researchers, policy makers, and leader of society:

1. The geographical background and natural resource base of tropical nations have different reasons to replace colonial ideology on the issues of population and development. The Malthusian beliefs should be tested on the ground of historical reality.
2. Prior to colonial rule most of the tropical countries, especially Indian subcontinent, considered population as an incentive for their agrarian development. This approach evolved through natural processes during the course of historical development. Its suitability was checked at various levels in a different mode of productions in its various stages of social formation.
3. Colonial intervention in most of Third World countries have diverted its genuine path of development and natural course of history, these processes have made dent on historical times and social formation. It helped the expansion of colonial ideology in a bigger way and irrelevant Malthus in Europe became relevant in most of the colonized nations of Third World.
4. In contemporary time population as natural resources have proved its utility in globally competitive economic development. The new emerging phenomenon has further justified for independent policies on population, which stands for genuiness, local suitability, pro-people and against misconception posed by colonial past.

Lastly, we may conclude that the issues of population should be formulated with geographical background, cultural heritage, and naturally evolved productive system without any colonial exploitative nexus.

REFERENCES

Harvey, D. (1974). Population, resources and the ideology of science. *Economic Geography*, 50, 256-77.

Zimmerman, E.W. (1944). Resources—an evolving concept. *Transaction*, Academy of Sciences, 157-165.

Grove, R. (1990). Colonial conservation, ecological hegemony and popular resistance: towards global synthesis. In J.M. Mckenzie (ed.), *Imperialism and Natural World*, Manchester University Press: Manchester, 15-50.

Halter, J. (1975). *Outcasts from Evolution: Scientific Attitudes of Racial Inferiority, 1859-1990*. New York: McGraw-Hill.

Hudson, J.C. (1969). A location theory for rural settlement. *Annals*, A.A.G. 59: June 1969, 366.

Jain (1972). Cited from P.K. Chaubey, *Population Policy for India* (2001). Knishika Publication: New Delhi, p. 31.

James, P.E. (1952). Towards a further understanding of the regional concept. *Annals of the Association of American Geographers*: 4, 197-198.

Marsh, G.P. (1965). *Man and Nature or Physical Geography*. As modified by Human action (reprinted), Harvard University Press, p. 36.

Marx, K. (1984). *Capital.* Vol. 1, Moscow: Progress Publisher, p. 200.

Bradshaw, M. (2005). Resources and Development. In *Human Geography*, Issues for 21st Century, Peter Daniel *et al.* (ed.), Delhi: Pearson, p. 244.

Peach, W.N. and Constantin, J.A. (1972). *Zimmerman's World Resources and Industries*, New York: Harper and Row Publisher, p. 60.

Powar, T.M. (1996). *Lost Landscape and Failed Economies*, Washington: Island Press.

Pratt, M.L. (1992). *Imperial Eyes: Travel Writing and Transculturalism*, London: Routledge, p. 39.

Qureshi, M.H. (2010). Keynote Address in national seminar on "Population Development and Environment: Issues and Challenges", Deptt. of Geography, B.H.U., Varanasi, dated 18-20 February 2010.

Rees, J. (1985). *Natural Resources: Allocation, Economics and Policy*. New York: Routledge.

Said, E. (1979). *Orientalism*, New York: Panthn.

Simmons, I.G. (1996), *Changing to Face of Earth: Culture, Environment, History*. Blackwell, p. 28.

Singh, R.L. (1973). *India: A Regional Geography*, USB Publishers, 129-130.

Smith, N. (1998). Nature at the millennium: production and re-enchantment. In B. Braun and N. Castree (eds.), *Remaking Reality*, New York: Routledge, 271-85.

Steers, A. and Litz, E. (1993). Measuring environmentally sustainable development. *Finance Development* 30(4), 20-23

Champion, T. (2005). Demographic transactions, In J. Sidaway, P. Daniels, M. Bradshaw and D. Shaw (eds.), *Human Geography, Issues for 21st Century*, New Delhi: Pearson, p. 213.

United Nations (1994). *United Nations International Conference on Population and Development* (ICPD), Cairo.

World Bank (1998, 1999, 2008). *World Development Report,* New York: Oxford University Press.

United Nations (2007, 2009, 2010). *Human Development Report*, New York: Oxford University Press.

Ehrlich, P.R., Ehrlich, A.H. and Holdren, J.P. (1977). *Eco-Science-Population, Resource and Environment*, San Francisco, USA: W.H. Freeman and Company.

Finch, J.C. and Trewartha, G.T. (1949). *Elements of Weather and Climate*, London: Prentice-Hall.

5

Climate Change and Sustainable Development: Issues and Challenges in India

Ajay Kumar, Bijay Kumar Das, Rajneesh Kumar and *Fulena Rajak*

Introduction

The climate change issue is part of the larger challenge of sustainable development. Sustainable development comprises social, economic, and environmental dimensions. Climate change modifies all these dimensions and therefore, alters the potential development pathways. Climate change exacerbates concerns about agricultural production and food security worldwide. At global and regional scales, food security is prominent among the human activities. According to the IPCC Third Assessment Report (IPCC, 2001), climate change is already happening, and will continue to happen even if global greenhouse gas emissions are curtailed. Many studies document the implications of climate change for agriculture and pose a reasonable concern that climate change is a threat to poverty and sustainable development, especially in developing countries. The greatest impact is upon the poor in developing countries—the people most dependent on natural resources for food, shelter, medicine, income and employment. This climate change challenges sustainable development in India who will be the most populated nation by the year 2030 surpassing China and by then in the era of climate change and degraded environment making India a food secure nation will be a very tough task. The concept of sustainable development is based on an awareness of environmental risk particularly the phenomenon of global warming. The sustainable development seeks to reconcile environmental protection, economic development and social development. It also goes hand in hand with the basic principles of environmental laws which include precaution, prevention, remedy at source, use of best available technology and polluter pays.

Issues in Climate Change and Sustainable Development

Climate Change and Global Warming

One of the greatest challenges that the world is facing today is climate change. Climate change is the variation in the earth's global climates over time. It involves changes in the variability or average state of the atmosphere over durations ranging from decades to millions of years. These changes can be caused by dynamic process on earth, external forces including variations in sunlight intensity and more recently by human activities. Human influences can be by increase in CO_2 level due to combustion of fossil fuels, aerosols, cement manufacture etc. Other factors like ozone depletion, animal agriculture and deforestation also change climate. The effect of climate change can be found on among other things, on rising sea level that may accelerate coastal erosion, on increasing temperature, on increase in intensity of natural disaster, and very importantly on vector borne diseases. There has been an increasing trend in the annual mean temperature in India. In recent decades the east coast has been experiencing fewer rainy days while the north-west has been experiencing heavy summer monsoon. There have also been some extreme climatic events like heat wave, intense rain, floods and droughts in India. Researchers have documented the increase in frequency of hot days and multiple-day heat waves in the past century. There has been record rainfall in Mumbai, India on 26 to 27 July, 2005, which led to loss of large numbers of lives. Consecutive droughts between 2000 and 2002 caused crop failures, mass starvation and affected millions of people in Odisha. Also, increased water stress poses to be a major problem for India. Accelerated glacier melt is likely to cause increase in the number and severity of glacial melt-related floods, slope destabilisation and a decrease in river flows as glaciers recede. The researchers have predicted that with the current trend in the melt of glaciers, the Ganga, Indus, Brahmaputra and other rivers could likely become seasonal rivers in the near future and affect the lives of people residing around them. It is likely that climate change will hamper sustainable development of India as it increases the pressures on natural resources and the environment associated with rapid urbanization, industrialization and economic development. The Intergovernmental Panel on Climate Change (IPCC) projects that the global mean temperature may increase between 1.4 and 5.8°C by 2100. The Intergovernmental Panel on Climate Change (IPCC) has projected that, if current greenhouse gas emission trends continue, the mean global surface temperature will rise from 1 to 3.5 degrees Celsius in the next century. The panel's best estimate scenario projects a sea-level rise of 15 to 95cm by 2100. The climate change issue is part of the larger challenge of sustainable development.

Results of Climate Change, including:

- increased air and water temperatures;
- degradation of freshwater fish habitat;
- diminished terrestrial biodiversity;
- increased bleaching and die-off of coral reefs;
- increased frequency and intensity of heavy downpours;
- a rise in sea level;
- reduced snow cover, glaciers, permafrost, and sea ice;
- reduced water supply in some regions;
- a longer ice-free period on lakes and rivers;
- lengthening of the growing season; and
- increased water vapour in the atmosphere.

The Greenhouse Effects

The earth's atmosphere is a gaseous layer consisting mainly of N_2 (78 per cent by volume) and O_2 (21 per cent). Various other gases present only in small quantities, however important due to the contribution most of them make to the greenhouse effect. Solar radiation reaching the earth is partly re-emitting by the earth as infrared rays is absorbed by greenhouse gases (greenhouse effect), which is main cause for global warming. Of the six GHGs, CO_2 accounted for 63 per cent, CH_4 24 per cent, N_2O 10 per cent, and the other gases the remaining 3 per cent of the carbon equivalent emissions in 2000. Thus in addition to CO_2, global mitigation efforts need to focus on the two largest and rapidly increasing GHGs. The combustion of fossil fuels and other human activities are the primary reasons for increased concentrations of CO_2 and other greenhouse gases. Between 1990 and 1999, an estimated 6.3 GtC/year was released due to the combustion of fossil fuels, and another 1.6 GtC/year was released due to the burning of forest vegetation. This was offset by the absorption of 2.3 GtC/year each by growing vegetation and the oceans. This left a balance of 3.3 GtC/year in the atmosphere. Controlling the release of greenhouse gases from fossil fuel combustion, land-use change and the burning of vegetation are therefore obvious opportunities for reducing greenhouse gas emissions. Reducing greenhouse gas emissions can lessen the projected rate and magnitude of warming and sea level rise. The greater the reductions in emissions and the earlier they are introduced, the smaller and slower the projected warming and the rise in sea levels. Future climate change is thus determined by historic, current and future emissions.

Sustainable Development

In 1987, a report entitled "Our Common Future" produced by the then Norwegian Prime Minister, Gro Harlem Brundtland for discussion at the 42nd UN Congress introduced the notion of sustainable development. The definition of sustainable development given by Brundtland was: "Sustainable development is development which meets the needs of the present without compromising the ability of future generation to meet own needs." It emphasized that global environmental problems are rooted to a large extent in the impoverishment of the greater part of the world's population. At the 1992 Rio Earth Summit, heads of states committed their notion to exploring ways of achieving "development which fulfils current needs without compromising the capacity of future generations to fulfil theirs." This concept of sustainable development is based on the three principles:

- Consideration of the "whole life cycle" of material.
- Development of use of natural raw material and renewable energy sources.
- Reduction in the materials and energy used in raw material extraction, product use and destruction or recycling of waste.

Challenges in Climate Change and Sustainable Development

Population Growth

An exploding human population burning more and more fossil fuels now has a greater effect on the climate than natural mechanisms. Greenhouse gas emissions due to burning of fossil fuels are closely linked to both population growth and development. The world's population has grown from around 1.5 billion in 1900 to 6 billion in 2000. The India's population has grown from around 238 million in 1901 to 1027 billion in 2001. The 2050 estimate of India' population is 1.5-1.8 billion. According to current rate by year 2030 India will surpass China in population statistics. By then India's greatest challenge will be to make it a food secure nation. The population growth has resulted in pressure on land, fragmentation of land holding, collapsing fisheries, shrinking forests, rising temperatures, loss of plant and animal species. Global warming due to increasing use of fossil fuels could have serious effects on the populous coastal regions in developing countries, their food production and essential water supplies. The United Nations Conference on Environment and Development (1992) acknowledged population growth, rising income levels, changing technologies, increasing consumption pattern will all have adverse impact on environment. This is one of the major challenges for sustainable development.

Poverty Eradication and Sustainable Livelihoods

Poverty and a degraded environment are closely inter-related, especially where people depend for their livelihoods primarily on the natural resource base of their immediate environment. Over seventy per cent of India's population still live in rural areas. Agriculture is the largest and one of the most important sectors of the rural economy and contributes both to economic growth and employment. Its contribution to the Gross Domestic Product has declined over the last five decades but agriculture still remains the source of livelihood for over 70 per cent of the country's population. A large proportion of the rural workforce is small and consists of marginal farmers and landless agricultural labourers. There is substantial under-employment among these people; both wages and productivity are low. These in turn result in poverty; it is estimated that 320 million people are still living below the poverty line in rural India. The survival needs of the poor force them to continue to degrade an already degraded environment. Removal of poverty is therefore a prerequisite for the protection of the environment. National Rural Employment Guarantee Scheme (NREGS)—A flagship programme of Government of India is the most ambitious poverty reduction programme designed and linked with the most critical problems of this time, namely extreme poverty and climate change. It provides guaranteed 100 days of wage employment in a financial year to every rural household and creates sustainable rural livelihood through rejuvenating natural resources base—*Jal* (water), *Jamin* (land) and *Jungle* (Forests) and creation of durable assets through water conservation, land development and afforestation. Rural Development Ministry of India has taken initiative for allowing farm ponds to be created on the lands of farmers using the labour supplied from the National Rural Employment Guarantee Scheme (NREGS). The revised NREGA guidelines have provision of introduction of 'A pond in every farm'. This concept is seen as an extension of earlier concept of 'A pond in every village' practiced in the rural settlement. This concept will harvest the rain water in the farm, use it for watering the crop after monsoon, and recharge the ground water with fishery as the other benefit. Some of the lists of activities/works that can be done under NREGA guidelines are listed below:

- Water conservation activities like digging of pond/percolation tank and desilting of pond.
- Small check dams and other water harvesting structures.
- Drainage to drain out extra water in water logged/flood prone areas.

Agriculture and Food Security

Agriculture is strongly dependent on water resources and climatic conditions, particularly in regions of the world that are particularly sensitive to climatic hazards, such as Africa, South and Central Asia. India is extremely vulnerable to changes in environmental factors. It is especially the case in countries where technological buffering to droughts and floods is less advanced, and where the main physical factors affecting production (soils, terrain, and climate) are less suited to farming. Crop production is consequently extremely sensitive to large year-to-year weather fluctuations. Crop diseases or pest infestations are also weather-dependent, and tend to cause more damages in countries with lower technological levels. Simulations using dynamic crop models indicate a decrease in yield of crops as temperature increases in different parts of India. However, this is offset by an increase in CO_2 at moderate rise in temperature and at higher warming; negative impact on crop productivity is projected due to reduced crop durations. Estimates of the effects of climate change on crop yields are predominantly negative for the tropics, even when adaptation and direct effects of CO_2 on plant processes are taken into consideration. This is one of the major challenges for sustainable development.

Increasing Consumption Patterns

The increasing purchasing power and supply driven market is stressing the resource base of India further. In several areas, desirable limits and standards for consumption need to be established and applied through appropriate mechanisms including education, incentives and legislation. Several traditional practices that are sustainable and environment friendly continue to be a regular part of the lives of people in developing countries. These need to be encouraged rather than replaced by more 'modern' but unsustainable practices and technologies. Development decisions regarding technology and infrastructure are a major determinant of consumption patterns. It is, therefore, important to evaluate and make development decisions which structurally lead to a more sustainable society. Technologies exist through which substantial reduction in consumption of resources is possible. Efforts to identify, evaluate, introduce and use these technologies must be made. Subsidies often lead to wasteful and unsustainable consumption by distorting the value of a resource. All pricing mechanisms must be evaluated from a sustainable development point of view. It is important to counter this through education and public awareness. This is one of the major challenges for sustainable development.

Providing Access to Water to Meet Basic Needs

Access to safe drinking water remains an urgent necessity, as 30 per cent of urban and 90 per cent of rural households still depend completely on untreated surface or groundwater. In India, water demand substantially exceeds sustainable water supply. In India, water withdrawal is estimated to be twice the rate of aquifer recharge; as a result water tables are falling by one to three metres every year; tapping deeper aquifers have resulted in larger population groups being exposed to newer health hazards such as high fluoride or arsenic content in drinking water. The hydrological cycle is likely to be altered and the severity of droughts and intensity of floods in various parts of India is likely to increase. Further, a general reduction in the quantity of available run-off is predicted. The widespread retreat of glaciers and icecaps in the twenty-first century will also lead to higher surface temperatures on land and increasing water stress. By 2025, as much as two-thirds of the world population, much of it in the developing world, may be subjected to moderate to high water stress. This is one of the major challenges for sustainable development.

Protection and Management of Natural Resource Base

The higher population density and rate of population growth has created high burden to the natural resource base. The climate change and global warming also depleted the natural resource base. This is one of the major challenges for sustainable development for India. The integration of agriculture with land and water management, and with ecosystem conservation is essential for both environmental sustainability and agricultural production. An environmental perspective must guide the evaluation of all development projects, recognizing the role of natural resources in local livelihoods. This recognition must be informed by a comprehensive understanding of the perceptions and opinions of local people about their stakes in the resource base. To ensure the sustainability of the natural resource base, the recognition of all stakeholders in it and their roles in its protection and management is essential. There is need to establish well-defined and enforceable rights (including customary rights) and security of tenure, and to ensure equal access to land, water and other natural and biological resources. It should be ensured that this applies, in particular, to indigenous communities, women and other disadvantaged groups living in poverty. Water governance arrangements should protect ecosystems and preserve or restore the ecological integrity of all natural water bodies and their catchments. This will maintain the wide range of ecological services that healthy ecosystems provide and the livelihoods that depend upon them. Biomass is, and will continue for a long time to be, a major source of fuel and energy, especially for the rural poor. Recognizing this fact, appropriate mechanisms must be evolved to make such consumption of biomass sustainable, through both resource management and the promotion of efficient and minimally polluting technologies, and technologies which will progressively reduce the pressures on biomass, which cause environmental degradation. The traditional approaches to natural resource management, such as sacred groves and ponds, water harvesting and management systems, etc., should be revived by creating institutional mechanisms which recapture the ecological wisdom and the spirit of community management inherent in those systems (Table 5.1).

Table 5.1: Energy Demand Projections (in Mtoe) for 8 per cent GDP Growth

Year	*Mtoe*
2011-12	677
2016-17	861
2021-22	1,082
2026-27	1,417
2031-32	1,818

Source: Energy Policy Report, Planning Commission, India, 2006.

Provision of Energy without Environmental Degradation

Climate change is a major challenge where world energy needs could grow threefold over the next 50 years. In order to maintain an average GDP growth of 8 per cent growth rate, proportionate growth in Energy sector is required. Under present conditions, 6 per cent increase in India's Gross Domestic Product (GDP) would impose an increased demand of 9 per cent dependency on its energy sector. For a growth rate of more than 8 per cent, a higher proportionate growth in Energy sector is required. Coal contributes 63 per cent of total primary energy production in India. Energy sector alone contributes 61 per cent of GHG emissions approximately. The high percentage of the use of fossil fuels rapidly diminishes their available resources, and cause environmental and health problems. The requirement of growth in further energy sector is major challenges for sustainable development. Renewable energy could mitigate global warming and climate change because it have the potential to reduce carbon emissions that will reduce and control the limit of atmospheric concentration of carbon dioxide as well as fulfil the requirement of energy to maintain the development spirit without damaging our environment.

Ensure Healthy Environments

Human health in its broadest sense of physical, mental and spiritual well-being mainly depends on the access to a healthy environment. For a healthy, productive and fulfilling life every individual should have the physical and economic access to a balanced diet, safe drinking water, clean air, sanitation, environmental hygiene, primary health care and education. Traditional diseases such as malaria and cholera, caused by unsafe drinking water and lack of environmental hygiene, have not yet been controlled. The strong relationship between health and the state of the environment in developing countries is becoming increasingly evident. This calls for greater emphasis on preventive and social medicine, and on research in both occupational health and epidemiology. Because of the close link between health and environment, there is need greater integration between the ministries of Health and Environment, and effective coordination and cooperation between them. Basic health and educational facilities in developing countries need to be strengthened. The role of public health services must give preventive health care equal emphasis as curative health care. People should be empowered through education and awareness to participate in managing preventive health care related to environmental sanitation and hygiene. India is repositories of a rich tradition of natural resource-based health care. This is under threat, on the one hand from modern mainstream medicine, and on the other from the degradation of the natural resource base. Traditional medicine in combination with modern medicine must be promoted while ensuring conservation of the resource base.

Conclusion

India is a large developing country with nearly two-thirds of the population depending directly on the climate sensitive sectors such as agriculture, fisheries and forests. The projected climate change under various scenarios is likely to have implications on food production, water supply, biodiversity and livelihoods. The integration of agriculture with land and water management, and with ecosystem conservation is essential for both environmental sustainability and agricultural production. Environmental education and community participation are crucial for the implementation of climate change policies within an agenda for sustainable development. International systems of compensatory taxes could be explored as sources of funds to support sustainable development objectives in the developing countries, in synergy with the climate change response strategies. An environmental perspectives must guide the evaluation of all development projects, recognizing the role of natural resources in local livelihoods. India should have significant stake in scientific advancement as well as an international understanding to promote mitigation and adaptation. This requires improved scientific understanding, capacity building, networking and broad consultation processes. Strengthening governance for effective management of resources requires participation by all stakeholders at the local, national and international levels and it will be keys of success for sustainable development. Lastly, we have to decide whether development means material comfort or whether development means peace, prosperity and happiness.

REFERENCES

IPCC (2001). *Climate Change 2001: Impacts, Adaptation and Vulnerability.* Contribution of Working Group II to the Third Assessment Report of the Intergovernmental Panel on Climate Change, Cambridge University Press: Cambridge.

Our Common Future, Report of the World Commission on Environment and Development (WCED), Oxford University Press, New York, 1987.

United Nations Conference on Environment and Development (UNCED), Rio de Janeiro, Brazil, 3-14 June 1992

Govt. of India (2006). *Energy Policy Report.* Planning Commission, New Delhi: GoI.

water supply, sewerage connection and door to door collection of solid waste should be ensured. Private agencies may be invited who may provide above services on 'pay and use' basis.

(b) A hundred per cent households be connected with sewerage network and have toilet facility with regular water supply to avoid defecating in the open and 100 per cent waste water generated be treated.

(c) Regarding solid waste management, waste collection should cover 100 per cent in closed trucks and be collected mechanically instead of 70 to 90 per cent collection in open trucks and manually.

8. The problem of transport and roads be tackled by making provision of higher dependence on public transport (presently only 22 per cent use) and use of private vehicles in the central areas and on the busy thoroughfares be restricted. This will create a condition of lifestyle dependent on sharing and sustenance. (Bhatia, 2011). To avoid traffic jams, road network and road width be expanded and separate lanes for footpath and slow and fast moving vehicles be made. Encroachments be strictly prohibited/penalized. Sufficient provisions be made for flyovers, bridges etc. at bottlenecks.
9. As an alternative mode of transport opening of sea in Mumbai and River Ganges in Kolkata along with multiple lane bridges be considered.
10. The provision of multistoreyed car parks, road-on-road and flyovers on intersection/bottlenecks/congested roads and extensions of railway track wherever possible be made.
11. To fight pollution, provision of lead free petrol/CNG/LPG and privatisation of infrastructure and services on pay and use policy will improve the situation.

Keeping in view the vital role played by cities in socio-economic transformation of the country and in global market Government of India has launched several programmes to overcome some of the problems raised above in the study to achieve balanced and sustainable development. The earlier Government scheme for infrastructure development in the mega cities introduced in 1993-94 to cover a wide

Part II
Migration and Urbanization

6
Migration in India: Patterns, Poverty, Underlying Phenomena, Spatial-Economic Processes, and Development Policies

Shekhar Mukherji

"Imagine! There are hundreds of millions of very poor, landless peasants, workers and unemployed, denied of any education, hope and aspirations for betterment, chained to fatalism, submissiveness, and traditionalism, designedly and deliberately pauperized through centuries of exploitation, murderous invasions, colonization and induced underdevelopment, ceaselessly struggling throughout their lives for barest sustenance! Imagine! Their emaciated faces, moist eyes, and tattered clothes! And, you will begin to hear the roaring sound of their struggle for survival and silent suffering! Perhaps then, you may begin to understand their real situations. No sordid statistics can tell their poignantly sad stories of dismal existence!!" (Mukherji, 1975)

Introduction

In spite of massive explosion in the number of the poor in India, about 430 million in 2005 (NSSO, 2006; World Bank, 2008), and still increasing, yet the Government of India do not consider the task of liquidating poverty as the topmost priority. Nor are the geographers, demographers, and population specialists yet socially committed to poverty reduction, nor even interested in unravelling the causal links between "poverty" and "migration-urbanization". Unless we soulfully do so, neither can we develop the correct theoretical perspective of this complex system, nor can develop right planning prescriptions. This study attempts to do just these. Unlike in the West, where rural-to-urban migration and urbanization are associated with a vertical shift in labour force from agricultural sector to urbanized-industrial sector, the migration in India mainly occurs from poor rural "peasant" sector to poor urban "informal" sectors. This is a typical spatial manifestation of poverty, underdevelopment, and spatial disorganization of the economy of India, which occurred due to past colonization and its adverse consequences on its space economy (Mukherji, 1981). These are symptoms of underdevelopment, not development, and lead to further underdevelopment. Traditional perspectives of urbanization and migration, as developed in the West, and religiously followed by conventional social scientists in the country, cannot explain true processes of migration and urbanization in India. Hence, this study presents a different story.

Purposes

- To provide an overall scenario of ongoing trends of migration and urbanization in India to highlight main problems;
- To unfold the underlying socio-economic-spatial processes of such poverty-induced migration and of lopsided urbanization in India, in terms of spatial disorganization, polarized industrialization, and related factors to evolve clear thinking about ongoing causal processes and their consequences; and
- To recommend development policies and strategies for relieving such problems of distress migration and lopsided urbanization in India.

Thus, this paper deals with the following five Ps:

1. Poverty and migration,
2. Patterns of migration,
3. Phenomena of the migration-urbanization system,
4. Processes underlying, and
5. Policies required.

PART I: POVERTY AND MIGRATION IN INDIA

Poverty-induced Migration and Urban Crisis in India

Crucial problems of massive distressed migration from villages to metropolises, and associated problems of unbalanced urbanization and extreme urban decay in India need the urgent attention of scholars and planners, before such maladies reach to a catastrophe. Several factors complicate the scenario.

Distressed Migration

First, in the rural sector, factors such as demographic explosion, increasing land inequality, continuous fragmentation of land holdings, and lack of rural industries and rural non-agricultural employment have virtually compelled the small farmers and landless agricultural labourers to leave poor villages and crowd in the cities. They gravitate towards larger metropolises, skipping smaller towns and cities, as they know that employment is more available in larger cities (Mukherji, 1975). But, their dreams remain unfulfilled; their lives end up in city slums. Slums, squatter settlements, and shanty towns in largest metros demonstrate the truth. Thus, masses of poor, landless, illiterate and unskilled agricultural labourers and petty farmers from backward states of UP, Bihar, Odisha, and Andhra Pradesh move to Kolkata, Mumbai, Delhi and Chennai, bypassing small towns, small cities, and district centres (Mukherji, 1979a). Such massive rural to metropolitan migration is a typical characteristic of migration in India. This is termed as '*poverty-induced migration*' (Mukherji, 1985a; Mukherji, 2002, 2005).

Urban Decay

Second, in the urban sector, economic decline has occurred due to a weak urban economic base, a rapid urban population growth, and heavy emphasis only upon foreign export activities (of raw materials). While the development of internal domestic markets is neglected, the growth of only light branches of industries, and low levels of technology resulted. Urban growth occurred, without internal economic strength and without means to provide adequate employment to the poor (Mukherji, 1977). Class I cities and metros serve more as focal points of collection of raw materials for export, rather than as poles of development to diffuse growth impulses to their respective hinterlands (Berry, 1966). Since they have limited employment generating capacities, due to capital-intensive industrialization, and a limited absorptive capacity in organized sector, the incoming labour find salvage only in bazaar-type activities, e.g. porters, servants, hawkers, vendors, construction labour, etc. Such informal sectors are characterized by very low wages, low productivity, cut-throat competition, insecurity and exploitation. *Although such migration helps to avoid starvation in the villages (hence desirable, and must not be restricted), but it does not improve their economic conditions, nor permit their upward social mobility.* Migrants are moving from rural poverty to urban poverty (Mukherji, 2005, 2006).

Urban Involution

Third, as a result, it led to acute urban congestion, urban decay, urban involution and proliferation of filthy urban slums, extreme urban squalor and poor levels of living (Mukherji, 1993). So, Indian metropolises became involuted, rather than evolved, i.e. they grew merely in population, not in prosperity. Cities are becoming merely over-blown villages, without an urban culture or urban functional characteristics. So, this phenomena is called as *"urban involution"* (Mukherji, 1979b, 1985b).

Urban Slums and Squalor

Fourth, Indian metropolitan areas and large cities are fast degenerating into extreme filth and indescribable squalor. Congestion in over-populated areas leads to shortages of even the most basic shelter space, water, sanitation and electricity in the slums.

Very Unjust Urban Society

Fifth, Indian metropolitan areas are very fast becoming the scenes of extreme social and economic inequalities where a small minority live in abundant affluence in contrast with abject poverty of the masses down below (Mukherji, 1993, 2001a).

These situations may create a dangerously eruptive situation, which may unleash extreme social disorder, severe class conflicts, crimes, widespread violence and even urban civil war, and therefore, warrant in-depth studies and planning. In short, if the problem of massive poverty-driven migration and the problems of acute urban congestion, urban failure, and urban inequalities in India are not attended and effective action not taken, then the entire process will fast reach the maximum entropy level, when the entire urban system might break open to a catastrophe (Mukherji, 2006).

These conditions gained momentum since 1981, showing typical traits of cancerous development of metropolises and urban involution in India (Mukherji, 1981, 1985a, 2001b, 2006). No cold statistics can adequately describe sadness, agony and pain associated with such massive migration of illiterate/semiliterate peasants and labourers to slums, squatter settlements, and mining zones—saturated with unemployment and declining basic human values. Perhaps the following lines may describe a bit.

Human Dimensions of Migration Problems in India

Now refer back to the quotation given in the beginning. Briefly, this encapsulates the poignantly very sad picture of migration in India. This, I had written in 1975 but I am forced to repeat the same lines even in February 2010 (during this conference). Even after 35 years! So much water has gone down the Holy river Ganga, but situation has not improved over all these years! Alas, India's so-called economic growth has not yet touched the lives of hundreds of millions of such poor migrants and citizens! Alas! Happiness is a very unfortunate lady, who has not seen the faces of such hundreds of millions of very poor people of India!!

PART II: PATTERNS OF MIGRATION IN INDIA

Analysis of Indian Migration data done for 1991-2001 period at six levels: (1) All-India, (2) Districts, (3) States, (4) Class I Cities, (5) Million Cities, and (6) Mega Cities. Very brief discussion is presented.

1. All-India Analysis

A. Broad Components of Migration

(a) Total Migration

According to Census 2001, there were a total of 314.5 million migrants, in a country of 1016 millions, i.e. almost every third Indian was a migrant (Table 6.1). *By 2011, about 450 million migrants are expected, which may rise to 785 million by 2051* (Mukherji, 2011 b). *Then, almost every other person will be migrant in India (44 per cent). To provide jobs and shelter to such huge masses will require a Himalayan task* (Mukherji, 2011a). Other studies also alarmed about India's this burgeoning problem (UNDP, 2007; United Nations, 2007).

Table 6.1: Total Migration in India, All Durations, Males and Females, Census 2001

Components	*Total migrants, all-India, all durations, 2001*	
	Persons	*Per cent*
Total Migration	314,541,350	100.0
Total Internal Migration	309,385,525	98.4
Intra-district migration	193,592,938	61.5
Inter-district migration	74,626,322	23.7
Inter-state migration	41,166,265	13.1
From Other countries	5,155,423	1.6

Source: Computed from Census of India (2001), Migration Tables, Series D, Census Office, New Delhi.

(b) Components of Migration

During 1991-2001, about 32.9 million males and 65.4 million females were migrants. Of them, mostly were short-distance intra-district migrants (males 43 per cent and females 61 per cent). Inter-state migrants were only 26 per cent among males, and 12.7 per cent among females (Table 6.2).

(c) Total Population Migration, Persons, All-India, 1991-2001 and its Parts

During 1991-2001, 98.3 million migrated, of which majority were intra-district migrants (51 per cent), followed by inter-district (26 per cent), inter-state (20 per cent) and international migration (3 per cent). (Table 6.2).

(d) Total Male Migration, All-India, 1991-2001 and its Parts

However, for the males, those components changed considerably: out of total 32.9 million male migrants, only 43 per cent moved intra-district, 29 per cent inter-district, and another 26 per cent moved for inter-state migration.

(e) Total Female Migration, All-India, 1991-2001 and its Parts

In comparison, among the females, such components changed substantially: out of a total 65.4 million female migrants, 60 per cent moved for short-distance intra-district migration, whereas 26 per cent did for inter-district, and only 13 per cent moved inter-state migration. Thus, majority of the females moved short distances.

B. Reasons for Migration

According to 2001 Census, 315 million migrants moved between different parts of India. Of them, 30 per cent were males, and 70 per cent females. The majority came from within the state (268 million, 85 per cent) and 41.2 million (13 per cent) from other states. There are *seven* reasons for migration: for the males, the most important reason was employment (28 per cent), followed by miscellaneous

Table 6.2: Total Migration by Types, All-India, Males and Females, 1991-2001

Total Migration	*Migrants for 0-9 years, All-India, 1991-2001*				
	Persons	*Males*	*Females*	*Males %*	*Females %*
Total Migration	98,301,342	32,896,986	65,404,356	100.0	100.0
Total Internal Migration	97,560,320	32,510,444	65,049,876	98.8	99.5
Intra-district migration	54,108,439	14,147,939	39,960,500	43.0	61.1
Inter-district migration	26,625,002	9,850,344	16,774,658	29.9	25.6
Inter-state migration	16,826,879	8,512,161	8,314,718	25.9	12.7
From Other countries	740,867	386,461	354,406	1.2	0.5

Source: Computed from Census of India (2001), Migration Tables, Series D, Census Office, New Delhi.

(34 per cent) and family moved (19 per cent), business (3 per cent), and education (3 per cent). For the females, marriage (70 per cent) was the most important reason, followed by family moved (11 per cent) and miscellaneous (14 per cent). Among them, education, or business was still not prominent, 0.2 per cent and 0.3 per cent, respectively. Indeed, the patterns have not changed, in spite of recent economic development.

C. Inter-state Migration Flows, 1991-2001

Figures 6.1 and 6.2 present inter-state migration streams of the males and females, respectively, who moved during 1991-2001. Only the major streams are illustrated (Mukherji, 2011b). Noteworthy features are:

(a) Uttar Pradesh, Bihar, Tamil Nadu, Kerala, Madhya Pradesh, and Rajasthan are the main migration-source regions for the males and females. Karnataka and Andhra Pradesh are secondary sending regions. These are less economically developed states from which males and females moved out.

(b) Maharashtra, Delhi, Gujarat, Karnataka, Madhya Pradesh, Punjab, Haryana, and West Bengal are main migration-destination states for both the sexes. These are economically more developed states towards which they move for employment, subsistence, or shelter. Kerala, Tamil Nadu, and Andhra Pradesh also get some migrants from neighbouring states. Like the males, the females are also drawn towards the same urban nodes, to the centres of production and trade-commerce. Exceptions are rural parts of Haryana, Kerala, Karnataka, and Gujarat where agricultural development has progressed recently.

(c) In sum, migration occurs as a response to regional disparities in the levels of economic development over India: moves arise from the less economically developed states (Uttar Pradesh, Bihar, etc.) to comparatively more developed states (Maharashtra, Delhi, Punjab, Haryana, and West Bengal). Maps of inter-state migration for females are almost identical to that of males. Only difference is that volumes are much

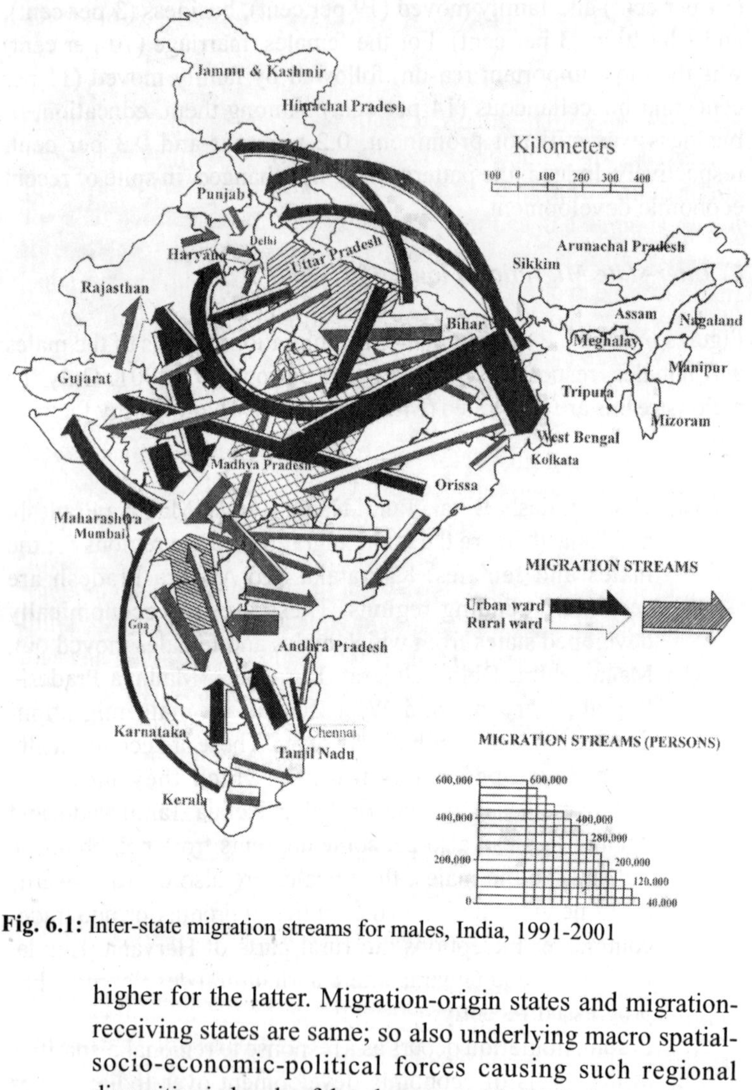

Fig. 6.1: Inter-state migration streams for males, India, 1991-2001

higher for the latter. Migration-origin states and migration-receiving states are same; so also underlying macro spatial-socio-economic-political forces causing such regional disparities in levels of development and which, in turn, induce migration to occur.

(d) Migration flows towards four mega cities (Mumbai, Kolkata, Delhi, and Chennai) (not shown), are similar, for both sexes, and also for inter-state streams. Difference is only in volume, patterns remain the same: showing same spatial responses

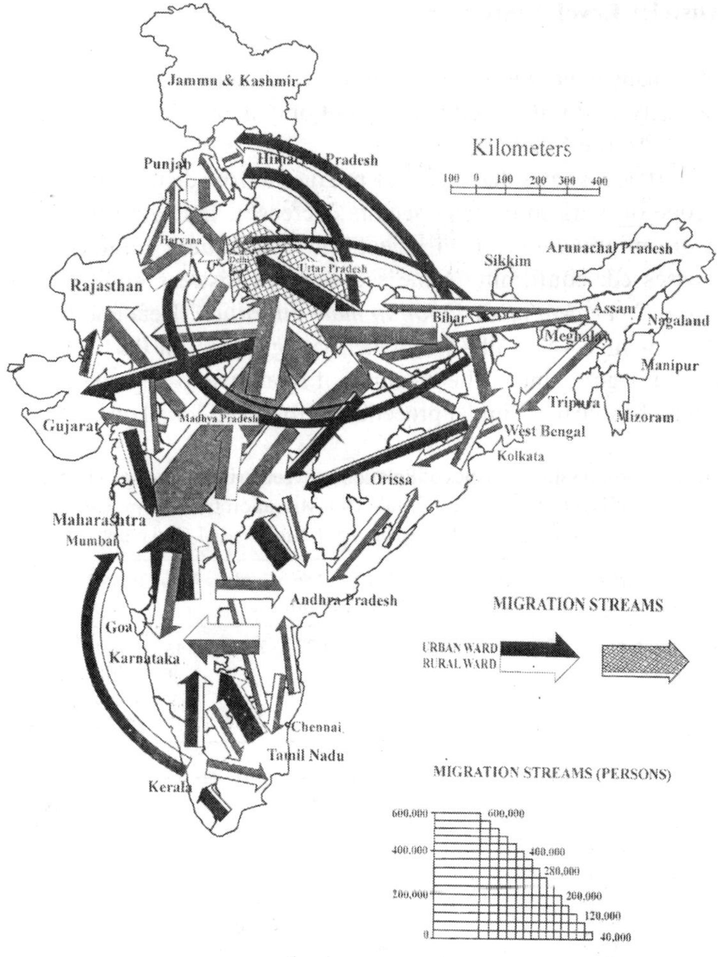

Fig. 6.2: Inter-state Migration Streams for Females, India, 1991-2001

to regional disparities. Same patterns can also be seen, if compared with that of last five decades (Mukherji, 1981). In fact, since the first Indian Census, 1872; it did not change since last 130 years, because spatial disorganization persisted all through (Mukherji, 1995). Sadly, it will continue many more decades, until we develop small towns/cities and re-configure India's spatial organization in favour of development of internal markets (Mukherji, 2011b).

2. District-Level Analysis

In fact, many hundreds of such analyses were made, at districts, states, class I city, and million city levels, but only a very few shown here (for details, see Mukherji, 2011b).

(a) Below shows that male out-migration from districts increases as scope of work in primary sectors decreases. Indicates uprooting and out-migration from villages, and consequent migration of dispossessed, confirming same sad story of distress migration. *It explains 7 per cent of variation in male migration.* Beta coefficient is 0.27, with T value of –5.9, which is significant at 0.01 level of statistical significance. The meaning of relationship is given in last column of Table 6.3. Similar procedure is followed for all other tables.

Table 6.3: Regression Statistics of Relation between Primary Sector of Economy of Districts and Vol. of Male Out-Migration from them, India, 2001 (N = 470) (Beta = –0.28, T = –5.9, Sig. T = .01, Intercept = 98807)

Dependent Variable	*Indep. Variable*	*Mult. R.*	*Adj R Sq*	*Expln.*	*Meaning of Relation*
Vol. of male out-migration	Primary sectors	0.27	0.07	7 %	Less primary sectors in district, larger volume of male out-migration

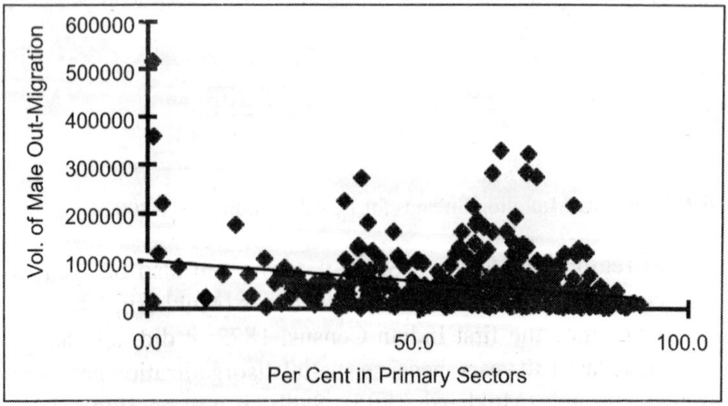

Fig. 6.3: Less Workers in Primary Sectors in Districts, the Larger Volume of Out-Migration, 2001

Source: Computed from Census of India, 2001.

(b) Positive links between tertiary sectors and male in migration in districts:

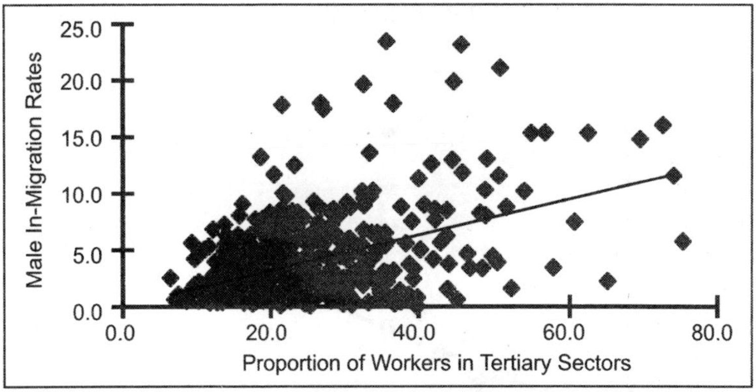

Fig. 6.4: Links between Tertiary Sectors and Male in-migration in Districts

(c) Positive causal links between total population and volume of male out-migration from districts:

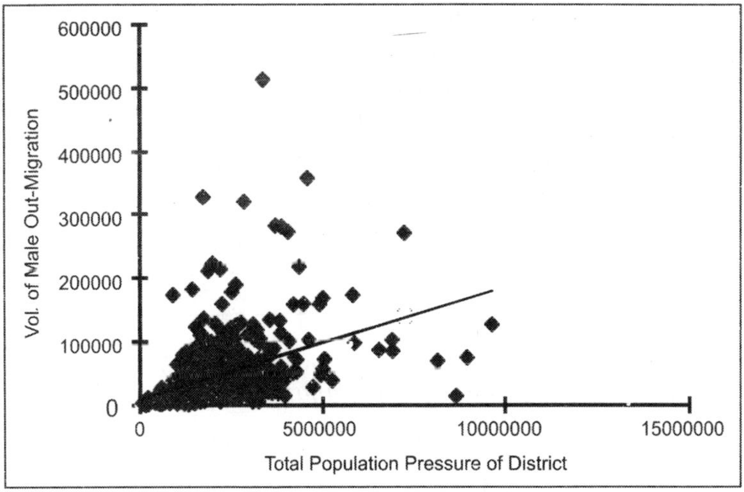

Fig. 6.5: Regression Line between Total Population and Vol. of Male Out-Migration from Districts, 2001

3. State-Level Analysis

(a) Causal relationships between urbanization in states and vol. of male in-migration (explains 36 per cent):

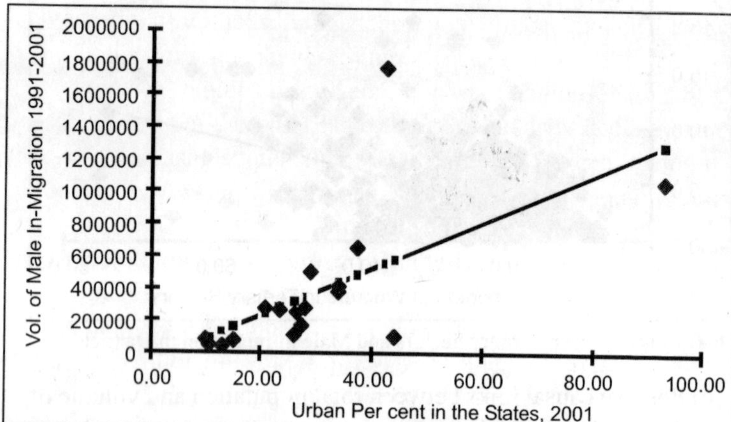

Fig. 6.6: Causal Relations between Urbanization in States and Vol. of Male In-Migration 1991-2001

Table 6.4: Relation between Urbanization in States and Vol. of Male in-migration in States, 1991-2001 (N = 17 states) (Beta = 0.67, T=3.1, Sig. T = .0001, Intercept = -488103)

Dependent Variable	*Indep. Variable*	*Mult. R.*	*Adj R Sq*	*Expln.*	*Meaning of Relation*
Vol. of Male In-migration	Urban per cent in States	0.62	0.39	36%	More urbanization in states, more male in-migration into states

(b) Very strong causal relationships between vol. of marginal workers in states and vol. of male out-migration, 1991-2001 (explains 72 per cent) (See Fig. 6.7 and Table 6.5).

4. Class-I City-Level Analysis

Causal links between secondary sectors of class I cities and male migration for work (1991-2001) (explains 13 per cent) (See Fig. 6.8).

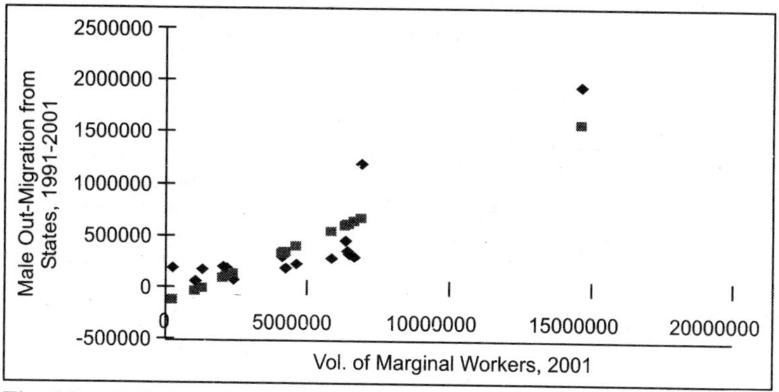

Fig. 6.7: Relation between Marginal Workers and Male Out-Migration (explains 70%)

Table 6.5: Relation between Vol. of Marginal Workers in States and Vol. of Male Out-Migration, 1991-2001 (N = 18 States) (Beta = .085, T = 12.5, Sig. T = .0001, Intercept = –163111)

Dependent variable	*Indep. variable*	*Mult. R.*	*Adj R Sq*	*Expln.*	*Meaning of relation*
Vol. of Male Out-Migration from States	Vol. of Marginal workers in State	0.85	0.72		More marginal workers in the states, the more male out-migration from

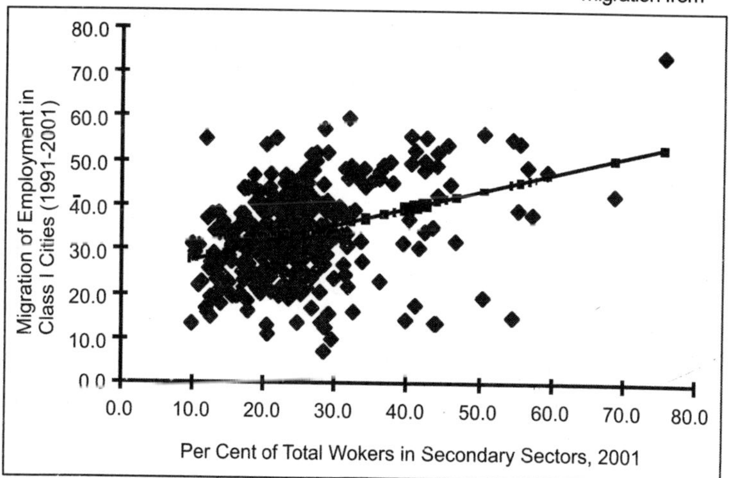

Fig. 6.8: Causal Links between Secondary Sectors of Economy in Class I Cities and Migration for Employment therein (1991-2001)

Table 6.6: Causal Links between Secondary Sectors of Class I Cities and Male Migration for Work into them, India, 2001 (N = Class I Cities Studied 388) (Beta = 0.38, T = 6.9, Sig. T = .0001, Intercept = 23.9)

Dependent variable	*Indep. variable*	*Mult. R.*	*Adj R Sq*	*Expln*	*Meaning of relation*
Male Migration for work	Secondary Sectors of cities	0.36	0.13	13 %	More secondary sectors in class I cities, more male migration for work

5. Million City-Level Analysis

(a) Exceptionally strong causal links between total population of metros and million cities and volume of male in-migration to them, 1991-2001 (explains 85 per cent).

Table 6.7: Causal Links between Total Population of Metros and Million Cities Male In-migration to them, India, 2001 (N = 19 Million cities) (Beta = 0.92, T = 10.4, Sig. T = .00001, Intercept = –17291.6)

Dependent variable	*Indep. variable*	*Mult. R.*	*Adj R Sq*	*Expln.*	*Meaning of relation*
Male In-migration to metros and million cities	Total population of metros and million cities	0.93	0.86	85 %	Larger population size of million cities, greater vol. of male migration to them

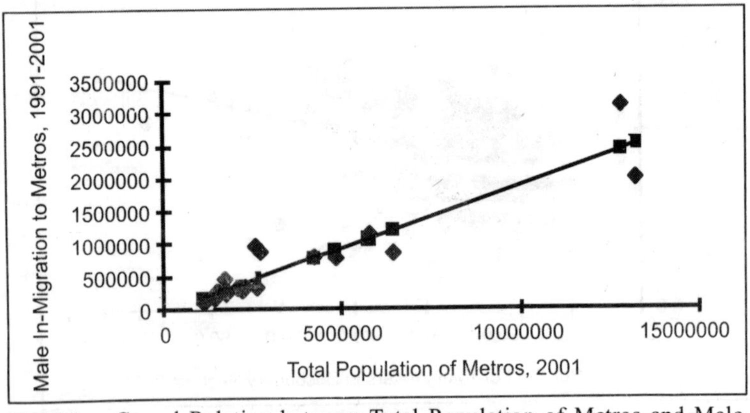

Fig. 6.9: Causal Relation between Total Population of Metros and Male In-Migration to Metros, 1991-2001

(b) Strong links between service sectors in million cities and female migration for work, 1991-2001 (explains 18 per cent) regression table not shown

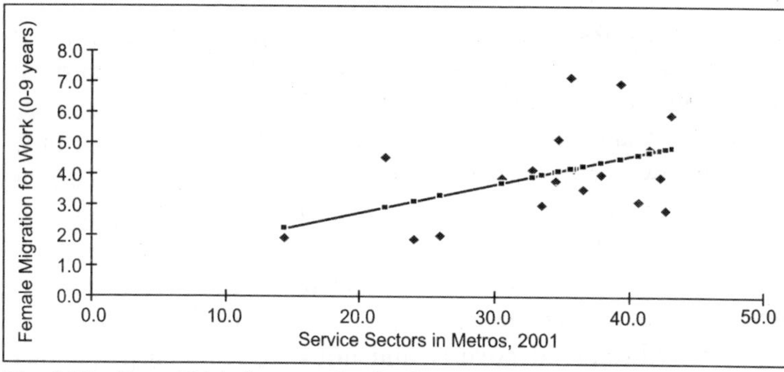

Fig. 6.10: Causal Links between Service Sectors and Female Migration for Work, 2001

6. Mega City Level Analysis

Components of Male Migration to Mumbai, 1991-2001

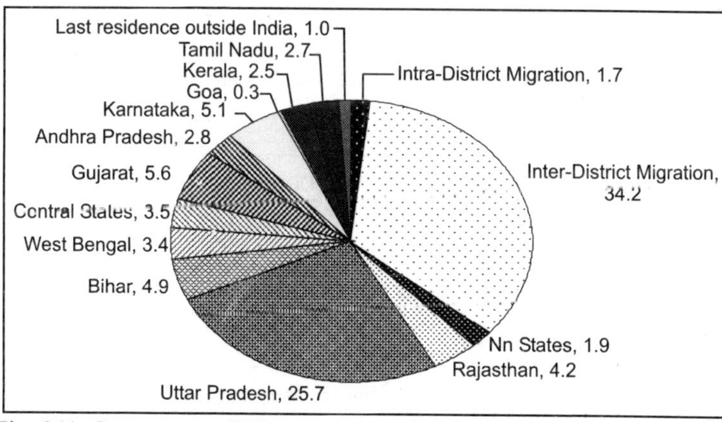

Fig. **6.11:** Components of Migration to Mumbai, during 1991-2001

In Mumbai, intra-state migration forms 34 per cent, inter-state migration, 60 per cent; of which migrants from UP (26 per cent) and Bihar (10 per cent) take large segments. Patterns remain same over time. Similar picture appears for other mega cities.

PART III: SIX PHENOMENA OF MIGRATION-URBANIZATION SYSTEM

1. Phenomenon of Poverty and Under-development

India was colonized by British and partly by French, Dutch, and Portuguese for over 200 years. Before that, India was foreign ruled and exploited for 750 years. No wonder that India manifests deep-rooted poverty and under-development even today, and consequently, generated low quality migration and urbanization. Indeed, current distressed migration and urban decay are just spatial symptoms of deep-seated nexus of poverty and underdevelopment, which, in turn, caused by historic adversaries, and which left legacy even in present uneven spatial landscape as regional disparities, uneven social inequities, widespread poverty and injustice (Mukherji, 2000a) (Fig. 6.12).

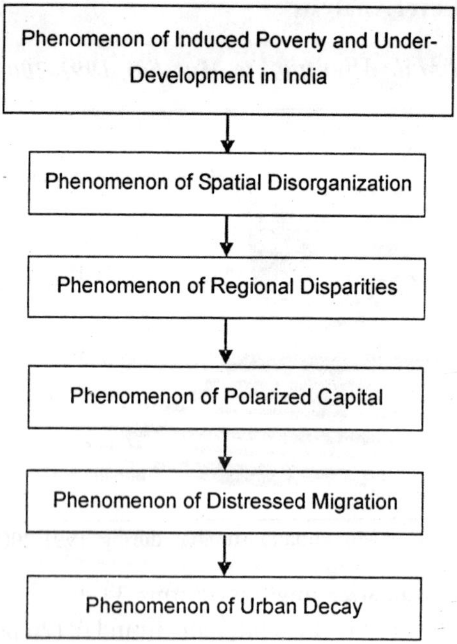

Fig. 6.12: Processes of Poverty, Underdevelopment and Spatial Disorganization and Migration and Urbanization in India

2. Phenomenon of Spatial Disorganization

Spatial dimension of unequal socio-economic development, in once-colonized country, have created and persisted disorganization and disintegration of spatial structure of economy (Mukherji, 2004).

3. Phenomenon of Regional Disparities

In this former colony, certain new ports/cities emerged (Bombay, Calcutta, Madras, New Delhi etc.) and grew up at the expense of other towns/cities; and certain specific foreign enclaves (mines, plantations, cantonments, hill resorts) became favoured and prosperous, neglecting interior regions. Consequently, marked regional disparities sprung up and continued, even today.

4. Phenomenon of Polarized Capital

One phenomenon of greatest causes is polarized capital. Investments, growth efforts, and location of industries have markedly uneven spatial concentration around primate city/large metros, utterly neglecting development of smaller towns/cities (Mukherji, 1991).

5. Phenomenon of Distressed Migration

Masses of poor and illiterate or semi-literate peasants/workers making quantum jump from marginalized villages directly to largest metropolises, causing severe urban congestion, decay, proliferation of slums and urban informal sectors, bazar economy, acute unemployment, and breakdown of essential services (water, transport). Such distressed migration and urban decay occur severely in all large cities/metros. This is termed, "distressed migration".

6. Phenomenon of Urban Involution

Massive population growth is occurring in mega cities/large cities, but without prosperity and structural changes. They burst in their seams merely as over-blown villages, without urban functions, infrastructure, and strong economic base. This phenomenon is termed, "urban decay and urban involution".

Such Six inter-related aspects termed as main phenomena of involutionary migration-urbanization system of India. Without de-linking them, such problems of low quality migration and urbanization will continue many decades, even centuries. Thus, needs correct perspectives, correct theory and explanations, based only upon which, correct policy prescriptions can be made for alleviating human problems of low quality migration and urban decay in India. For, these problems involve lives and livelihoods of hundreds of millions of poor Indian migrants and slum dwellers, and poor peasants and workers (Mukherji, 2000b).

PART IV: UNDERLYING SPATIAL, ECONOMIC AND POLITICAL PROCESSES BEHIND NEXUS OF POVERTY AND MIGRATION-URBANIZATION IN INDIA

A. Poverty, Underdevelopment, and Spatial Disorganization

1. Poverty and Regional Imbalances

These are spatial manifestations of poverty and regional imbalances of a developing economy, like India. Under prolonged colonization, dependent nature of economy of India had generated certain distortions in the spatial structure of economy and regional imbalances within the country, which did not permit a strong and self-generating economic base to emerge (Mukherji, 1981) (see Fig. 6.13).

2. Urban Development Bias

Persistent development bias only towards metropolises and large cities continued unabated since long time. All urban services, capital, and development efforts have been concentrated only at a few favoured centres and foreign enclaves (Kolkata, Mumbai, Delhi and Chennai), neglecting growth of other settlements (Mukherji, 1985b, 1995).

3. Induced Spatial Disorganization and Regional Disparities

These processes have created sharp regional imbalances, which, in turn, have led to prolonged rural neglect and consequent labour migration. Tremendous urban bias in the development process has led to further concentration of urban services, facilities, credits,

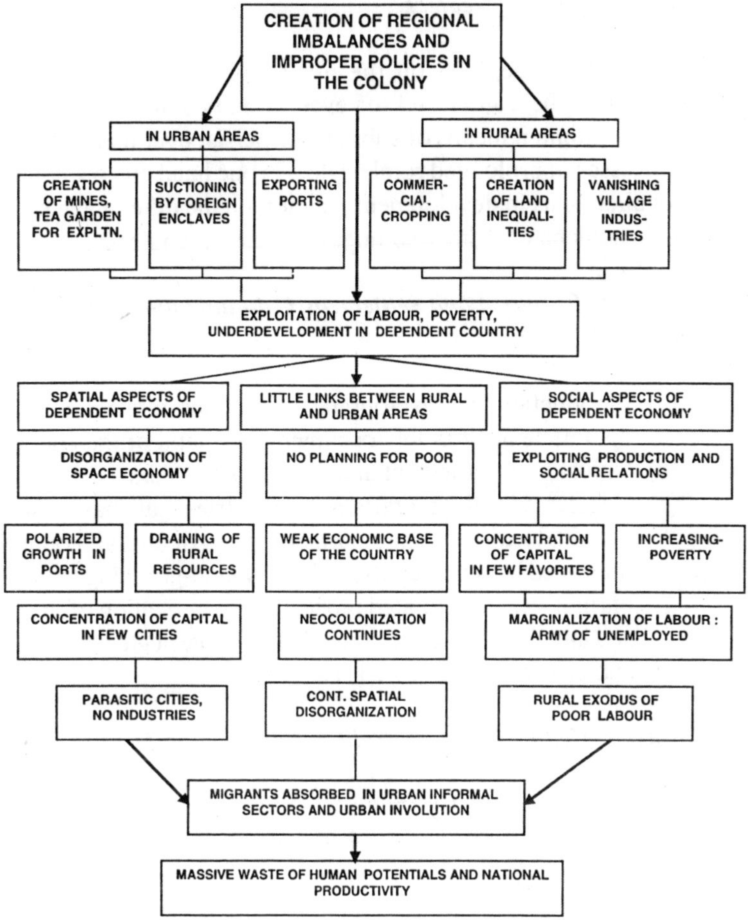

Source: Mukherji (2011b).

Fig. 6.13: Spatial-Economic Processes of poverty-induced migration and urban involution in developing countries

investments and infrastructural development only to those few privileged urban centres and regions (favoured by the colonizers and later continued by their local substitutes), at the expense of others. Evidently, if all the blood in a human body rushes only towards head, then that is not a sign of health; rather it might cause death. Exactly similar things happening in space economy of India. This implies acute spatial disorganization (*Ibid.*).

4. Prolonged Rural Neglect

Prolonged rural neglect, on the other hand, have stagnated the countryside; where utter lack of even primary school, or drinking water, or minimum means of subsistence persist over many decades. So, processes of prolonged rural neglect in one hand and pronounced metropolitan bias in development policies, on the other, have caused such phenomena of distressed migration and urban decay.

5. Creation of Unjust Land Settlement, Colonial Land Use and Land Inequality

(a) Introduction of cash cropping (plantations, tobacco, tea, jute, opium, hemp) forcibly occupying best land of the country,
(b) Unjust land settlement and tenure systems, and
(c) Neglect of village crafts and industries—all have caused marginalization of small farmers and agricultural labourers. Increased land inequality, caused by land-grabbing tendency of rural rich, accelerated process of pauperization of farmers and labourers, who are compelled to leave neglected villages and crowd into large metros.

6. Parasitic Mega Cities and Proliferation of Poor Tertiary Sectors

Excessive export-oriented economy (mainly of raw materials at throw-away prices) and distortion in urban economy, on the other hand, resulted in the growth of only parasitic cities and proliferation of low-grade trade and services, without strong manufacturing base. As Fig. 6.13 shows, therefore, we have parasitic urban centres without proper industrialization, on the one hand, and marginalization of peasants, on the other; with weak economic base of the country in between (Mukherji, 1997).

7. Resultant Rural Exodus and Labour Migration of Unemployed

Consequently, massive and continuous rural exodus of refugee labourers occur from rural to urban centres. This is economic-historical context of such low quality labour migration and urbanization in a developing economy (Mukherji, 1993, 2008).

B. Continuing Underdevelopment during Post-Independent and Planning Era

8. Past Unjust Forces Continue Determining Migration-Urbanization under Post-Independent Period

The forces set in motion in historic past still largely determine labour migration and urbanization in India, even now, because the development measures undertaken during Five Year Plans were inadequate to ameliorate the condition of acute poverty and spatial disorganization, and widespread unemployment. So, there has been a great rural exodus of uprooted labourers from backward regions to a few parasitic urban centres (Kolkata, Delhi, Mumbai, Chennai, and others). So, even after 64 years of planning, economy of India continue to be oriented towards export of mainly raw materials. Railways and roadways mainly feed export centres, or primate cities, instead of the interior areas within the country (Mukherji, 1981, 1993, 2008) (Fig. 6.13).

9. Continued Spatial Disorganization even under Planning Era

Thus, spatial disorganization persists even now, in spite of a few attempts to correct it. So, internal trade circuits remain stultified. Breaks in settlement hierarchy continue, impeding free flow of goods, commodities and services across all parts of India. It is very symbolic that no big city has ever emerged around Kolkata, Mumbai, Delhi or Chennai during last 300 years. Investments, capital, and industries continue to be polarized as in the past, only in a few selected urban nodes. Rarely there was any real attempt for rural development; nor development of hitherto neglected regions.

10. Persistent Poverty and Rural Neglect and Continuing Migration Streams of the Dispossessed

Numerous villages are, therefore, reeling under poverty, unemployment, and under-employment. No wonder that these masses of illiterate, unskilled, and poor labourers keep crowding to the Mega cities to find any job, no matter how unskilled or low pay, nor how unbearable the filthy urban slums.

11. Parasitic Urban Centres and Proliferation of Urban Informal Sectors

Naturally, rural migrants have to make quantum jumps to Mega cities or coastal metropolises, bypassing small local towns, because, like flow of capitals and commodities to such nodes, the human labour are also compelled to move under the pull of the same centrifugal forces. They somehow get absorbed in the urban informal sectors, because these are the only means available to these masses of illiterate or semi-literate migrants for survival.

12. Migration only as Survival Strategy

Although they somehow survive in the Mega city at the most minimum level of living, but there is very little scope for upward social mobility. They continue to live at the same low level of existence in the cruel city, even after 30 or 40 years. This results in massive waste of human resources and national potentials. Briefly this is the central idea of underlying processes. Various planning strategies are illustrated in another figure (Fig. 6.14), and discussed in last section.

C. Globalization and Its Adverse Effects on Migration

13. Current Processes

What is the present and the future scenario? This section highlights these processes. Under the current trends of liberalization, privatization, and globalization, evidently the more and more resources, capital, investments, production units, and growth efforts are polarized only around mega cities/ports (mainly in elitist-consumption oriented industries), thus neglecting all-round development of interior countryside, villages, small towns, and hinterland of India. As a result, much more skewed forms of afore-mentioned types of distressed migration and urban decay will increasingly manifest in now. True nature of such processes are never truly unfolded. We really need more socially-committed social scientists, who will undertake such research and will help the poor.

14. Global Processes and Further Pauperization of Labour

Process of liberalization in fact has placed common people at the mercy of international market forces, buried issues of equity and social justice, and widened the gaps between the rich and the poor. Secondly, process of privatization, that involves disinvestment of Public Sector Units (PSUs) at throw away prices, has actually usurped huge amount of public money, benefiting a handful of privileged industrialists, at the cost of an ocean of the poor. Third, ongoing processes of globalization is hurting the poor most, as prices of daily necessities sky-rocketing, beyond reach of common people. New capital or investments are mainly concentrated in mega cities, and to benefit only a few privileged elites. These capital and investments are not available for agricultural regeneration or rural development. Rural areas, thus devoid of any development efforts and capital, are reeling under abject poverty, acute unemployment and widespread illiteracy (Mukherji, 2011b, and 2011c).

15. Globalization Adversely Affecting Migration

The masses of dispossessed farmers and labourers are thus further compelled to swell in urban slums and construction sites around large cities for minimum sustenance. Recently, a study of such migrant labourers is made in Gujarat which amply testified such agonizing process of migration under globalization and privatization (Breman *et al.,* 2000).

PART V: POLICIES REQUIRED FOR REDUCING PROBLEMS OF LOW QUALITY MIGRATION AND URBANIZATION

Discussion here is very brief (Fig. 6.14). All details of such policies and strategies are given elsewhere (Mukherji, 2011b).

A. Rural Development Strategies

1. Need more Effective Rural and Regional Development

To save the metros, very necessary is that all-round socio-economic development of all rural interior and neglected parts of India are urgently made, from which massive migration of poor people occur to metros. (For details see Mukherji 2011b) (see Fig. 6.14).

Unless done, problems of distress migration, over-congestion, and urban involution of metros and larger cities, will continue unabated. Problem must be attacked at its roots, where it arises.

2. Eliminate Rural Poverty

- Eliminating rural poverty be given maximum focus in "Planning for Poor", without which, low quality migration will remain in India, for many more decades!
- In recent book on "Poverty and Fertility in India", Mukherji (2007: 250-280) much focused on strategies for reduction of rural poverty for fertility decline and low quality migration.
- Generate more non-agricultural employment in rural and neglected areas, which will reduce poverty-induced migration, also high fertility.
- Also, ensure that benefits of economic development not eaten up through income inequality and unjust social development, without which difficult to break vicious cycle of poverty and fertility and poverty and migration.

3. Generate Massive Rural Employment for Eliminating Rural Poverty

(a) Tremendous efforts needed for massive generation of employment in villages, through

- rural industrialization,
- irrigation development,
- land conservation,
- land redistribution,
- multiple cropping, and
- development of village crafts and cottage industries (see Fig. 6.14).

Source: Mukherji (2011b).

Fig. 6.14: Migration of Labourers and Required Planning Strategies and Development Policies

(b) Rural poverty can only be reduced

- If purchasing power of rural people are sufficiently enhanced by generating massive rural employment,

- By proper distribution of benefits to village society, specially for weak and under-privileged.
- Unless attempts made for upliftment of pauperized rural people, nothing really can be done for controlling current migration from rural poverty to urban poverty.
- Will require very strong political will and determined economic discipline (Mukherji, 2006).

4. Irrigation Development

- Need massive irrigation development programmes in villages for:
  - canal,
  - river,
  - well irrigation,
  - reconstructing field drains, and
  - irrigation channels, etc.
- Will considerably add to total land capable of cultivation and to productivity.
- Will create jobs for landless labourers and seasonally unemployed.
- Irrigation is main panacea for rural development and agricultural development. Two-third variations in agro-productivity depends solely upon this.
- Create irrigation channels, deep tube-wells, pond irrigation, and water harvesting.
- Steps will greatly raise agricultural productivity, create massive rural employment, raise levels of living and purchasing power, and rejuvenate villages.

5. Rural Industrialization and Economic Value-Enhancing Process

- Great efforts required that raw materials processed within villages, in finished/semi-finished form.
- So that economic values enhanced to products locally before transferred to cities.
- Set up many rural industries/cottage industries (rice-milling, *gur*-making, oil-pressing, weaving, garments) (with subsidies

and co-operatives) at properly chosen, centrally located villages, each serving 20 villages (Mukherji, 2001b).

- Such value-enhancing process will raise rural productivity, improve levels of living and motivate them to work more productively as they will derive benefits themselves (Mukherji, 2006).
- When more rural jobs will be created, then, it will reduce current mass migration from rural poverty to urban poverty. This is most mandatory step for migration-influencing strategy for India.

6. Integration of Industry and Agriculture

- Must dispense with two great disarticulations between towns and villages and between industry and agriculture: bases upon which contradictions between growing and lagging regions founded can be achieved by following steps:
    - Creation of new division of agricultural labour: transferring part of food-producing labour to industrial crop-growing, livestock-breeding, fishing, forestry, horticulture, and another to agro-industries, small-scale industries, transport and communications (Mukherji, 2007).
    - Need conversion of huge masses of unskilled labour to semi-skilled and skilled labour through short effective training, and raising per capita labour productivity to higher levels.
    - Industrialization of agriculture, implementation of industry into villages, and for general integration of industry and agriculture.

7. To Change Unequal Rural-Urban Linkages and Relations

- Thrust of rural transformation programme: accelerate rapid development of villages and reduce rural-urban disparities in wages, amenities, income, growth efforts, investment, education, nutrition, and well-being.
- Need definite pro-rural development policy and all-pronged

attack at roots of rural poverty and underdevelopment. Corrective steps needed for changing existing unjust rural-urban relations.

- Dispense away with false notion of downward-filtration of development from towns to villages, and adopt direct programmes which benefit rural poor directly.
- Free villages from excessive dependence upon cities, and from expropriation of rural agricultural surpluses by urbanites. Fix proper pricing policy in favour of peasants so that they derive due share of their labour. Strengthen rural-urban relations in favour of villages, by marketing agricultural goods, roads, electrification.

8. Develop Agri-Pole Centres

- One crucial step for rural reconstruction should be to create "*Agricultural Growth Pole Centres*" in central locations in villages itself, where agricultural goods/commodities from surrounding villages may be marketed.
- And, the farmers may derive their benefits, without being looted by middlemen and village money lenders. Existing agricultural collecting centres may be developed to perform these roles.

B. Urban Development Strategies

9. Development of Small Towns—Intermediate Cities

- Restructuring of the space economy of the hinterland of large cities is thus urgently necessary. That is, the need for providing much greater capital investment, growth efforts, industrial licences in villages, smaller towns, and intermediate cities (Mukherji, 2011b).
- And, also making them as poles of development that will diffuse growth waves to surrounding villages, instead of the present day excessive reliance only upon a few primate cities (Fig. 6.14).

10. Massive Employment Generation in Cities

- Objective should be to generate massive employment, self-employment, and gainful employment, both for migrants and residents, by strengthening productive linkages between urban formal and informal sectors, and increasing their efficiency and technology.
- Make all-pronged effort for creation of employment for rural masses in villages themselves. Irrigation development, transfer of technology, and rural industrialization of villages are the means which are greatest generators of employment in villages. These are difficult tasks, but many Asian nations are succeeding in such efforts; and there could be no reason why India cannot succeed.

11. Eliminate Urban Poverty

- Urban poor are very rapidly swelling in number everywhere. To eliminate poverty among urban poor, ensure that value received by them must enhance. They derive benefits of their only capital, labour, and they should be made to pay less for goods and services they receive from big city.
- One World Bank Policy study (1977) suggested two interventions: one, we should increase the value of labour by increasing the demand for labour and by upgrading their quality of labour; and other, improve quality and quantity of public services available to urban poor.
- Providing quick but useful training for their skill development and providing cheap and subsidized school and health services to urban poor, especially slum dwellers will reduce urban poverty and human resource development. Skill development will provide more self-employment or upgraded employment both in formal and informal sectors.

12. More Humanistic Urban Planning and Housing for Slum

- Objective should be to give a better deal to slum-dwellers and ensure a more humanistic approach to urban planning for welfare. Specifically, concrete programmes be made for construction of inexpensive, heavily-subsidized, low-rise apartments for slum dwellers within the slums; each apartment with 150-200 sq. ft and designed to accommodate a 6-7 person family.

- Toilets and bathrooms can be centrally located and commonly operated. Free labour of slum-dwellers for construction of own houses at present sites can be ensured. Expenses can be met by subsidies from city corporations, state government, voluntary organizations, and monthly levy from not-poor segments of metro residents (as slum development tax).
- Goal will be to provide cheap housing and sanitary facilities to slum people within slums; instead of perpetuation of such problems. So, cities and metropolises hopefully will become self-sustained, more habitable and more egalitarian. A more humanistic approach to this entire problem of slum development and shelter of urban poor is very urgently needed.

13. Need Efficient Metropolitan Management

- All mega cities of developing countries must learn planning lessons from very efficient Tokyo Metropolitan Government (TMG, 1998) and Japanese Urban management experiences. In extremely strenuous world, efficiency has no alternative. TMG has a very large, organized, very efficient workforce, looking after efficient management of Tokyo's all aspects of development.
- It includes: right from urban traffic, transport, communications, administration, resources, planning, implementation, housing, health care, social welfare. Their efficiency is manifest in concrete plans, updated every two years, and their implementation in right earnest as well as in city life which flows smoothly in Tokyo, in spite of big population (36 M in 2001) (Mukherji, 2000a).
- Efficient management of TMG should be replicated in metropolitan management of Indian mega cities, Mumbai,

Kolkata, etc. Better management skills of Japanese people, discipline, orderliness, work ethic, even innovativeness to elicit financial resources from urban residents can be applied.

14. Develop Urban Infrastructure

Many cities have inadequate essential urban infrastructure like roads, electricity, water, utilities, transport-communications. Frequent breakdown of essential urban services acutely hamper urban lives and productivity. Developing essential urban infrastructure must be given top priority.

15. Redirect Migration to Growth Centres

- This avowed goal may not be possible, unless we make a paradigm shift towards development of small cities/towns, and re-orient current iniquitous bias towards only mega city development.
- Must promote new growth centres and redirect migration flow to district headquarters/small cities.

Next *five* steps elaborate various aspects of programmes for restructuring of spatial organization.

C. Regional Development Strategies

16. Spatial Restructuring

- To relieve acute problems of such migration-urbanization and for achieving economic development, a major task is to re-orient India's spatial-economic structure and spatial linkages to internal markets (Mukherji, 1997, 2006) (Fig. 6.14).
- Spatial restructuring implies development of and use of vast unutilized domestic markets within country and internal orientation, rather than excessive dependence upon exporting.
- Because, development means internal development, not external dependency.

17. Evening out Spatial and Socio-Economic Disparities

- Strenuous attempts to be made for achieving a more equitable distribution, over whole national space, of productive forces and means of production, employment, consumption, investments, and services, e.g. schools, hospitals, vocational training, and basic amenities of life (Mukherji, 2002).
- That is, even distribution for all people over the nation of "per capita civilization". It requires an evening-out of social and economic disparities among social classes and regions throughout India.

18. Focus upon Neglected Regions and Enhance their Potentials

- Balance between production and consumption must be achieved at all levels. So, we need to develop latent possibilities and expand each small region's productive base according to its potentials and basic human needs (Mukherji, 2006).
- Current very accentuated concentration of wealth and consumption only among privileged few and widening gap between growing and lagging regions must be given now a final burial.

19. Fostering Inter-Regional Developmental Bonds

- Attention must be given to develop inter-regional economic bonds through proper extension and re-orientation of transport and communication links and forging mutually beneficial inter-local and inter-regional trade links.
- This re-orientation should be away from current excessive focusing of rails-roads and trade links towards exporting ports/large metros, and must be towards small, local headquarters, and micro-regional and meso-regional centres, so that future migration streams can be attracted to them.

20. Create Service Centres and Central Agricultural Market Centres

- This step will be similar to creation of agri-pole centres, but has wider "sphere of influence" than former. These may be set up at district level or regional level, at central locations to serve a broader rural community. So, generate "*Agricultural Market Centres*" in central locations in rural and regional setting itself, where agricultural goods from surroundings may be marketed and linked with agro-industries (Mukherji, 2011b).

- Likewise, setting up new "Service Centres" and growth of existing "Service Centres" must be encouraged at regional level, located at central position, where villagers may get urban services, agricultural repairs, rural bank, micro-finance, communications and marketing information.
- Such steps will make farmers to derive their own benefits, without deceived by conmen, bureaucrats/rural money lenders. Existing service centres may be promoted to achieve these goals.

Conclusion

The paper has discussed acute problems of distressed migration and urban involution in India, and has specifically focused on related five Ps: (a) poverty and migration in India, (b) patterns of migration in India, (c) phenomena of migration-urbanization system, (d) processes, and (e) policies required.

- In sum, I must state that such problems cannot be fully redressed unless we realize that these are consequences of spatial imbalances in economy, which, in turn, are consequence of excessively polarized economic activities and resources at a few nodes and wrong policies (Mukherji, 2011b).
- So, great efforts required to restructure spatial organization of economy over national space. Means more emphasis on development of internal markets, rather than only upon external one.
- Will require reorientation of trade-transport links, resource-use, growth and investment, transport links and functions of urban centres, away from current excessive concentration at

only larger metropolises, rather towards development of hitherto neglected villages and small urban centres.

- Requires greater welfare for the poor and to raise their purchasing power. In sum, requires more equitable distribution of income, wealth, means of production and consumption, among classes.
- Implies total elimination of poverty from India and making all people, especially poor, humiliated, and downtrodden, free from all human bondages and fully liberate them for their sojourn towards higher human and spiritual fulfilment!

REFERENCES

Berry, B.J.L. (1966). *Essays on Commodity Flows and the Spatial Structure of the Indian Economy.* University of Chicago, Research Paper No. 111, Chicago: University of Chicago.

Breman, J. and Das, A. (2000). *Down and Out: Labouring under Global Capitalism,* Delhi: Oxford University Press,12-48.

Census of India (2001). *All-India, Migration Tables, Series D,* Census Commissioner and Registrar General's Office, New Delhi, CDs, Various Tables, 2005-6.

Mukherji, S. (1975). The Mobility Field Theory of Human Spatial Behaviour: A Spatial Behavioural Approach to the Study of Migration and Circulation in the Indian Situation, *Ph.D. dissertation,* Department of Geog., Univ. of Hawaii (Xeroxed Univ. Microfilms, Ann Arbor, Michigan): 1-378.

Mukherji, S. (1977). Spatial Disorganization and Internal Migration in India: Alternative Strategies for Restructuring the Space Economy and Development, *Paper, ESCAP Expert Group Meeting on Population and Economic Development in Sub-National Areas,* UN-ESCAP, Bangkok: 1-56. *Summary in Asian Population Study,* Series No. 40, Bangkok, 1977: 17-18.

Mukherji, S. (1979a). Spatial Disorganization and Internal Migration in India: Probable Strategies for Restructuring of Space Economy and Development, *Canadian Studies in Population*, 6, 45-61.

Mukherji, S. (1979b), *The Mobility Field Theory: A Spatial-Behavioural Foundation to the Movement Dynamics, with Application to Indian Situation.* International Geog. Union's and National Geog. Society of India's Res. Monog No. 25, Dept. of Geog., Banaras Hindu University, Varanasi: 1-80.

Mukherji, S. (1981). *Mechanisms of Underdevelopment, Labour Migration and Planning Strategies in India,* Calcutta: Prajna, 1-190.

Mukherji, S. (1985a). The Process of Wage Labour Circulation in Northern India, in Guy Standing (ed.), *Labour Circulation and Labour Process* by Croom Helm, London: International Labour Office, 252-289.

Mukherji, S. (1985b). The Syndrome of Poverty and Wage Labour Circulation: The Indian Scene, in R.M. Prothero and Murray Chapman, (eds.), *Circulation in Third World Countries*. London: Routledge and Kegan Paul, 279-299.

Mukherji, S. (1991). The Nature of Migration and Urbanization in India: The Search for Alternative Planning Strategies, by S. Srinivasan and K.B. Pathak (eds.) in *Dynamics of Population and Family Welfare*, Bombay: Himalaya Pub. Co., 203-249.

Mukherji, S. (1993). *Poverty-Induced Migration and Urban Involution in India: Causes and Consequences*. I.I.P.S., Mumbai, Res. Monog: 1-135.

Mukherji, S. (1995). Poverty-Induced Migration and Urban Involution in ESCAP Region. *Background Paper for UN-ESCAP Expert Meeting on Population and Poverty, Bangkok, Sept. 1995:* 1-65. Summary in *UN-ESCAP Expert Group Meeting's Proceedings,* June 1996.

Mukherji, S. (1997). *Poverty-Induced Migration and Urban Involution in ESCAP Region: and Urban and Regional Planning Lessons from Japan*. Inst. of Developing Economies, Tokyo, Res. Monog. No. 300, 1-81.

Mukherji, S. (2000a). Urban Planning of Tokyo: Metropolitan Management Lessons for Asian Mega Cities, *Nagarlok*: *Journal of Urban Studies,* New Delhi: Indian Inst. of Public Adm. 1-12.

Mukherji, S. (2000b). Urban Decay in India: A Plea for Sustainable Planning, *International Sem. on Land-use and technology in Developing Countries*, in O. Singh, J. Singh and V.K. Kumra (eds.), *Changing Land and Environment Systems,* Varanasi: Nat. Geog. Soc of India, 171-82.

Mukherji, S. (2001a). *Causal Linkages between Migration, Urbanization, and Regional Disparities in India: Required Planning Strategies*, IIPS, Mumbai: Research Monograph No. 29, 1-226.

Mukherji, S. (2001b). *Low Quality Migration and Urban Decay in India: Processes and Planning Strategies*, International Union for Scientific Study of Pop (IUSSP), Salvadore, Brazil, 1-26.

Mukherji, S. (2002). Urbanization and migration in India: a different scene. In H.S. Geyer (ed.), *International Handbook of Urban Systems: Studies of Urbanization and Migration in Advanced and Developing Countries*. Cheltenham, U.K. and Massachusetts, USA: Edward Elgar, 525-559.

Mukherji, S. (2004). Poverty-Induced Migration and Urban Decay in India. in S. Banerjee-Guha (ed.), *Space, Society, and Geography*. Jaipur and New Delhi: Rawat Publications: 156-165.

Mukherji, S. (2005). Distressed Migration and Urban Decay in India: Required Planning Strategies, in *Poverty and Population in India*. In S. Mukherji, *Prof. R.N. Dubey Memorial Lectures*, 2004, Allahabad: Prof. R.N. Dubey Foundation, 75-95.

Mukherji, S. (2006). *Migration and Urban Decay: Asian Experiences*. Jaipur and New Delhi: Rawat Publications, 1-330.

Mukherji, S. (2007). *Poverty and Fertility in India: The Demographic Field Theory Perspectives*. Jaipur and New Delhi: Rawat Publications, 1-278

Mukherji, S. (2008). Distressed Migration and Urban Decay in Some Asian Metropolises: Phenomena, Processes, and Planning. In A.K. Dutt, H.N. Misra and M. Chatterjee (eds.) *Explorations in Applied Geography*. New Delhi: Prentice-Hall, 308-328.

Mukherji, S. (2011a). Migration and Urbanization in India: Patterns, Phenomena, and Required Development Policies. Lucknow Management Convention 2010, *Proceed,* Lucknow, 1-20.

Mukherji, S. (2011b). *Migration in India: Links to Urbanization, Regional Disparities, and Development Policies.* Jaipur and New Delhi: Rawat Publications, 1-550 (forthcoming).

Mukherji, Shekhar (2011c). Urban-Rural Inequalities in South and SE Asia: Colonial Impacts and Post Colonial Spatial-Economic Disparities. in H.S. Geyer (ed.), *International Handbook of Urban Policy*, Vol. III, London/Cheltenham, U.K. and Massachusetts, USA: Edward Elgar Pub, Ch. 7.

National Sample Survey Organization (NSSO), 2006, *Consumer Expenditure Survey, 2004-05* (61st round), New Delhi: NSSO, 1-125.

Tokyo Metropolitan Government (1998). *Planning of Tokyo, 1998*. Tokyo: Bureau of City Planning, 1-5.

UNDP (2007). *The State of World Population 2007: Unleashing the Potentials of Urban Growth.* New York: United Nations.

United Nations (2007*). World Urbanization Prospects, 2006 Revision*, New York: United Nations.

World Bank (1975). *The Task Ahead for the Cities of the Developing Countries*. Bank Staff Working Paper No. 209, Washington, D.C.: 47-56.

World Bank (2008). *World Development Report, 2008*, Washington D.C., 1-35.

7

Globalisation Induced Migration of Labour and their Exploitation

Shrikrishna Singh and *Shiv Shankar Mishra*

Introduction

We are witnessing a peculiar problem of jobless growth after liberalization, privatization and globalization (LPG) era. Despite a comfortable growth rate, the pace of creation of work opportunity has not kept pace with the growing requirement. There is substantial decline in employment elasticity (e.g. increase in employment for every unit rise in GDP) in almost all the major productive sectors, except for transport and finance. In agriculture, the employment elasticity has dropped to near zero. The reason for the phenomenon of jobless growth could be that growth in India has essentially been capital intensive. In the name of LPG many public sector enterprises are also engaged in trimming the labour size. This trend of rising unemployment is aggravated by the historical developmental imbalances in our country. After 1991 government has withdrawn several incentives given to industries located in backward region.

India is a country of baffling paradoxes, i.e. regional developmental imbalances, huge population majority of which are illiterates with few skills. Nearly 65 to 70 per cent of population still dependent upon agriculture which is also dependent on the mercy of raingod. Agriculture the traditional mainstay of majority of Indians, is contributing less and less in GDP, which is barely 18 per cent. In the nutshell, agriculture economy contributes only 18 per cent in GDP still support 70 per cent of the population. This itself tells the story of miserable condition of Indian people,. There are many groups which were traditionally dependent upon agricultural economy, namely rural artisan, wooden and iron equipment makers so on and so forth. These people engaged in unorganized sector of our economy, like handloom weavers, *zari* workers, utensils makers are integrated traditionally with agriculture economy.

A Brief Historical Perspective of Underlying Causes of Migration

The process of decline in agriculture and associated traditional economy started simultaneously with the decline of Mughal empire in Seventeenth century. The weakening of power at Delhi led to the emergence of several Zamindars and Tallukedars declaring themselves as kings. Their ruthless exploitation of rural peasantry and total neglect of agriculture and allied industry made people pauper and powerless. The poor and helpless Indian peasantry devoid of any self-respect and morale did not present the united force against colonials in Eighteenth century. A poor and devastated country with huge but faceless population disgusted at the core proved to be an easy prey to the British merchants and colonials of every hue and cry.

The arrival of British at this juncture added fuel to the fire. Their ruthless exploitation of economy spell doom to the unorganized sector which rendered millions of people jobless. The 200 years of British rule in India aggravated the already neglected agriculture sector and dependent industries. The occurrence of drought and flood became routine phenomenon. All these lead to unemployment and penury to Indian masses. Therefore, we witnessed the most tragic phenomenon of migration of our people in the form of indentured labour in several Caribbean islands. Migration of our people in Nineteenth century was not out of will but out of despair and all-round hopelessness pervaded in Indian subcontinent.

After independence in 1947, our leaders initiated planned development of our economy. Pt. Jawaharlal Nehru, our first Prime Minister, was very much fascinated by the Soviet model of development. The USSR model of planned industrial developments was appreciated because USSR has the biggest landmass with low population base. They need modern industrial activity for speedy development of the backward but resourceful country. But the case of India is different. Here we have huge numbers with limited land area. Our socio-economic parameters are different. We are not religiously and socially homogenous people. Majority of people are landless with no economic means at all. Almost all business resources are controlled by few and the rest are daily wage earners. Even after 60 years of independence nearly 40 per cent of our population lives below poverty line, which means they have to manage their affairs in less than one dollar a day.

So planned development of our economy which was started by Nehru did not make considerable dent in revival of agriculture and associated industries. The hope of the planners was that due to industrialization a large number of people will be taken out of agriculture but it did not materialize. The capital intensive industrial growth model adopted, where most work is being performed by machines so there was little scope for large scale employment of people from agriculture to industries. Slowly and slowly relentless pressure of population on limited amount of cultivable land has further enhanced land inequality. This has led to the process of marginalization of vast majority of cultivators, who eventually were relegated to share-croppers, tenant-farmers or landless agricultural labourers. The neglect of the indigenous crafts and industries in villages has rendered many skilled workers jobless.

Globalisation Encourages Private Investment in Richer States

Nearly 40 years of plan development witnessed heavy investment in core industrial activity. Power, steel, coal, hydro-electric and metal sector was identified as prime industrial mover. And there were very few private sector investment in these sectors. Therefore, it was thoughts that state investment in these sectors will galvanise industrial activity. And these industries came up in very backward region of UP, Bihar, Odisha, Chhattisgarh, MP and West Bengal.

In agriculture sector, irrigation and flood control measures were initiated by constructing big multipurpose dams on rivers. Several agricultural universities and specialized agricultural institutes started with state support to supply quality seeds, pesticides, on affordable cost to the farmers. District level and block level agricultural extension services were started in big way. Several state owned or co-operative sector sugar mills opened in sugarcane producing zone of UP, Bihar and Maharashtra. The cumulative activity of the last forty years have made a real dent in age old poverty of Indian people. We were able to produce more to feed the stomach of poorest of the poor through public distribution system. The state intervention in industrial and agricultural activity has surely reduced the poverty driven migration of people from BIMARU states to prosperous state.

But the whole scenario of slow but self-sufficient growth have few takers in globalization era where high growth rate of other country is lauded. Now people view government sector a liability which always needs budgetary support. They judge public sector performance *vis-à-vis* capitalist business group. They forget the role played by these companies is different than purely commercial activity of a business company. Many sugar mills in UP and Bihar which were directly related with the farmers are sick. Almost 70 per cent state and cooperative sugar mills have stopped operation for want of state support. These mills were not only providing job opportunities to less skilled villagers but providing much needed cash to farmers. The old industrial cities, i.e. Kanpur, Rohtas, Moradabad, Varanasi, Allahabad, Mau located in backward regions look like ghost town. The power generating companies in these states are unable to meet the existing or growing demand of entrepreneurs. The cumulative effect of above mentioned factors are the real cause of economic migration to greener pastures.

In the wake of economic reforms initiated in 1991, the role of private investment has acquired a special significance in the context of economic development of various states of India. Indeed, there has been an element of competition among states for attracting private investment, both domestic and foreign. Some of the states have been offering various tax concessions and other special facilities to new investors on a competitive basis. We present state-wise data on investment proposals, assistance by all India financial institutions and assistance by state financial corporations in Table 7.1.

Table 7.1: Investment Proposals and Disbursal of Financial Assistance for Investment

Sl. No.	*State*	*% of Investment Proposals Aug. 1991-Dec. 1998*	*Cumulative share of financial assistance disbursed by all India financial institutions (up to March end 1997)*	*Cumulative financial assistance disbursed by State financial corporations (up to March end 1997)*
	1	2	3	4
1.	Andhra Pradesh	8.3	7.2	7.8
2.	Gujarat	18.7	13.5	9.3
3.	Haryana	3.6	2.5	4.8
4.	Karnataka	5.6	6.1	15.5
5.	Kerala	1.1	1.7	4.4
6.	Maharashtra	18.0	21.0	11.5
7.	Punjab	3.4	2.4	3.6
8.	Tamil Nadu	7.2	9.0	10.6
	Sub-Total (1 to 8)	**65.9**	**63.4**	**67.5**
9.	Assam	0.7	0.5	0.5
10.	Bihar	1.2	1.4	2.0
11.	Madhya Pradesh	7.4	5.1	3.2
12.	Odisha	2.2	1.8	3.7
13.	Rajasthan	3.9	4.5	6.1
14.	Uttar Pradesh	9.4	7.9	11.1
15.	West Bengal	3.3	3.9	2.5
	Sub-total (9 to 15)	**28.1**	**25.1**	**29.1**
16.	Other states	6.0	11.5	3.4
	All India	**100.0 (757316.0)**	**100.0 (312502.0)**	**100.0 (20896.0)**

Note: All India Financial Institutions include IDBI, ICICI, LIC, UTI, GIC, IRBI and SIDBI.

Source: 1. Annual Report 1998-99, Ministry of Industry, Govt. of India.

2. RBI, Report on Currency and Finance 1997-98, Vol. 1.

The percentage share of different states in these investment proposals are given in column 2 of Table 7.1. The disparities are obvious. The group of forward States accounted for about two-third of the amount while the group of backward states accounted for just about 28 per cent of the amount. Indeed, Gujarat and Maharashtra received together 36.7 per cent of the investment proposals which is significantly more than the total investment proposals received by all the States in the second group. While Gujarat which accounted for less than 5 per cent of the population of the country received about

19 per cent of the private investment proposals, Bihar which accounts for more than 10 per cent of the population of the country received just little over one per cent of such proposals. This is a clear pointer to the direction of private investment in the coming years.

The cumulative share of financial assistance disbursed by All India Financial Institutions upto March end 1997, State-wise are given in column 3 of Table 7.1. The great divide between forward and backward States is clear. Maharashtra alone received almost as much financial assistance as all the States in the second group put together. It may, however, be noted that Mumbai being the headquarters of large number of private companies in the country, it is possible that some of the financial assistance accounted for as Maharashtra's may be actually flowing into other States for actual investment. Here again, the share of States like Assam, Bihar and Odisha are far below their respective population shares. These figures give a clear indication as to where the resources mobilized through the all-India financial institutions are flowing into.

The last column of Table 7.1 gives the share of cumulative financial assistance provided by the State financial corporations during 1991-96. The pattern is not any different from the other sources of private investment as far as the State-wise distribution is concerned. While over two-thirds of such assistance is provided by the financial corporations in the forward States, just 29.1 per cent is accounted for by all the States in the backward group.

The State-wise details of banking operations in the country as on the last Friday of March 1998 are presented in Table 7.2. Column two gives the State-wise distribution of bank branches in the country. It is obvious that, by and large, the bank branches are fairly distributed across the States without any major bias towards the group of forward States. It may need mention that this could be attributed to the banking sector policies pursued after nationalization of the major commercial banks in the country in 1969.

The distribution of bank credit across the States given in column 4 of Table 7.2 shows that bank credit distribution is even more skewed than bank deposit distribution. This implies that a part of the deposits mobilized in the backward States is getting transferred to the advanced States. Indeed, Maharashtra alone accounted for more bank credit than all the seven states in the second group put together. Similarly,

Table 7.2: Bank Branches, Deposits and Bank Credit (as on last Friday of March, 1998)

Sl. No. State	Bank branches	Share of bank deposits	Share of bank credit	Credit-deposit ratio
1	2	3	4	5
1. Andhra Pradesh	4956	5.3	6.9	72.1
2. Gujarat	3562	5.9	5.1	48.2
3. Haryana	1423	2.1	1.6	42.9
4. Karnataka	4572	5.4	6.6	68.2
5. Kerala	3165	4.6	3.7	44.3
6. Maharashtra	6071	19.9	25.9	72.3
7. Punjab	2427	4.6	3.2	38.6
8. Tamil Nadu	4676	6.6	11.5	96.1
Sub-Total (1 to 8)	**30861 (48.0)**	**54.4**	**64.5**	
9. Assam	1241	1.0	0.6	32.8
10. Bihar	4981	4.3	2.1	27.5
11. Madhya Pradesh	4449	3.9	3.6	51.4
12. Odisha	2170	1.5	1.2	45.2
13. Rajasthan	3259	2.8	2.4	47.4
14. Uttar Pradesh	8810	9.9	5.1	28.6
15. West Bengal	4352	7.5	6.2	46.1
Sub-total (9 to 15)	**29262 (45.5)**	**30.9**	**21.2**	
All India	**64267**	**100.00**	**100.0**	**55.5**

Source: RBI, Report on Currency and Finance, 1997-98, Volume I.

all the states in the second group, except Uttar Pradesh and West Bengal, put together received less bank credit than Tamil Nadu. The implications of such skewed distribution of bank credit across the States on economic growth and income distribution in the coming years are obvious. The fact that Maharashtra and Tamil Nadu have major metros in them might have helped them to get higher share of bank credit. Having Calcutta as the State capital might have helped West Bengal also somewhat. In this connection, it may be of interest to note that all the 15 states considered together which account for 96.5 per cent of the population of the country accounted for only around 85 per cent of bank deposit of bank credit. The fact that the remaining 15 per cent has gone to the minor States and UTS may be somewhat surprising. This, however, is because of NCT of Delhi accounting for over 10 per cent of bank deposits and bank credit. In Table 7.3, region-wise number of credit accounts per 100 persons of scheduled commercial banks also shows the skewed figure in favour of southern and western region.

Table 7.3: Region-wise Number of Credit Accounts per 100 Persons—Scheduled Commercial Banks in India

Region	*Rural*		*Urban*		*Total*	
	1991	*2005*	*1991*	*2005*	*1991*	*2005*
Northern	6.6	5.1	5.9	5.7	6.4	5.7
North-East	4.4	3.2	4.4	3.9	4.4	3.3
Eastern	7.2	4.2	4.3	4.3	6.6	4.2
Central	5.8	4.2	4.4	4.5	5.5	4.3
Western	6.2	4.2	4.8	12.2	5.7	7.5
Southern	13.6	12.7	7.6	17.4	11.8	14.2
All-India	7.9	6.0	5.5	9.8	7.3	7.0

Source: RBI, Report on Currency and Finance, 2007-08.

This trend has further accelerated the divide during ten years. Our big cities, and well-off states obviously attract large chunk of investment. This further fuels the already existed regional imbalances. Very few foreign investors have shown any interest in investing in backward region or underdeveloped states. Most of the F.D.I. in India has gone in service and financial sector, very little money is invested in power or basic industries where less skilled Indian people can be accommodated. The F.D.I. inflows in sector specific allocation from 2000 to 2007 show the direction of investment (Table 7.4).

Table 7.4: Sectors Attracting Highest FDI Flows

(Rs. crore)

Sector	*Cumulative (inflows (April 2000 to Nov. 2007)*	*Share of inflows (per cent) (April 2000 to Nov. 2007)*
Services sector[a]	38228	19.86
Computer software and hardware	30760	15.98
Telecommunications[b]	15607	8.11
Construction[c]	9989	5.19
Automobile industry	8350	4.34
Power	5958	3.09
Chemicals except fertilizers	5956	3.09
Housing and real estate	7573	3.93
Drugs and pharmaceuticals	4633	2.41
Metallurgical industries	4572	2.37

[a] financial and non-financial services; [b] radio paging, cellular mobile, basic telephone services; [c] Construction including roads and highways.

Source: Department of Industrial Policy and Promotion.

Table 7.5: Region-wise Break-up of FDI Received (April 2000 to November 2007)

Regional Office of the RBI	*States covered*	*Share in FDI inflows (%)*
Mumbai	Maharashtra, Dadra Nagar Haveli, Daman and Diu	25.14
New Delhi	Delhi, parts of UP and Haryana	22.68
Bangalore	Karnataka	7.03
Chennai	Tamil Nadu, Pondicherry	6.69
Hyderabad	Andhra Pradesh	4.12
Ahmedabad	Gujarat	2.84

Source: Department of Industrial Policy Promotion.

Table 7.5 gives region-wise break up of F.D.I. received between April 2000 to November 2007 which also demonstrates that Mumbai, New Delhi, Bangalore, Chennai, Ahmedabad region received almost 70 per cent of inflow and the Eastern, Central and backward region of Rajasthan received little money. Since we have already said good-bye to controlled and regulated growth, private investment will surely flows where higher return is expected. No foreign country or F.D.I. inflow is interested in balance development of our country, it is the prime responsibility of our leader to regulate the flow of investment in such fashion that it did not aggravate the already existed skewed development.

Globalization is Pushing Regional Disparities Further

With the early nineties, we have adopted globalization as the guiding mantra. The people at the power centre now-a-days, see globalization as the final remedy to the ills of Indian economy. Several plan era restrictions and reservations, which were somehow able to protect the small sector industries, where majority of Indians find jobs, were thrown open to big business houses. Several reserved categories products are now in the free-zone. After 1991 we are living in LPG era where government's intervention in economic activity is not welcome. The licence-permit system which was able to protect the traditional skill based industries, have gone. Now these skill-based people without having capital, modern management skill, are thrown into competition with big capitalist business houses. This open market economy operates as, Marx famously wrote, "functions essentially on ever more concentration of capital in fewer and fewer hands. There is a need within the market economy to reduce real wages to withstand competition and this leads to the ever more marginalization of labour." Our globalization led growth process is not taking place in manufacturing or agricultural sector but in the service sector. The service sector economy operates by excluding vast majority of people as they don't become factory workers nor modern avenues are available to them. And given the poor literacy rate wider participation of these people in service growth is less. And this is the paradox: faster the Indian economy grows, the more exclusionary it becomes. Vast agriculture sector lumbers on the margins as planners and commentators hope and pray for better rains and fewer floods, to keep it going and achieving four per cent growth target.

Globalization and Income Inequality

Although globalization has led to the economic development in the country, it has not really helped to minimize income inequalities across the country. Recent surveys say that lots of people across the country still live below the poverty line and their standard of living has gone down considerably. In fact, the poverty level in India is much more than that of China. One of the main causes of poverty in India is income inequality. While the effects of globalization, economic liberalization and market growth are being felt in the cities and urban sectors across the country, most of the rural areas are still not so developed and the condition has also not significantly improved.

A significant portion of the rural mass does not have access to the basic amenities such as electricity, education, sanitation, drinking water, and infrastructure. The wealth distribution pattern is also uneven in the country. Recent survey has shown that the top 10 per cent of the income groups share around 33 per cent of the total income of the country. Even after globalization and economic progress, around a 40 per cent of the population of the country has a earning less than the minimum level of one dollar per day.

Globalization has also not done much good to reduce disparities between various regions and states across the country. While some states have a high earning, there are also some states which have a very low growth rate. Some of the states which have a high income are Gujarat with 8.8 per cent, Delhi with 7.4 and Haryana with 8.7 per cent. In comparison to these, Bihar with 5.1 per cent, Uttar Pradesh with 4.4 per cent and Madhya Pradesh with 3.5 per cent are some of the lowest ranking states with very little socio-economic growth.

India saw an annual growth rate of GDP (Gross Domestic Product) of only 3.6 during the three decades, post-impendence. It is a growth of only 1.4 per cent in terms of per capita. Then it saw a growth rate of 5.4 per cent. The growth rate of GDP soared up to 7.3 per cent by 2007-08, whereas the per capita GDP saw a growth of 5.7 per cent. Due to sufficient foreign exchange reserves, moderate inflation, stable exchange rates and adequate food stocks, the country achieved a sustainable growth rate. However, the gross state domestic product growth rates largely varied between states in India. The states like Assam, Uttar Pradesh, Madhya Pradesh and Bihar saw a growth rate of less than 5 per cent in pre and post-reform period. On the other hand, states like Andhra Pradesh, Tamil Nadu, Gujarat, Karnataka and Haryana saw a growth rate of less than 5 per cent during 1980-1993, which improved considerably during 1993-2006.

Table 7.6: Growth of Gross State Domestic Product from 1980 to 2005-06

States	*GSDP*		*Per capita GSDP*	
	1980-81 to 1992-93	*1993-94 to 2005-06*	*1980-81 to 1992-93*	*1993-94 to 2005-06*
Andhra Pradesh	5.72	6.53	3.50	5.41
Gujarat	6.77	7.10	4.74	5.30
Haryana	5.75	6.98	3.24	4.81
Karnataka	5.57	6.88	3.59	5.50
Kerala	3.82	6.28	2.44	5.51
Maharashtra	6.32	6.17	3.97	4.42
Punjab	5.21	4.53	3.23	2.93
Tamil Nadu	5.46	6.12	4.03	5.16
Assam	3.98	3.54	1.75	2.14
Bihar	3.41	4.73	1.23	5.26
Madhya Pradesh	4.49	4.91	2.07	6.00
Odisha	3.67	5.53	1.80	4.31
Rajasthan	7.37	5.44	4.74	3.29
Uttar Pradesh	4.47	4.19	2.17	2.59
West Bengal	4.57	6.61	2.35	5.23

Source: CSO, 2007-08.

Even after 60 years of planning, the country's economy continues to be oriented towards export, mainly of raw materials. Railways and roadways still feed export centres, instead of the interior areas within the country. Investments and industries continue to be polarized, as in the past, only in a few selected cities. Rarely there has been vigorous attempt at rural development or attempts for development of hitherto neglected states and regions. Numerous villages are, therefore, reeling under-poverty, unemployment and under-employment. This is the reason that these masses of illiterate, unskilled, semi-skilled poor labours keep crowding to the big cities to find any job, no matter how low the pay or unbearable conditions prevail. In a recent statistics, it was revealed that more than 10 crores of Indian people have migrated since 1991. Out of which nearly 40 per cent people live in West Bengal, Maharashtra, Punjab and Delhi region. Delhi has become the heaven of migrants from poor states of India. For every ten resident four belong to interior poverty stricken hinterland. As Desai stated "the state instead of becoming a funnal to pump out resources from the rich for distributing them to the poor, in fact works in the opposite direction through an elaborate system of deficit financing, loans (foreign, internal) and indirect taxation—a system which hurts the poor most" (Desai, 1974).

As a result, whatever developmental planning has been attempted has actually increased the gulf between the rich and the poor, and whatever regional developmental programmes have been taken are characterized by *ad-hocism* and maintenance of the *status quo*. In fact, increased regional inequalities, have increased instead of narrowing down,

The ever increasing population and consequential unemployment is compounded by the existence of regional imbalances in development within the country which have collectively accelerated the phenomenon of migration. All theories of migration concede that migration occurs when the region of origin lacks the opportunities which the destination promises. It is inherently a combination of push and pull factors. The rural poor are concentrated in eastern India, and in the rainfall-dependent parts of central and western India, which continue to have low agricultural productivity, while bulk of jobs are being created in western and southern India. Invariably, migrant labour is paid lower wages compared to local labour and the implementation of the Interstate Migrant Workman (Regulation of Employment and

Conditions of Service) Act, 1979 is largely on paper. Migrants of backward regions are willing to accept any distress wages that are offered as long as they have access to employment. In the bargain they undercut the employment prospects of local labour. This uncontrolled supply also contributes in reducing the wage rate. This is the chief cause of social unrest and opposition in Maharashtra, Punjab, Assam against migrant labourers.

The Mahatma Gandhi National Rural Employment Guarantee Scheme Falters

One of the significant objectives of the MGNREGA is to arrest out-migration of unskilled, landless labour force from the rural areas to urban areas by ensuring up to 100 days of wage employment within their native jurisdiction so that these 100 days guaranteed wage employment can be judiciously and rationally utilized by the landless peasants during lean and distress seasons.

A recent Planning Commission study illustrated the mass migration scenario in all the 20 districts surveyed. In Malda, and South Garo Hills, almost half of the rural folks expressed that there is a mass migration from their places. In Medak (A.P.) which is adjoining the fast growing Hyderabad metropolitan is experiencing mass migration to the extent of 40 per cent. In Dahod of western region it is almost one-third of the rural population. In this study an attempt is made to know the reasons for migration purely in terms of whether there exists enhanced wages upon migration in other places compared to the same kind of work in their own native villages. It is surprising to note that contrary to the general perception of better wages upon migration, 70 per cent of the beneficiaries revealed that the migration is only for just wages and not for any better wages. This implies that there is a distress migration for just minimum wages to earn the livelihood and for survival rather than for better wages.

After 1991 government has started divesting the public sector in big way and the proceeds were to be utilized in rural development so that distressed migration can be checked. After almost 10 years of the implementation of this scheme, objective evaluation have confirmed that this much hyped scheme is mired in corruption and seepage of funds is reported in big way. Many poor states have not been able to provide more than 30 to 40 days of work to the needy. In a recent newspaper report job card holders of Bihar burnt their job cards in protest, as they are neither getting jobs on demand nor unemployment allowance as required in the scheme. The all-India employment rate of job card holders has not crossed 50 per cent under this scheme. However, government is trying to stop the pilferage by signing a memorandum of understanding with the postal department in November last year to resolve the procedural delay in getting the wage money to the panchayat post offices, the major roadblock yet to overcome is the delay in making the pay advice by the MGNREGA department.

Mechanisation of Agricultural Pushing Migration

LPG era policy has made import cheaper due to our commitment to WTO. This has facilitated easy import of different kinds of agricultural implements which are used in labour deficit European countries. These machineries, like harvester, tractor, sowing weeding machine are replacing village landless labourers at a very fast rate. The bargaining capacity of village people has seriously compromised due to these machines. In a labour surplus economy uncontrolled use of machines only adds the unemployment of unskilled or semi-skilled people which ultimately leads to large scale migration towards big cities.

Several kinds of weaving machines are being imported which is making competition tougher to the handloom sector. These machine operated looms are much faster *vis-à-vis* to handloom which is playing havoc to the livelihood of poor weavers in Uttar Pradesh, Bihar, Andhra Pradesh. Free flow of capital and technology is bulldozing the poor farmers and weavers of poor states. That is why more than 2 lacs of weavers have already migrated from Varanasi and surrounding area to big cities as their livelihood is destroyed.

Conclusion

Migration especially force compulsive have been old phenomenon. Historical and political dynamics, war and famines also increased the rate of displacement. After independence Governments at centre and states initiated several measures to arrest this phenomenon. But the New Economic Policy adopted in 1991 again pushing the jobless villagers towards cities. The old economy is shrinking and new investment is coming in selected pockets. The widespread mechanisation in agriculture and destruction of cottage industries are rendering thousands of people jobless. This is causing distress migration. The much talked MNREGA scheme which was started to arrest migration is mired in corruption.

REFERENCES

Desai, A.R. (1974). An alternative approach to development. In A. Bose, P.B. Desai, A. Mitra, and J.N. Sharma (eds.) *Population in India's Development, 1947-2000*. New Delhi: Vikas Pub.

Government of India. (1999). *Annual Report 1998-99*. Ministry of Commerce and Industry, Government of India.

Government of India (2008). *CSO Reports, 2007-08*, Ministry of Statistics and Programme Implementation, Goernment of India.

RBI, *Report on Currency and Finance, 1997-98*, Vol. 1.

RBI, *Report on Currency and Finance, 2007-08*.

8

The Mega Cities of India: A Question of their Sustainability

Tara Devi Singh

Introduction

Third World urbanization especially the metropolisation variant of it has not been involved in the process of spatio-economic transformation and scholars have most often emphasized the need for pursuing a positive policy of urbanization and human settlement distribution in order to be benefited from the process of polarization and modify the resultant phenomena of spatial dualism. It should be noted that in 1951 urban sector accounted for 29 per cent of the total GDP which grew to 47 per cent in 1980-81 and 62-63 per cent in 2009-10, while the population contained in them grew from 17 per cent to 23 per cent and 31.16 per cent (Census of India, 2011, Prov.) respectively. With the enhancement in economic growth through economic reforms linked to liberalization these proportions are likely to increase, GDP to 75 per cent and population to become 40 per cent by 2021 (estimates by Planning Commission, GoI, 2011).

In the Eleventh Five Year Plan Report of Planning Commission, it has been remarked, that 'Indian cities will be the locus and engine of economic growth over the next two decades and the realization of 9-10 per cent growth in GDP depends fundamentally on making Indian cities much more livable, inclusive, bankable and competitive

(Planning Commission, GoI, 2011). India has a young and rapidly growing population and by 2030, 180 million persons will enter India's workforce and thriving cities will account for 70 per cent of the 170 million net new jobs created by 2030 to provide more than 70 per cent of the GDP (with a modest estimate of annual GDP growth of 7.4 per cent) and drive a fourfold increase in per capita income across the nation. Besides, the urban economy is also vital for the overall development as it provides 80 per cent of total tax revenue for the benefit of 200 million rural Indians living in the proximity of 68 future large cities in the country (Mckinsey, 2010). It will also be important for eventual increase in agricultural productivity and income. In this way, as urbanization and economic development go hand in hand, it will lead to an overall economic development of the country.

Thus, a great part of economic potential of the nation lies in these cities. The development of the hinterland (rural areas) depends on the development of the core (cities). As such, the efficiency and sustainability of our cities become a prerequisite for the progress of our nation. A corresponding increase in investment on physical and social infrastructure renewal and addition, city planning and urban development augment the process. It has rightly been examined that issues arising out of such urbanization deserve special attention as the urban future is 'inevitable and it should not be feared. It carries many risks ... such as the possibility of collapse of basic services, intolerable environmental degradation and escalating social conflicts ... but the very nature of large cities as concentration of human creativity and the highest forms of social organizations opens up new avenues for human development' (Chakrabarti, 1996). Correa (1988) observes, 'urbanization is the most massive phenomena of our time. It is going to happen anyway. Are we going to be passive receptors of its effects or are we going to rise to its challenges and make the best of it?'

Trend of Urbanization

The history of urbanization reveals that at the onset of Twentieth century, five per cent of the world population lived in cities which shot up to roughly 50.6 per cent after 110 years in 2010. Proportion of population living in urban areas of the world registered an increase of 15 per cent from 30 to 45 per cent during 1950 to 2000, developed nations 21 per cent from 55 to 76 per cent and developing nations 22 per cent from 18 to 40 per cent in the same duration (Brockerhoff, 2000). That is, in recent years the most explosive growth has been recorded in the developing world, whereas between 1950 to 1995 the number of one million plus cities more than doubled from 49 to 112 in the developed world, during the same period these multiplied to a staggering almost six times from 34 to 213 in the developing countries. In 2010, of the 414 million plus cities 75 per cent are found in developing world, showing that while urban growth on the one hand has more or less stabilized in the developed world, it is leaping fast in developing world.

In India, although the proportion of urban population (31.80 per cent in 2011) is far behind as compared to even other Asian countries like Malaysia (72.21 per cent), Indonesia (53.7 per cent), China (44.9 per cent), Pakistan (37 per cent) and Thailand (34 per cent) in 2007 (UNDP, 2007), its aggregate 377 million urban population (Census of India, Prov. 2011) is the second largest only after China (535 million, 2005) and, is the size of total population of many countries including U.S.A.

Growth of Urban Population

In contrast to slow growth of urbanization the population of class I cities and more so of million plus cities is growing continuously and regularly by leaps and bounds, i.e. major share of increased urban population is taken up by class I and metropolitan cities. In 2011 the 7935 urban centres in the country housed 31.16 per cent (377.10 million) population of the country's total population. The 468 class I cities and 53 million plus cities witnessed 70 per cent (264.9 million persons) and 42.6 per cent of the total urban population respectively.

The Mega Cities

For this paper, the identification of mega cities is done on the basis of cities having population more than 10 million in 2001 census, i.e. Mumbai UA (16.49), Delhi UA (13.20) and Kolkata UA (12.87)

Table 8.1: India's Total and Urban Population (1951-2011)

Census Year	Total Population (million)	Urban Population (million)	No. of UAs % Towns	Urban Popn. as % of total popn.	% of Urban Popn. in Class I	Popn. of million+ as % of total urban popn.
1951	361.08	62.44	2843	17.29	44.63	18.81
1961	439.23	78.93	2365	17.97	51.42	22.93
1971	548.13	109.11	2590	19.91	57.24	25.51
1981	683.32	159.46	3378	23.34	60.42	26.41
1991	844.32	217.17	3768	25.72	65.20	32.54
2001	1028.61	285.35	5161	27.78	61.48	37.77
2011	1210.19	377.10	7935	31.16	70.24	42.60

Source: Census of India, 1991. Series-1, Paper 2 of 1991; Census of India, 2011, Provisional Population Totals.

million. Now, in 2011 their population has shot up to Mumbai UA 18.4 million, Delhi UA 16.3 million and Kolkata UA 14.1 million. They stood 5th, 6th and 8th rank in world's top 10 largest metropolises in context of population in 2005 and are expected to rise. Further, Mumbai UA (21.9 million) is expected to hold 2nd rank after Tokyo (35.5 million), Delhi UA 6th rank (18.6 million) and Kolkata UA 8th rank (17.0 million) by the year 2015 (UNDP, 2007).

It has been found that mega cities have reached to saturation points which is evident from the fact that other metropolitan and class I cities are growing faster as their aggregate share declined gradually in the populations of total metropolitan as well as class I cities. In a total of five metropolitan cities in 1951, the present three cities accounted for 90.37 per cent, while in 2011 out of 53, these accounted for only 30.36 per cent. Similarly, the share of the three mega cities in total class I cities' population declined from 38.5 per cent in 1951 to 18.42 per cent in 2011, which is still too high indicating that urban system of India lacks proper articulation as class I cities are growing faster than towns, metropolitan faster than class I and mega cities faster than metropolitan cities.

Total population growth in all the three mega cities exhibits declining trend from 1981 as is pointed out by their mean growth rate (Table 8.2). In 1951, Mumbai M.C. had 2.9 million persons which rose to 8.24 million in 1981. For the same duration Kolkata M.C. started with 4.66 million, reached to 9.19 million holding the first position among the most populous urban centres of the country

Table 8.2: Growth of Population in Mega Cities (1951-2011)

Mega cities	*1951*			*1961*			*1971*			*1981*			*1991*			*2001*			*2011*	
	Popn. (m)	*Growth Rate %*	*Share in class I %*	*Popn. (m)*	*Growth Rate %*	*Share in class I %*	*Popn. (m)*	*Growth Rate %*	*Share in class I %*	*Popn. (m)*	*Growth Rate %*	*Share in class I %*	*Popn. (m)*	*Growth Rate %*	*Share in class I %*	*Popn. (m)*	*Growth Rate %*	*Share in class I %*	*Popn. (m)*	*Share in class I %*
Greater Mumbai	2.9	39.94	10.78	4.15	43.79	10.41	5.97	38.06	10.00	8.24	52.50	8.8	12.57	30.72	8.99	16.49	12.05	9.39	18.4	6.94
								←Areal Expansion→												
Kolkata UA	4.6	28.14	16.97	5.98	24.01	15.00	7.42	23.90	11.60	9.19	18.73	9.8	10.91	20.97	7.81	13.20	6.81	7.52	14.1	5.32
								←Areal Expansion→												
Delhi UA	1.4	64.17	5.22	2.35	54.57	5.91	3.64	51.09	6.10	5.72	46.18	6.1	8.37	53.75	5.99	12.87	26.69	7.33	16.3	6.15
								←Areal Expansion→												
Total Popn.	8.90			12.48			17.03			23.15			31.50			42.56			48.80	
Mean Growth Rate		40.22			36.45			35.93			36.06			35.11			14.66			
Total Metro. Popn. (m)	11.74			18.10			27.83			42.12			70.66			107.93			160.70	
No. of Metros	5			7			9			12			23			35			53	
% share of Mega popn. in Metro popn.	90.37			79.76			72.60			65.18			52.62			39.43			30.36	
No. of class I cities	76			102			148			216			296			394			468	
% of mega popn. in total class I popn.	38.56			36.16			36.00			29.30			26.62			24.25			18.42	

Source: Census of India, 2001, Census of India, 2011 Provisional Population Totals.

because of being the capital of British India for long. Inclusion of Kalyan and Thane M.C., Ulhasnagar, New Bombay and Mirabhayander municipalities to Bombay M.C. raised Greater Mumbai to 1st position in 1991 with 12.57 million persons, closely followed by Kolkata UA and Delhi MC with 10.91 and 8.37 million persons respectively.

As the data in Table 8.2 reveal because of the shift of capital from Kolkata in 1911, *Delhi* became the most rapidly growing mega city in the country and maintained consistently high growth of population especially since Independence. It has been estimated that migration from neighbouring states especially from the east contributed an additional 2.7 per cent growth of population while natural growth added another 2 per cent. The large number of migrants constitute construction workers who came here to work during Asiad and Olympics besides the Bangladeshi migrants. Later on, they set up their shanty towns which have become unauthroised settlements made of *jhuggi* clusters. *Kolkata* was most populous city from the beginning of the century. Bangladesh liberation war in 1971 led to a mass influx of thousands of refugees in Kolkata resulting in a massive strain on its infrastructure. In mid-1980, due to faster economic growth of Mumbai, Kolkata slipped to second place. After economic reforms in the year 2000 and growth of the city, its population also is growing faster.

The growth of population in the mega cities has slowed down considerably during the last decade. Greater Mumbai UA, which had witnessed 30.47 per cent growth in population during 1991-2001 has recorded 12.05 per cent during 2001-2011. Similarly, Delhi UA (from 52.24 per cent to 26.69 per cent in 2001-2011) and Kolkata UA (from 19.60 per cent to 6.87 in 2001-2011) have also slowed down considerably (Census of India, 2011). However, the annual growth rate of individual mega cities in all the decades may not be very high rather declining also, still the base itself is so wide that even the slightest rise in percentage terms translates into several thousands. Hence, the base fact remains, Mumbai UA multiplied its population to staggering 6.34 times from about 2.9 m to 18.4 m in 60 years duration from 1951 to 2011, Kolkata UA about 3 times from 4.6 to 14.1 m and Delhi UA, surpassing all, to a 11.64 times from 1.4 to 16.3 millions.

Density of Population

The worst result of the population explosion in these cities is the concentration of people leading to unbearable density. However, one can have a sigh of relief with the decreasing rate of urban population growth during 2001-11. There were 29,850 persons/km^2 on *Mumbai* city space during 1994 which got reduced to 20,694/km^2 during 2011 due to slackened population growth rate of 12.05 per cent during 2001-2011 against 30.47 per cent of the previous decade (Census of India, Prov. 2011). One of the world's highest density levels which is getting literally too close to breathe is found in the island city of South Mumbai which has only 67 km^2 area and holds 40 per cent of the city's population leaving the northern Suburban District spanning 370 km^2. The millions who come in to work daily further enhance the problem. In *Delhi* NCT, density of population escalated from 4200 persons/km^2 in 1981 to 11,297 persons/km^2 in 2011. In fact, the urban zone of the Municipal Corporation has density much higher than this figure. Similarly, *Kolkata* city has the highest density of population among the three mega cities to the scale of 24,252/km^2 making it one of the most populous urban areas in the world (Census of India, Prov. 2011).

Causes of Urban Explosion in the Mega Cities

The causes of the urban explosion in the mega cities are not far away to seek. It continues unabated partly due to incessant migration from the countryside as a result of social and economic revolution and partly due to natural growth within it. In recent years migration trends have changed so that more women have accompanied their spouses to the cities, birth rate balancing out decline in numbers which keep out growth rate from declining more rapidly. In fact, reproduction took over migration about two decades age.

The three mega cities, two being state capitals and the one being national capital, are no longer dominated merely by bureaucrats, professionals, academicians and traders, rather a large proportion of population in these cities belong to the informal economy. In fact, these cities now have witnessed phenomenal expansions of large-scale industries, besides mushrooming of small scale squalid industries, wholesale and distributive trade and proliferation of service sector, public and private both entrepreneurs, all these function as pull factors for the migrants. On the other hand, push factors of massive migrations include the deepening agrarian crisis that led to over 2,40,000 farmers suicide between 1995 and 2009, most of them mired in debt and no job avenue for the youth in agriculture sector or the closure of powerloom units in Gujarat during economic crisis of 2008 (Sainath, 2011). Most of the migrations from U.P., Bihar, Odisha, Andhra Pradesh, Kerala and other parts of the country to the mega cities are distress laden and footloose[1] with no particular skill. So they are bound to be absorbed as cheap labour in construction works, brick kilns or pulling *rickshaws* etc. (Sainath, 2011).

A brief description of the various avenues/opportunities offered by the three individual mega cities, Mumbai, Delhi and Kolkata follows here:

Mumbai, developed from an ancient fishing community and a colonial centre of trade, has become South Asia's largest city. It serves as an economic hub of India contributing 10 per cent of factory employment, 25 per cent of industrial output, 33 per cent of income tax collection, 60 per cent of customs duty collection, 20 per cent of central excise tax collection, 40 per cent of India's foreign trade and Rs. 4000 crore (US$ 892 million) as corporate taxes. It accounts for 70 per cent of maritime trade in India and 70 per cent of capital transaction to Indian economy. It is commercial capital of India, one of the world's top 10 centres of commerce in terms of global finance flow. Mumbai is home to important financial institutions like RBI, BSE, NSE, SEBI and Headquarters of numerous Indian companies and Multinational Corporations. India's premier scientific and nuclear institutes like BARC, NPCL, IREL, TIFA, AERB, AECI and DoAE have found place there. Also, it is entertainment capital of India as Hindi and Marathi TV industry exist here. Many technological institutes, several independent and autonomous universities, management schools, law colleges and other institutions/organizations of international repute are located here. Large number of State and Central Government employees, skilled professionals and high percentage of unskilled and semi-skilled population earn their livelihood as hawkers, drivers, mechanics in the above sectors. Mumbai port and around 15,000 industrial units run in single room in Dharavi employ large number of workers. All these provide sufficient reasons for it to act as magnet for migrants from all over India resulting into a mix of many communities and cultures.

Similar is the case with ***Delhi***. Being the seat of Federal Government of India along with its local Government Owing to migration of people from across the country (mostly from northern and eastern states of India) Delhi has grown to be a multicultural, cosmopolitan mega city. Its rapid development and urbanization coupled with high average income has transformed Delhi into a major cultural, political and commercial centre of India. Delhi's key service industries, backed by strong and well laid out infrastructure including IT, telecommunication, hotels, banking, media and tourism has put it next to Mumbai. In recent times, manufacturing industry has grown considerably especially manufacturing units of consumer goods industry have been established in and around the capital. The backbone of Delhi's economy is formed by construction, power, real estate, health and community services. These are the key areas where migrants especially from eastern, western and surrounding states come to seek employment.

Kolkata's contribution to the country's GDP (PPP[2]) is third among the three mega cities. As earlier stated, Kolkata suffered from economic stagnation in post-independence period. However, after economic reforms in the year 2000 an economic rejuvenation has led to an acceleration in the development of Kolkata due to growths in IT sector, services and manufacturing sectors. Kolkata is also an important centre for banking and finance. Besides being the Headquarters of three Nationalised banks, Kolkata houses 17 universities or autonomous institutions and numerous affiliated colleges. The opening of Nathu La pass in Sikkim as a border trade route and the immense interest of South East Asian Countries to enter Indian labour market is further going to add to the burgeoning population of Kolkata (rediff.com, 2003). Also, Kolkata is home to many industrial units operated by large corporations with products ranging from electronics to jute. Its 40 per cent of labour force range from skilled/semi-skilled in informal sector to Government employees and blue collar/knowledge workers.

Besides diverse and innumerable employment opportunities available to migrants in these cities, *women* in large numbers also enter the employment stream as they have more chances to exercise autonomy and generally face less inequality in key areas such as access to education, reproductive health services and waged jobs. Some people especially from lower strata migrate to escape caste problem. Here, in the face of survival, caste problems submerge.

Thus, despite the various handicaps of living in these cities, diverse employment opportunities and infrastructural facilities offered here are better than those of neighbouring regional centres (step migration)/rural areas. Hence, urban population is growing faster than total population and mega cities are experiencing fastest aggregate population growth ever seen.

The Problem

Fast growth of population in mega cities forces migrants to live in such *bustees* which are devoid of basic amenities. Life in such areas are worse than the rural areas. The influx of population has augmented the unplanned and uncontrolled development of the city leading to proliferation of unauthorisd construction, mushrooming of slums amidst highrise residential buildings. The condition of urban infrastructure and utility services is deteriorated and the institutions/ organisations providing civic amenities are finding themselves unable to cope with satisfying the minimum needs. The roads clogged with unauthorised pavement shops, overcrowded public/private transport result in traffic congestion, inadequate water supply, frequent breakdown of power supply and use of noise creating generators, uncleared garbage, sewage system heavily loaded and untreated sewage flowing into water bodies, all lead to rapid environmental degradation. Similarly, heavy vehicular traffic releasing fume and rising level of noise pollution in market and residential areas lead them to breaking point. Tuberculosis, asthma and other respiratory/ allergic ailments spread like epidemics in these mega cities.

Slums

One of the fallouts of uncontrolled urbanization is the proliferation of slums which are housed and occupied by the poor, deprived and destitute urban dwellers. The condition of mega cities as regards slum condition is even worse. Slum dwellers along with those residing in congested *chawls* occupy only as much as 16 to 17 per cent of space that a family living in middle class flat would occupy. *Mumbai* slum population has already crossed the 50 per cent mark (Table 8.3). Dharavi, Asia's second largest slum, alone houses about 8,00,000 people. In Mumbai 2,335 pockets having over 9 lakh huts and 1,110 slums with over 4.5 lakh huts are on lands owned by Central and State Governments. Total housing demand in Mumbai in the next few years would be about 2.5 million units. *Delhi* has the second largest slum population in India with 1.8 million people (18.74 per cent of total population in 2001) living in it. These people are mostly unemployed or daily wage workers who cannot even afford basic necessities of life. Here, every year 8000 to 10000 people have to be resettled for the next 10 years. Around 230 clusters were identified for providing basic facilities like toilets, water supply, electricity etc. In *Kolkata*, over the past 10 years expansion of slum in municipal limits has recorded nearly 5 per cent increase each year. Here, 34.48 per cent persons (1.48 million in 2001), higher proportion than Delhi, live in 2011 registered and 3500 unregistered (occupied by squatters) slums. More so, increasing social and economic tensions in mega cities as a result of widening gap between rich and poor and declining moral and cultural values result in raising crime rate.

Table 8.3: Slum Population in Mega Cities, 2001

Mega cities	*T. Popn. (million)*	*Slum Popn. (million)*	*% of Slum to T. Popn.*
Greater Mumbai	11.97	6.47	54.06
Delhi M. Corp.	9.97	1.85	18.74
Kolkata	4.57	1.48	32.48

Source: Census of India, 2001.

Crimes in Mega Cities

Crime rate is increasing unabated in these cities. Only a general idea can be had from the reported cases as thousands of cases go unreported. In 2007, *Delhi* accounted for the highest percentage (15.1 per cent) of the crimes reported in the 35 metropolitan cities in India which declined to 13.2 per cent in the year 2009. Crime rate in metropolitan cities *versus* domain state is also higher in Delhi both

Table 8.4: IPC Crime Rate Metropolitan Cities vs Domain State

Cities	*Crime rate*		*% of total crime in mega cities*	
	2007	*2009*	*2007*	*2009*
Mumbai City	186.2	191.0	9.0	9.1
Delhi City	397.9	353.7	15.1	13.2
Kolkata City	82.5	103.0	–	–
Avg. rate of crimes in 35 metropolitan cities	312.3	318.6	–	–
National rate of crime	175.1	181.4	–	–

Source: Crime in mega cities: http://ncrib.nic.in/cii_2009/chapter 2

in 2007 (397.9) and 2009 (353.7) against the average rate of crime in metropolitan cities (312.3 and 318.6) and national rate of crime (175.1 and 181.4, Table 8.4). Here, women and children are the most vulnerable group. The city has the highest rate of crime against women usually manifested in different forms be it dowry deaths, rape or molestation/sexual harassment. Crime against children and juvenile delinquents are quite common. While the crime rate in *Mumbai* is not as high as in Delhi, underworld dons are notorious for extortion and harassment of the rich (Windo2India, 2011). The crime rate in Mumbai has gone up slightly from 186 per lakh population in 2007 to 191 per lakh in 2009 (accounting for 9.0 per cent and 9.1 per cent of the total crimes in metropolitan cities). In *Kolkata* it was 82.5 against the national rate of 175.1 in 2007 and 103.0 against the national rate of 181.4 in 2009 which is lowest among the three mega cities in India (GoI, 2009). No doubt, all these in aggregate deteriorate the condition of ideal urban life. The vasteness of the urban problems can be further assessed from an indepth analysis of various aspects of environmental condition of these mega cities.

Environmental Condition

In the face of increasing urbanization, industrialization and changing life style maintenance of environmental quality has emerged as the most challenging task before the mega cities. The huge quantities of pollutants—solid, liquid and gaseous—are being let into the air, water and land systems worsening the relationship between man and nature with new complexities. Life support systems in built in the ecosystem are being strained almost to the point of no return.

Vehicular Pollution

The increase in vehicular pollution in the mega cities during the past few decades has contributed significantly towards increased levels of air pollution. The air quality has worsened eightfold over the levels of twenty years ago since 1991 due to vehicular emission and four times due to industrial pollution during the same period. Air quality of the mega cities—Kolkata, Delhi and Mumbai—have been found to be the worst. The levels of respirable suspended particulate matter (RSPM) above permissible limits have grown significantly over the past three years. People living near the enormous waste dumping and landfill sites at Deonar (since 1927, festering pile covers more than 120 ha and eight storeys high) and Mulund complained of the foul stench and authorities had to purchase 42000 litres of perfume to spray on these dumps. These are responsible for about 50 per cent of air pollution. Automobile exhaust of both petrol and diesel vehicles cause air pollution by way of releasing excessive smoke which contribute to the concentration of higher quantities of particulate matter in the form of oxides of nitrogen and sulphur, invisible carbon monoxide, unburnt hydrocarbons and lead in the atmosphere.

In *Mumbai* an estimated more than 8 million people pouring into the city everyday, play havoc in areas of transport, traffic and pollution. Although, public transport network is efficient and 88 per cent of the city's commuters travel by public transport. Mumbai still continues to struggle with traffic congestion. People are packed everywhere. The suburban railway carried a total of 6.3 million passengers everyday in 2007 (Mumbai Rail Vikas Corp., 2008). The bus services of BEST (run on CNG) and MSRTC carried over 5.5 million passengers per day in 2008 (BEST, 2009). Besides, total transport and non-transport vehicles, which grew from about 13 lakh to 14 lakh, one lakh in a single year from 2005 to 2006, spew over 650 tonnes per day of vehiculars pollution load (1994) into Mumbai air leading to serious health hazards (Tables 8.5 and 8.6). They are substantially responsible for its being one of the most polluted and noisiest cities in the world.

Delhi's high population growth rate coupled with high economic growth rate has resulted in an ever increasing demand for transport creating excess pressure on the city's infrastructure. In Delhi also,

Table 8.5: Registered Transport and non-Transport Motor Vehicles in Mega Cities (million)

Cities	2005	2006
Delhi	4.18	4.48
Kolkata	0.91	0.94
Mumbai	1.29	1.39

Source: Road Transport/morth.nic.in/2006-7-Book 292768426

Table 8.6: Vehicular Pollution Load in Mega Cities (Estimated) tonnes/day, 1994

Cities	Particulates	SO_2	NO_x	HC	CO	Total
Delhi	10.30	8.96	126.46	249.57	651.01	1046.30
Mumbai	5.59	4.03	70.82	108.21	469.92	659.57
Kolkata	3.25	3.65	54.69	43.88	188.24	293.71

Source: CPCB, 1995, envfor.nic.in/soer/2001/ind_air GIT4U.com

buses mostly run by DTC using CNG as fuel, being the most popular means of public transport catering to about 60 per cent of the demand. Auto-rickshaws, use CNG fuel and as they charge lower fare, have become an integral part of public transport in Delhi. Delhi Metro plays significant role in public transport. Private vehicles account for 30 per cent of the total demand for transport, is higher than the other two mega cities. In 2008, there were 85 cars per 1000 of its residents (*Economic Times*, 2008). As of 2006, there were nearly 45 lakh transport and non-transport vehicles, which show an increase of three lakh in a single year from 2005 to 2006. Consonantly, estimated vehicular pollution load (Table 8.6) of 1046.30 tonnes/day is highest among the mega cities which is more than the human endurance level creating serious health problems among its residents. Owing to huge number of new cars and other diesel vehicles releasing toxic smog, pollution levels are up to 350 micrograms per metre in 2006-07 and the levels of nitrogen oxides are increasing in the city to dangerous levels. Delhi's air is choking with pollutant pm 2.5 (Gits4u.com, 2011). Inhaling this air deeper leads to lead deposit, asthma and lung cancer.

Buses are the preferred mode of transport in *Kolkata* run by both government agencies and private operators. Private owned vehicles are less in number and usage compared to other major cities due to state intervention and the abundance in both variety and number of public vehicles. However, the 2006 data show an increase of 36,000 in transport and non-transport vehicles from 9.11 lakh in 2005 to 9.47 lakh in 2006 (Table 8.5) which is the least in comparison to other two mega cities. That too create major traffic problems. As such among the three mega cities Kolkata is least polluted with vehicular pollution load being only 293.71 tonnes/day (Table 8.6).

A comparative analysis of the vehicular pollution load causing air pollution in the mega-cities in 1995 reveals certain eye-raising factors. Analysis of vehicular pollution data for 1995 (Table 8.4) states that the level of particulates, SO_2, NO_x, HC and CO highest in Delhi followed by Mumbai and Kolkata. Buses and auto-rickshaws in the three mega cities use CNG as fuel in compliance with order of Supreme Court of India but study says that burning CNG releases highest rate of methane emission which is key contributor to climate change (Gits4u.com, 2011). According to CPCB Report (2007) based on National Ambient Air Quality Standards (NAAQS) in 2005, Delhi holding 22nd position in the country as regards air pollution, is most polluted among the mega cities followed by Kolkata (25th) and Mumbai (40th) in the country.

Urban Service Delivery

In studying the urban services delivery, the biggest challenge is the paucity and inconsistency of data in general. However, in the ensuing study a clear picture of deficiency and neglect emerges regarding water supply, sewerage and sanitation, solid waste management and transport system based on whatever data is available from various agencies.

The state of *public utilities and services* is made worse because of rapid population growth, excessive influx to the cities, haphazard and uncontrolled growth of the residential colonies in the surrounding villages of the city due to skyrocketing urban land prices and commuters' pressures on the infrastructural facilities etc. Physical expansion of the cities without corresponding increase in services has aggravated problems related to unhygienic living quarters, public transport, health, water supply, solid waste disposal and pollution etc. Besides, people living in slums, *chawls* and dilapidated buildings and those surviving on pavements, footpaths and public gardens are compelled to defecate in the open as only 30 per cent have community toilets. In the overcrowded slums *jhuggis*, *jhopadis* and *chawls* of Mumbai, Kolkata and Delhi various epidemics, e.g. malaria, gastroenteritis, cholera, bronchitis, tuberculosis, asthma etc. are teeming with migrants. Poor public health is essentially a function of higher men/basic infrastructural facilities.

The *transport system* can no longer cope with the pressures that have been driving the cities to the brink over the last few decades. Because of overcrowding and extreme congestion in buses and railways and unusual delays in traversing small distances because of traffic jams at bottlenecks, air and noise pollution leave adverse impact on health and well-being of people. A variety of vehicles on the roads, moving at different speeds without any demarcated lanes for footpaths or cycle tracks and above all encroachment of commercial establishments, street vendors and on-street parking make these cities unsustainable. Also, the fast multiplying motor vehicles population of India and road networks/space not keeping pace with it pose serious threat to the sustainability of these cities. Urban planners recommend 20 to 30 per cent of a city's surface area to be devoted to roads. In the three mega cities only National Capital Delhi with 23 per cent road space fall within the norm. Mumbai with 17 per cent is trying hard to come closer and Kolkata has 6 per cent only.

Water Supply

Another factor pertaining to the environmental scenario is the problem of water, in terms of demand-supply gap, quality aspcct and related health problems. Rapid increase in population and continuing increase in city limits added with theft and leakages result in inadequate coverage, intermittent supplies, unreliable pressure and poor quality of water supply in the mega cities of India. In general, 30 to 35 per cent of the total water needed to large cities is never met. To keep up the quality of life of urban residents are required a standard of 135 lpcd[3] (GoI, 2011). At present, nearly 70 per cent of the available water in Indian cities is polluted. A variety of pathogen and bacteria, thus, reach the users resulting in most of the urban poor suffering from amoebiosis, typhoid, dysentry, cholera, jaundice and hepatitis. Over 73 million workdays are lost annually due to water related diseases.

In *Mumbai* out of a daily supply of 3500 million litres about 700 million litres of water is lost per day by way of water thefts, illegal connections and leakages. Mumbai flat owners get 140 lpcd but slum dwellers get only 40. The poor receive only the marginal share and that too is often contaminated because of the usual problem of sewerage lines. About 50 per cent of the population in Mumbai has no water supply because most of them live in slums, shanties or hutments. With falling ground water level and rising population density *Delhi* faces acute water shortage, as such, it supplied only 2465 million litres/day in the year 2000 while the demand was estimated to be 3650 million litres/day. Rest of the demand is met by private and public tubewells and hand pumps. *Kolkata* receives water from Hoogly river, purified and treated at Palta water pumping station in North 24 Parganas and then supplies potable water to its inhabitants. This is insufficient to cover the total requirement and coverage. As a result Kolkata residents on an average get only 90 lpcd, which is too less in quantity.

Sewerage and Sanitation

The sustainability of mega cities faces serious challenge of poor sewerage network leading to insanitary conditions. A large number of urban poor depend on public toilets which often have no water supply or are not connected to city's sewerage system. About 18 per cent of the urban households do not have access to any form of toilet facility and defecate in the open making the city environment dirty and unhealthy and the inhabitants themselves get infected giving rise to several stomach problems. In comparison to Delhi and Kolkata the sewerage treatment of *Mumbai* metropolis is better managed which is carried out at Worli and Bandra and disposed of by two marine outfalls 3.4 km away at Bandra and 3.7 km at Worli.

The daily domestic waste water production of *Delhi* is 1718 million litres/day and industrial waste is 265 million litres/day. A large proportion of sewerage flows untreated in river Yamuna which meets large portion of water requirement of the city. The septic condition of the river is shocking. About 30 per cent of the city's population is directly dependent on the untreated river or ground water made available from tubewells which is also contaminated. In July 1995, the concentration of BHC downstream of Mathura was eight times higher than that of upstream of Delhi. Similarly, the concentration of DDT was 9 times higher. Thus, what Haryana does to Delhi, Delhi then calmly does the same to the cities downstream. After Delhi, the river literally turns into a sewer (Agrawal and Narain, 1997). On the other hand, many parts of *Kolkata* metropolis still lack sewage facility leading to unsanitary method of waste disposal.

Many parts of the mega cities especially the slums and newly included outer areas or villages into the corporations are not connected with sewerage system, hence lack sewerage treatment. As a result, untreated domestic sewerage and industrial effluents are discharged into sea in Mumbai, in Yamuna in Delhi and in river Ganges in Kolkata. Many times the underground drains of sewerage system gets choked under pressure and the road is bathed in water and gets dotted with potholes leaving foul stench in the atmosphere and become breeding ground of waterborne diseases.

Solid Waste Management

The state of garbage collection in the mega cities is not satisfactory. Out of the total garbage generated by the city, 40 per cent in Delhi, 20 per cent in Kolkata and 15 per cent in Mumbai remain uncollected. In *Mumbai* heaps of garbage and squalor on the roadsides are a common sight. Though the authorities of BMC claim that garbage dumps are removed twice daily, the cleanliness drive is perceptibly sluggish. The situation is worse in slums. Almost all of 7800 metric tonnes of daily refuse of Mumbai, of which 40 metric tonnes is plastic waste, is transported to dumping grounds in Gorai in the northwest, Mulund in the northeast and Deonar in the east (World Bank, 2009). However, large covered plastic dustbins have been put at most of suburban railway stations of both the Central and Western Railways by voluntary Agencies. In *Delhi*, about 40 per cent of garbage generated remains uncollected and a foul stench fills the air. It produces daily 8000 tonnes of solid waste which is dumped at three landfill sites of M.C.D. (GoNCT, Delhi, 2005-06). In *Kolkata*, in the busy business district of Burra Bazar, the buildings are dilapidated and a foul stench fills the air. Keeping in view the worsening condition of garbage besides recently launched door-to-door waste collection

KMC has also introduced computerised weighing of garbage at the dumping ground of Dhapa lying on the city's eastern fringe. Nearly 40 per cent of the city's daily output of garbage is transported by private agencies while KMC's own disposal vans cart the rest (Chakrabarti, 1996). Almost all of daily refuse of 2500 tonnes of Kolkata, the least among mega cities, is transported to the dumping grounds at Dhapa in the east of the town. Here, agriculture is practised for natural recycling of garbage and sewer water (UNEP, 2006). Ragpickers clear parts of the city garbage by picking waste to the extent of 20 per cent each in Kolkata and Delhi and 10 per cent in Mumbai, which make them prone to various health hazards.

The proper disposal of solid waste and control of air pollution is the other challenging tasks before the authorities. Many parts of the mega cities still lack sewage facility giving rise to unsanitary methods of waste disposal worsening environmental condition resulting in health hazard of the inhabitants. Noxious emissions and callous dumping pose great threat to human health, soil fertility and ground water quality. The waste collection from dumpsites is infrequent, segregation of biodegradable waste from the rest and disposal rules are not followed. Waste collection coverage ranges from 70 to 90 per cent in major metropolitan areas (Planning Commission, GoI, 2011). Transport of waste often takes place in open vehicles with manual loading and to open dumping grounds. Besides, with the expansion of the city limits, the early relatively remotely lying dumpsites have become inner part of the city spreading foul stench into the atmosphere polluting the air.

The Future Prospects

The above indepth analysis of urbanization, population growth and density levels, causes of urban explosion in mega cities, slums, increasing crime rate, degrading environmental condition mainly air pollution generated because of extreme congestion due to large number of vehicular addition and road width not keeping pace with it, various aspects of urban service delivery especially water supply, sewerage and sanitation, solid waste disposal, all reveal the negative aspects of mega cities. But if we analyse the attraction/pull of multifaceted job opportunities in services, administration, industry and entertainment, educational and health services offered in these mega cities we can understand the reasons of uninterrupted/incessant migration from all parts of the country generated out of negative/push factors of limited job opportunities in farm sector.

The mega cities can become new engines of growth in the transformation of the country by providing employment to millions of people and give them an opportunity to create value for themselves and also for the country ensuring sustainable development. Sufficient provision for future growth and expansion of medium and large scale industries and service activities will further add to the employment opportunities. A report by McKinsey Global Institute (2010) states that India will have 6 mega cities by 2030. Cities, in general, will generate 70 per cent of new jobs, more than 70 per cent of India's total GDP and 85 per cent of the tax revenue that will finance development and drive a fourfold increase in per capita income. Mumbai and New Delhi will be among the five largest cities in the world and many largest cities will be bigger than many major countries in both population and development (McKinsey, 2010).

The future prospects of the three mega cities under study can be assessed from the projected figures of their present contribution to national economy. Presently, Mumbai the commercial capital of India accounts for about 1 per cent of total population (2011) and 6.1 per cent ($ 209 billion in 2008) GDP of the country (PPP based GDP $ 3380 billion) and has a per capita income which is almost three times that of India. Similarly, Delhi has the second highest GDP of $ 167 billion contributing 4.94 per cent to all India GDP. Its per capita income is more than double the national average. Kolkata shares 4.43 per cent ($ 150 GDP PPP 2008) of GDP of the country due to mainly the development of IT industry growing at 70 per cent yearly, twice that of national average and manufacturing and infrastructure sector. Thus, the three mega cities altogether contributed 15.47 per cent to the total GDP of the country in 2008 (Rediff.com Business, 2010).

According to projection of McKinsey Global Institute Mumbai Metropolitan Region (MMR)'s GDP is projected to reach $ 265 billion by 2030 from $209 billion (2008) which is larger than the GDP of many countries. While, Delhi presently on second position after Mumbai with $165 billion will surpass Mumbai and will generate $296 billion GDP with lesser population of 26 million than Mumbai's

33 million. Kolkata will maintain its modest growth in GDP and population to contribute $ 169 billion with 23 million population. Thus, the 83 million aggregate population of these cities will generate a huge sum of GDP $ 730 billion by 2030 (McKinsey, 2010). It will be highly vital vehicle for transforming the country by funding the development, being instrumental in raising the quality of life of its inhabitants, migrants and the people of their places of origin, its surrounding region and spread the fruit of India's economic growth to a wider section of the population of the country. Hence, the robust health of these cities is vital for the overall development of the country.

A few suggestions have been forwarded here to address the sustainability of mega cities, to make them a livable, pleasantly habitable place:

1. Measures be taken to accommodate the influx of migrants who come to these cities with modest expectations of employment. In the last 50 years the migrants were either increasing the densities or are settled on the outer limit of the city and boundary was to be extended. As about 70 to 75 per cent of them belong to lower middle to middle class, due to inflated land/building costs in all the three mega cities they cannot afford a descent home to live in. It is suggested here that the multistoreyed buildings of Mumbai, Delhi and Kolkata be raised high enough to match Singapore levels to reduce the cost and provide affordable housing to larger numbers (Bhatia, Gautam, 2011).
2. The multiple corridors/expressways being developed between the mega cities, Delhi-Mumbai and Kolkata-Delhi will include many new towns and villages on the way. These may be preplanned with proper roads and utility services to accommodate the migrants on the fringe areas. This will also ease out the social and cultural constraints of their lives besides being instrumental in developing the affected villages also.
3. Other small and medium towns lying in the vicinity of these mega cities may be planned/developed along with various economic activities as satellite towns to accommodate migrants coming from distantly located small and medium towns and rural hinterland.

4. The suggestions of working group of National Advisory Council (NAC) taken in October 2011 regarding the resettlement of slum dwellers where they live rather than displacing them so that they continue to remain close to their places of work, be welcomed and followed. Several projects like Rajiv Awas Yojana or Kanshi Ram Awas Yojana should be taken up to provide cheap housing facility to the poor deprived workers. However, caution should be taken up by administration against such nexus of illegal owners who rent the flats to migrants. It was reported recently that Dharavi, once Asia's largest slum may have 59,000 tenements but less than 20 per cent of families have any degree of permanence or ownership. Mere 100 families own Dharavi slum and only 17 per cent of those surveyed are born and brought up there. Rest are migrants (Lewis, 2011).
5. To avoid starvation, debt, diseases and suicides, the migrants' return to their villages be discouraged because of the limited possibility of employment in the agriculture sector. They themselves would not leave the villages if the possibilities and prospects of better life through transformation of villages into a new town arise.
6. Like western developed countries for proper planning of a city housing first road/electric/water/gas/drainage/sewer networks and other services be laid, only then housing process should begin.
7. As already discussed earlier the state of urban service delivery is very poor in these cities resulting in the pollution of air, water and land, which is instrumental in the proliferation of diseases like dengue, chikungunya, swine flu, diarrhoea, acute respiratory diseases etc. and wasting time in traffic jams, queuing up for water, all these hinder from exploiting the full potential of people involved in rapid economic growth.
   (a) The distribution of urban service delivery should cover equally the privileged class, poor or slum dwellers. Presently, per capita water supply ranges from 37 to 298 lpcd for a limited period with low pressure and only 64 per cent of the population is covered. Instead a 24 × 7

water supply, sewerage connection and door to door collection of solid waste should be ensured. Private agencies may be invited who may provide above services on 'pay and use' basis.

(b) A hundred per cent households be connected with sewerage network and have toilet facility with regular water supply to avoid defecating in the open and 100 per cent waste water generated be treated.

(c) Regarding solid waste management, waste collection should cover 100 per cent in closed trucks and be collected mechanically instead of 70 to 90 per cent collection in open trucks and manually.

8. The problem of transport and roads be tackled by making provision of higher dependence on public transport (presently only 22 per cent use) and use of private vehicles in the central areas and on the busy thoroughfares be restricted. This will create a condition of lifestyle dependent on sharing and sustenance (Bhatia, 2011). To avoid traffic jams, road network and road width be expanded and separate lanes for footpath and slow and fast moving vehicles be made. Encroachments be strictly prohibited/penalized. Sufficient provisions be made for flyovers, bridges etc. at bottlenecks.
9. As an alternative mode of transport opening of sea in Mumbai and River Ganges in Kolkata along with multiple lane bridges be considered.
10. The provision of multistoreyed car parks, road-on-road and flyovers on intersection/bottlenecks/congested roads and extensions of railway track wherever possible be made.
11. To fight pollution, provision of lead free petrol/CNG/LPG and privatisation of infrastructure and services on pay and use policy will improve the situation.

Keeping in view the vital role played by cities in socio-economic transformation of the country and in global market Government of India has launched several programmes to overcome some of the problems raised above in the study to achieve balanced and sustainable development. The earlier Government scheme for infrastructure development in the mega cities introduced in 1993-94 to cover a wide range of projects on water supply and sewerage, roads and bridges, city transport, solid waste management etc. has been subsumed in Jawaharlal Nehru National Urban Renewal Mission (JNNURM) taken up in December 2005 for integrated development of urban infrastructure and services. JNNURM, National Urban Transport Policy (2006b) and National Urban Sanitation Policy (2008a) will function under Ministry of Urban Development for their respective services to fulfil Millennium Development Goals.

All these efforts will definitely make null and void the question of sustainability of these mega cities. Rather, they will accelerate the degree of urbanization and quality of life to the aspiration levels of the people of the country directly and will function as engines of economic growth to raise the country from the present fourth largest economy to the second largest after China in the world indirectly.

Conclusion

The history of urbanization in India reveals that total population growth in urban areas is almost double (multiplied 6 times between 1951 to 2011) than that of total population (multiplied 3.32 times between the same period) and most of the urban population is concentrated in class one cities. It indicates that urban system of India lacks proper articulation as class one cities are growing faster than towns, metropolitan faster than class I and mega cities faster than metropolitan cities. Although, total population in mega cities exhibits a declining trend from 1981 onward, their aggregate population in themselves has grown to staggering more than 11 times in Delhi, 6 times in Mumbai and more than 3 times in Kolkata during 1951-81. They are bursting at the seams and the question of their sustainability in the present circumstances is raised. The quality of life in these mega cities is being degraded. The exodus of population augments the growth of slums devoid of urban infrastructure and utility services and the authorities are being unable to cope with satisfying the minimum needs. Inadequate water supply, uncleared garbage, sewage system heavily loaded and traffic congestion due to heavy vehicular traffic releasing fume, all lead to rapid environmental degradation giving rise to various ailments.

The mega cities can become new engines of growth in the

transformation of the country by providing multifaceted job opportunities to millions of additional people through which value for themselves and also for the country will be added by increasing production in every sphere leading to increased contribution in GDP of the country and fund for development by collection of various taxes. Hence, robust health of these cities is vital. To address the sustainability of mega cities and to make them a livable pleasantly habitable place investment in infrastructure and management skills of a very high order and provision of public utility services like housing, water supply, transport, medical and educational services be made. The first thing the mega cities need is an upgradation in public transport. To decongest the cities development of mini cities or townships surrounding the mega cities to fight pollution, provision of CNG/lead free petrol and privatization of infrastructure and services on pay and use policy may improve the situation, otherwise a total breakdown of even existing services will take place giving rise to anarchy and unsustainability.

ENDNOTES

1. Footloose is the desperate search for work that drives poorer people in multiple directions with no clear final destination.
2. PPP—Purchasing Power Parity
3. lpcd—litre per capita per day

REFERENCES

Agrawal, A. and Narain, S. (1997). The Yamuna is becoming Delhi's river of death. *Times of India*, March 7, p. 11.

BEST (2009). Bus transport profile. bestundertaking.com/trans_botright.asp.

Bhatia, G. (2011). City is going to village. *Sunday Times, Times of India*, Oct. 2.

Census of India (2001). *Slum Population of Million Plus Cities in India*. New Delhi: Office of the Registrar General and Census Commissioners.

Census of India (2011). Urban Agglomerations and Cities, Provisional Population Totals. New Delhi: Office of the Registrar General and Census Commissioners, p. 3.

Chakrabarti, C. (1996). Calcutta ready for population boom, *Times of India*, June 14, p. 10.

Correa, C. (1988). *Prospect of World Urbanization*, National Commission on Urbanization in United Nations, p. 2.

Economic Times (2008). Indiatimes.com/Every_12th_Delhiite_owns_a_car/ article show/2667484.cms, Nov.

CPCB (1995). Centre for Pollution Control Board, Ministry of Environment and Forest, Govt. of India. Retrieved from http://envfor.nic.in/soer/ 2001/ind_air

—— (2007). Report on National Ambient Air Quality Control in Indian Cities, 2005. Centre for Pollution Control Board, Ministry of Environment and Forest, Govt. of India.

Gits4u.com (2011). Environmental Pollution in India, http:// www.gits4u.com/envo/envo4.htm, accessed on 12/8/2011

Gupta, S. (2011). Slum dewellers must be resettled where they live: NAC, *The Hindu*, Oct. 22.

Lewis, C. (2011). Mere 100 families Own Dharavi Slum, *Sunday Times, Times of India,* p. 1.

Sankhe, S., Vittal, I., Dobbs, R., Mohan, A., Gulati, A., Ablett, J. (...) Sethy, G. (2010). Urbanization is critical to India's development. *India's Urban Awakening: Building Inclusive Cities, Sustaining Economic Growth*. McKinsey Global Institute, pp. 37-50. Retrieved from www.mckinsey.com/./urbanization/urban_awakening_in_India.

morth.nic.in (2007). *Road Transport 2006-7*. Retrieved from morth.nic.in/ ./Road Transport 2006_7_Book 292768426

Mumbai Rail Vikas Corp. (2008). Overview of existing Mumbai Suburban Railway, Retrieved from http://archieve.org/web/20080620033027// www.mrvc.indianrail.gov.in/overview.htm.

Planning Commission (2008). *Urban Infrastructure, Housing, Basic Services and Poverty Alleviation*, pp. 394, 396, 398, retrieved from http:// www.planning commission.nic.in/plans/planrel/fiveyr/11th/11./11v3 _ch11

GoNCT Delhi (2005-6). *Water supply and sewerage. Economic Survey of Delhi*. Planning Dept., GoNCT Delhi, Ch. 13, pp. 147-162

Brockerhoff, M.P. (2000). An Urbanizing World. *Population Bulletin*, PRB, 55(3).

Rediff.com (2003). Nathula trade may spur business in NE, Retrieved from http://www.rediff.com/money/2003/sep/09trading.htm. retrieved 2007-09-18.

Rediff.com Business (2010). GDP the top 10 cities in India, Retrieved from http://www.business. rediff.com/slide_show/2010/apr/23/ slide_show_1_the_top_10_ cities_in_India_b. 1/4/2002

Government of India (2009). *Crimes in Mega Cities*, National Crime Records

Bureau, Min. of Home Affairs, Retrieved from http://ncrib.nic.in/cii_2009/Ch II

Government of India (2011). Indian urban infrastructure and services, *The State of Urban Service Delivery*, Min. of Urban Development, Chapter II, pp. 44-47, 50-57 Retrieved from http://www.niua.org/projects/hpec/final reports_hpec.5, March

Sainath, P. (2011). Decadal Journeys: debt and despair spur urban growth, *The Hindu*, OP-ED, Sept. 27, p. 11.

UNDP (2007). World Urbanization Prospects, Revision, Population Data Base, Fact Sheet 7.

UNEP (2006). Sound Practices Composting, Retrieved from http://www.unep.or.jp/ietc/EST dir/Pub/MSW/SP/SP4/SP42.asp. Retrieved 2006-04-26.

Windo2India (2011). Crime rates in Indian cities—out of control? Retrieved from http://www.windo2india.com/cms/admin/article.jsp?aid=2165 accessed on 12/8/2011

World Bank (2009). *Bombay Sewage Disposal*. Retrieved from http://www.worlbank.org.in/default/main, Project id=P010480.

9

Urban Development and Involuntary Resettlement: Indian Experience

Mrutyunjaya Sahu

Introduction

Today almost one half of the world's population live in cities. Cities across the world are growing by one million persons every week. In India, there has been huge increase in urban population since independence, and the trend is likely to continue. Cities play a significant role in development and they continue to attract migrants from rural areas because they enable people to advance socially and economically. Cities also offer significant economies of scale in the provision of jobs, housing and services and are important centres of productivity and social development.

Widespread all pervading urbanization is a truly Twentieth century phenomenon. Although cities have always existed such as Memphis, Babylon, Thebes, Athens, Sparta, Mohen Jodaro and Anuradhapura in antiquity, there is little evidence of widespread urbanization in the early years of civilization. Rome was perhaps the first settlement to reach one million population; and in 1800 AD London became the second city to reach this population size. Virtually all this growth has taken place in developing countries. Developing countries have most of the world's largest cities and, in the more urbanized among them, the urban poor already outnumber the rural poor. Coping with such a population explosion is not an easy task for urban planners. What makes the situation particularly grim, is the general lack of planning, investment, and enforceable regulatory mechanisms. Most cities in developing countries are largely the products of massive and uncontrollable rural-urban inflows. Planners will need to find, in already such highly congested cities, space enough for housing, civic amenities, road networks, commercial activities, industrial estates, and so on. However, projects to promote development in urban areas necessitate forcible relocation of people, especially from slum areas, causing severe adverse social and economic impacts (Sharma, 2000). As involuntary resettlement occurs with greater frequency, raising major social, economic, political and financial issues, it obviously will demand effective policy and institutional mechanism to implement it (Cernea, 1995). Involuntary resettlement is often associated with the dams that are built to generate irrigation, power, and other benefits. Resettlement also occurs due to other kinds of projects—highways, ports, airports, thermal power stations, mining, gas pipeline, new industries and urban environment improvement. Urban resettlement has already emerged as a major problem, dislocating the lives of hundreds and thousands of people, especially the poor. This paper attempts at understanding the challenges posed by Indian cities on their path to sustainability. It seeks to highlight the rehabilitation and resettlement programmes for people affected by urbanization in India and suggests that the resettlement programme must centre on enhancing human capabilities and expanding social opportunities by addressing the social and personal constraints that restrict people's choices. Finally, it concludes that unless the displaced people are brought closer to decision making processes, the goal of inclusive development will remain ever more elusive.

Urbanisation: A Recent Phenomenon

Rapid urbanization, one of the most significant demographic and social changes in the Twentieth century, has continued into the Twenty-first century. In 1800, only two per cent of the world's population was urbanized. By the year 1900, out of total world population, close to 1.5 to 1.7 billion, only 15 per cent of the population, about 250 million, lived and worked in urban areas, a number lower than the total urban population of India alone today (Dasgupta and Mohan, 2005). The last 50 years have been truly remarkable in terms of the number of people who have been successfully absorbed in cities in an incredibly short time period by historical standards.

As in other countries of South Asia, India's urban population has grown relatively slowly over the last century. The world has about 20 megacities—urban areas with 10 million or more. Three are in India: Delhi, Kolkata and Mumbai. The pattern of urban growth across states is significantly different from that of the levels of urbanization. Since independence until 1991, developed states such as Tamil Nadu, Punjab and Karnataka (having a high percentage of urbanization) had shown medium or low growth of urban population. In contrast, high urban growth was registered in relatively backward states, namely Bihar, Uttar Pradesh, Rajasthan, Odisha and Madhya Pradesh, the states also having low percentages of urban population. This implies that the relationship between urban growth and economic development was generally negative. However, some of the developed states like Maharashtra and Haryana proved exceptions, as they had recorded urban growth higher than that of the country.

The urban population of India has rapidly increased in recent years. In 1961 about 79 million persons lived in urban areas, by 2001 their number had gone up to over 285 million, an increase of over 350 per cent in the last four decades, which will increase to over 400 million by the year 2011 and 533 million by the year 2021 (Kundu, 2003). The percentage of urban population to the total population of the country stands at 27.8 per cent (Table 9.1). In 1991 there were 23 metropolitan cities, which have increased to 35 in 2001. Among the major states, Tamil Nadu is the most urbanized state (43.9 per cent) followed by Maharashtra (42.4 per cent) and Gujarat (37.4 per cent). The trend is quite evident that more and more people are living in urban centres and we cannot ignore the issues in urban areas any longer. Among metropolitan cities Greater Mumbai is the largest (16.4 million) followed by Kolkata, Delhi, Chennai, Bangalore, Hyderabad, Ahmedabad, Pune, Surat, Kanpur, Jaipur, Lucknow, Nagpur etc. According to the UN study if urbanization continues at the present rate, then 46 per cent of the total population, i.e. 634 million people

Table 9.1: Urban Population Growth Rate in India

Census	Total population (million)	Urban population (million)	% of Urban population to total population	Decadal Urban growth rate (per cent)
1951	361.08	62.44	17.29	—
1961	439.23	78.93	17.97	26.41
1971	548.15	109.11	19.91	38.24
1981	683.32	159.46	23.34	46.15
1991	846.30	217.61	25.71	36.47
2001	1027.01	285.00	27.78	36.47

Source: Population Census Data 2001.

will be in urban regions of India by 2030 (UN, 1998). If such an exodus is not checked and corrected, it would lead to extreme urban decay where urban sustainability would then become one big utopian vision.

Growth of Urban Slums in India

Slums are an inescapable part of India's urban landscape. They have infested all parts of the city. Slums and more so the non-notified slums are often located on public/private lards near drains, dumping grounds, roadsides or railway lines making them prone to a number of vulnerabilities like displacement, disease prone or accidents (Verma, 2002). More than 40 million urban Indians live in areas classified as slums—a number roughly equal to the population of Spain. Slums (*jhuggis*) are defined as any area designated as such by a state or local government or any compact area of at least 300 population or about 60 to 70 households of poorly built, congested tenements in an unhygienic environment usually with inadequate infrastructure and lacking in proper sanitary and drinking water facilities.

India conducted a systematic enumeration of the urban slum population for the first time during the 2001 Census, pegging the slum population at 42.6 million, or about 14.9 per cent of the national urban population of which more than 96 per cent work in the informal sector with lack of legal protection and social security. The largest slum populations are in major cities like Mumbai (6.5 million slum dwellers), Delhi (1.9 million), Kolkata (1.5 million), Chennai (0.8

million), and Nagpur (0.7 million). Even the "hi-tech" cities of Bangalore and Hyderabad have one million slum population. By far the largest percentage of population living in slums is in Mumbai, a shocking 54 per cent. Next are Faridabad (46 per cent slum dwellers) and Meerut (44 per cent), both in the Delhi National Capital Region, followed by Kolkata (32 per cent) and the lowest in Patna (0.25 per cent).

Most inhabitants of slums come to the city with hopes of earning some income, no matter how meagre. Slums are by definition illegal, usually rising on a piece of empty government or private land in less desirable places near railway lines or drainage canals. However, large numbers of people living in the cities do not have any entitlements. These people are called as informal dwellers as they live in so called slums. "The existence of slum means the authorities have failed," says the World Bank. The slums are characterized by lack of basic services including water, sanitation, health, education etc. Slums located adjacent to large open drains have a greater incidence of diarrhoea and other water-borne diseases. Slums with inadequate housing are more likely to have worse health indicators. Improper housing in terms of lack of ventilation, overcrowding, lack of cleanliness, dampness leads to faster spread of many communicable diseases like measles and TB (Modi, 2009).

Urban Development—Induced Displacement

A new urban order has appeared since post-1970s coinciding with the crises of the Fordist-Keynesian accumulation regime and breakdown of the Bretton Woods system. It was a time when the world capitalist system became increasingly neo-liberalised, subsequently taking a dominant form in-turn impacting social and economic spaces of countries across the world (Banerjee and Guha, 2009). Many third-world countries are the product of massive, relatively rapid and spontaneous rural-urban inflows and natural population growth, combined with insufficient master planning, low investments and weak regulatory frameworks. The world today has been in the stream of extreme development, which is not well planned and egalitarian. India is no exception to this. Indian cities of various sizes are being remodelled in recent times as 'world class cities' to function as nodals of circulation of global finance and hi-tech activities of a diverse nature. Since independence, externally assisted urban sector projects in India have accounted for more than $230 million. In the mid-1990s, it marked the beginning of a new regime of regulation in the Indian urban sector. The process of economic reforms has cleared the ground for vigorous implementation of state-sponsored neo-liberal progammes in a number of cities in subsequent years, impacting municipal finance, infrastructure, basic services, the land and housing market, land use, urban form etc., and most importantly, the shelter and livelihoods of millions of urban poor. Improvements became imperative. They often require a certain amount of population relocation.

According to one projection, 'every year about 10 million people globally are displaced by projects that build dams, transportation systems, or other urban infrastructures, without counting those development projects in other sectors that also cause forced displacements'. In India alone, involuntary resettlement is estimated to have affected about 50 million people in the last five decades (Roy, 1999). The World Bank has estimated that 60 per cent of development-induced displacement every year. About 6 million people are a result of urban infrastructure and transportation projects. This same proportion is not reflected in World Bank-assisted projects—in 1993, only 22.6 per cent of displacement was caused by urban and transportation projects. Urban infrastructure and transportation projects that cause displacement include slum clearance and upgrading; the establishment of industrial and commercial estates; the building and upgrading of sewerage systems, schools, hospitals, ports, etc., and the construction of communication and transportation networks, including those connecting different urban centres. The various causal factors that account for the deliberately planned displacement in urban areas can be conceptually captured in taxonomy of the key causes of planned displacement (Doggan and Kassards, 1988). There are many reasons for displacement of people in cities but the four fundamental reasons that lead to urban displacement include:

Urban Economic Growth

Cities are becoming engines of national growth at a pace faster than anticipated earlier. Industrial manufacturing and service activities that benefit and diversify have led to displacement of people. Recent estimates indicate continuous economic expansion as the trend for the immediate future. Relocation is used to make room for new industrial estates, transportation corridors, and economic ancillary activities.

Environment Improvement

The anarchic spatial and demographic expansion of urban settlements has outgrown by far the capacities for supplying drinking water and energy, for waste disposal and sewerage, for sanitation, etc., leading to snowballing deterioration of the environment. Relocation, temporary or permanent, is caused by the need to make room for structural and infrastructural equipments for environmental services, health facilities, water supply systems, and to insert them into already densely inhabited downtown or residential areas.

Slum Upgrading

Shanty towns or slum areas are sometimes beyond rehabilitation or improvement. The accumulated needs for poverty alleviation and quality of life improvement reach the point when slum conditions must be changed. Sometimes, slum/squatter areas cannot be upgraded without some population movement. In certain situations, relocation of the entire slum is undertaken.

Non-Urban Programmes

Certain projects located outside the urban area, too, can trigger displacement affecting urban populations. This happens, for example, when a dam reservoir even though located at a distance submerges an existing town. A case in point is Tehri, a town that due to Tehri dam submergence had to be completely shifted to another location.

Approaches to Relocating Slums

Rapid urbanization, the concentration of the urban population in large cities into wider geographical areas and the rapid growth of mega cities are among the most significant transformation of human settlements (Mathur & Marsden, 1998). Over the years, approaches to slum relocation have seen many changes. There are three approaches to understand the relocation of slum dwellers:

(a) Earlier approaches to slums regarded them as illegal settlements and an eyesore to the whole city, and for that reason worthy of no consideration, illegal, human or any other. Slum people were expected to go back to their villages or wherever they came from. When that did not happen, it was common for public agencies to literally throw people out their own devices. The idea was to clearly convey that they would not be welcome if they dared to return.
(b) A broader approach towards urban development, somewhat more sensitive to the plight of the poor, later emerged. It did not rely on brutal evictions to reclaim valuable government land from slum dwellers; it went further to facilitate their relocation in specifically built housing colonies, sited on the fringes of the city, at a distance in satellite towns. This however did not work. People who lost their livelihoods due to relocation often failed to find any in the new place. Slowly, they began returning to the city, not exactly to where they were before, but as closer to that place as possible.
(c) The problems with insensitive resettlement, often encountered in the above approaches to urban planning, have now led to approaches in favour of slum up gradation/squatter improvement, with only limited relocation. Essentially, this approach calls for providing on-site improvements in slums, and site-and-services to support those relocated to peripheral areas. Slum upgradation component is evident in most current urban development policies. In a way, this policy concedes that though slums remain illegal settlements the residents can continue to stay without the fear of being arbitrarily evicted.

The approaches, described above, do not clearly correspond to any particular periods in the history of urban development. They broadly indicate the trends as they have occurred over time. It is possible to find examples of these approaches or some variations of them existing side by side at the same time. The forced eviction procedures of the early period have not been given up everywhere, and they recurrently flare up with terrible consequences in slum relocation in one or another country.

Displacement and Impoverishment

The Indian cities have been facing the wrath of unplanned development, due to which there are large number of displacements. The displaced people belong to the vulnerable groups; Scheduled Castes and Scheduled Tribes. While the overall economic effects of urban growth are positive, the parallel compulsory relocation that it often generates has serious negative effects, in many cases causing aggravating impoverishment. The displacement of households and economic units (artisans, workshops, commercial small producer's shops, units and food stalls) deprives those affected either of dwellings, or of employment, or of access to their customers, or of a combination of these loses. Besides disrupting the family life of the displaced and the social fabric of communities, the movement of people to non-camp, urban settings is further exacerbating the vulnerability of the already resident urban poor. The arrival of new displaced people further stresses already inadequate water and sanitation infrastructure, shelter and access to land. Competition for resources and livelihoods among the urban displaced and host population increases social tension and result in new conflict. The arrival of displaced people in a city or town may not only generate problems for the city but also jeopardise its ability to plan for its future.

People have become vulnerable in terms of bad resettlement and rehabilitation. The distance factor is critically important in urban relocation. If people are relocated at a short distance from their original places of living, that is not so hurtful. At least, that allows them to hold on to their jobs and retain the existing social and cultural bonds, though this costs them time to commute and money to pay for transport. However, sites nearby suitable to relocate large groups of affected people are not always easy to find. The distance of the relocation site from original place and jobs often become an insurmountable obstacle to maintaining prior employment. Furthermore, those who lack legal title to their shelters and house plots are often regarded as intelligible for compensation. For many, the standard of life deteriorates if policies and programmes do not provide for adequate re-establishment. Displacement practices that impoverish people are contrary to the stated policy goals of both national governments and donor agencies, yet they continue to occur frequently (Cernea, 1995).

Urban Resettlement: Lessons Learned

Urban resettlement has turned out to be far more complex issues than resettlement in rural areas. The urban environment offers challenges of a different kind. The review of a large number of urban projects reveal the wide differences in the quality and effectiveness of the approaches used by government agencies in various developing countries. Many fundamental issues facing the displaced groups continue to be disregarded or are resolved inequitably. The following are among some lessons learned from recent experiences with resettlement projects in urban areas:

Resettlement Policy

The presence or absence of a policy has a decisive influence on resettlement performance. Where there are no policies, the recovery does not receive the attention that they should. At present there is no comprehensive Resettlement and Rehabilitation policy regarding displacement in the urban context drawn up either by the World Bank or by the Central Government. India has not yet finalized its draft slums policy. The lack of an official policy is proving to be a hindrance in managing resettlement problems that arise in connection with urban development projects. The policy to resettle people that the authorities are mostly followed is by bulldozing their living spaces, removing them from the site and then leaving them to their own devices.

Democratise the Process of Participation

The concept of participation has become a virtual *sine qua non* for all development projects. Most resettlement programmes, in effect, become extremely large bureaucratic technical organizations operating with specific models of development progress. Progress in democratization and decentralization has brought new opportunities for improved participatory urban governance in many countries. This has increased the scope for community leadership in urban-improvement initiatives. Except in projects funded by international development agencies, the involvement of people in resettlement projects is virtually absent. Consultation with the people affected is considered a waste of time, and an avoidable hurdle in the process. Experience, however, shows that far from hindering the planning process, participation actually reduces tensions, conflicts and negative social impacts; builds trust and collaboration; avoids delays due to lack of arrangement at the community level; and improves the feasibility of the project. There is no evidence that participatory processes cause delays and cost overruns. Such difficulties are more due to planning inadequacy, implementation weakness, or insufficient resources that actually hinder participation (Mejia, 1999).

Economic Rehabilitation

Experience has repeatedly shown that if people are relocated far away from their sources of livelihood they are likely to get further impoverished. The availability of work is often largely a function of location rather than of opportunities for income generating schemes (Marsden, 1998). Project affected people do not see any employment schemes as a source of dependable livelihood, generating sufficient income. In resettling the urban poor it is therefore important to move them the least distance possible from their original location. Generating new opportunities at the sites that are too far from the city centres does not succeed. People do not like to change over to new kinds of employment, with self-employment schemes being the lowest in their order of performance.

On-site Resettlement

Because of the difficulties in providing economic recovery opportunities at distant relocation sites, on site resettlement is widely recommended. But this approach too has its implication. Moreover, some problems arise due to unavailability of land in areas large enough to accommodate the displaced people from one place to another place. It is sometimes possible to resettle people on site, but this usually involves building vertically and smaller plots—disadvantages offset by the economic advantages likely to accrue from a desirable central location and enhanced ability to rent, but involving major cultural changes in the mechanics of daily life.

Role of Civil Society Groups

Rehabilitation of people affected by development projects is a long-drawn process. It involves an understanding of people and requires planning. It needs the coordinated efforts of specialists from diverse disciplines, including those from voluntary and philanthropic associations, social workers, project officials, and affected persons themselves. The civil society and NGOs contribution to economic recovery efforts can be quite productive. Non-government organizations work both separately and alongside expert advisers, municipal managers, and private investors. They play a big part in monitoring and documenting the state of the local environment. National and international NGOs can bring technical and financial assistance to help improve local capacity. Local NGOs provide valuable resources for mobilising local expertise, developing local legitimacy for projects, improving local ownership and gaining knowledge of local conditions, needs and priorities. Experience indicates that enlisting the support of NGOs can be crucial to resettlement efforts. NGOs can be helpful in a variety of ways. The Mumbai Urban Transport Project provides one example of how successful NGO involvement can be with a major urban resettlement operation. NGOs are not monolith. Much depends on which ones are associated with the project.

Need for Fresh Initiative

Displacement due to 'development' in India is not new phenomena. The colonial period has produced a vast segment of displaced people. These segments, with the enactment of land settlement laws, forest laws and commercialization of forest products and minerals, have undergone a metamorphosis, where legally the access to the various natural resources are denied and these segments are treated as hostages within their environment. Another productive segment was also a part of displacement due to the process of industrialization and forced commercialization of agriculture—these comprise the differentiated peasantry, the artisanal groups and the traditional service groups (Fernandes, 1995). Any resistance to the displacement was treated as a 'law and order' problem, so no question of R & R policy. Land was acquired by the draconian provisions of Land Acquisition Act, 1894, which still continues, with some amendments in 1967 and 1984, to be a weapon in the hands of independent Indian state for acquiring land from its citizens.

Urban development authorities have, however, become so accustomed to insensitive methods in dealing with slum people that any other way to manage relocation is just unthinkable to them. The use of coercive methods to remove people from their existing settlements is a common practice in India. The literature on urban resettlement is full of cases where people have been ruthlessly evicted to keep the city clean and green, among other public purposes. Often, people are given no warning and bulldozers appear unannounced all of a sudden to raze to the ground their homes and living spaces in minutes, in most insensitive ways. The forced expulsion of people from their homes is making urban development an increasingly explosive phenomenon.

In 1997, the Government of India's Urban Development Ministry (GoI, 2001) initiated the process of preparing of a slum policy, ostensibly to improve the existing situation. Endorsed by NGOs at regional consultations, the Draft National Slum Policy (DNSP) was circulated in 1999. While emphasis in the draft policy is on slum up-gradation, it is its resettlement part that is particularly commendable. As the following extract shows, the aim of the policy guidelines is unequivocally to stop forced evictions. Alternative to resettlement should be explored before any decision is taken to move people. Relocation distances should be minimized to reduce the impact on livelihoods. Resident dwellers must be provided with some choice of alternative sites and where feasible, an alternative rehabilitation package. All resettlement sites should be adequately serviced and provision should be made for public transportation prior to resettlement. The livelihoods of affected people must be sufficiently compensated within a fixed period. Participation of primary stakeholders, particularly women, in planning and decision making is a prerequisite for any resettlement process. Women's particular needs and constraints must be specifically addressed. Any urban development projects that lead to the involuntary resettlement of communities must make provision to cover the costs of R&R. All stages of the resettlement process including the transition and follow-up periods, should be closely monitored and supervised by the ULB with community representatives (Draft National Slum Policy, 2001). However, urban development authorities have become so accustomed to insensitive methods in dealing with slum people, that any other way to manage relocation is just unthinkable. When they actually confront the task, especially the humane provisions of the DNSP, they forget their own guidelines on relocation.

Sustainable Human Resettlement

Sustainable human settlement implies to individual settlement in every scale, ranging from the smallest tribal settlement to individual settlements in large metropolis. Most of the urban resettlements are however, made by altering natural environment. These are the one which are threatened with unsuitable lifestyle and unstable settlements. The structure even varies from place to place within urban areas. The resettlement, in some places, range from 300 to 1000 families. At the most fundamental level, human settlements can be classified on the spatial significance. In the case of urban areas, the settlements have significance as to whether it is in the heart of the city or in the outskirts of the city. Most of the time the classification is according to the population and the sustainability is for the already existing community.

Sustainability of human settlement is vital part of human development. The human settlements today have become, most of the time to show just that settlements have been given to people or alternative settlements have been provided for the people who are displaced so that the political parties and civil society groups do not question the ruling party. The key issues that threaten human settlements sustainability are growing population, poverty and severely overexploited, disrupted and polluted natural environment (Monto, Ganesh & Varghese, 2005). Achieving sustainability in the urban spheres becomes crucial, as urban areas are growing everyday with additional population. The infrastructure and the facilities are not enough to handle the populations in the town as the populations have crossed the limit, which must be according to the capacity of the cities. Overcrowding, use of space and amenities for living that should be available for education or recreation, for example, and uncontrolled urban sprawl are a drain and a burden on the ability of a city and it is the duty of residents to see that conditions improve or at least do not deteriorate.

The Constitution of India grants right to life, which incorporates guaranteeing shelter and other basic services including the right to livelihood. Various fundamental rights and directive principles that define a framework of equality and distributive justice necessarily draw an agenda of providing adequate, habitable housing to every citizen. It is in this contextual framework of values and principles that any policy/legislation should be formulated (Patkar & Singh, 2007). The goal for a proper equitable housing policy can only be attained if the policy and the plans are made with the objective of utilising all our resources, land and human power effectively and efficiently. It also requires that the primacy to the poor and due recognition of value framework of equity and justice becomes the basis for development planning as a whole and housing projects specifically. Sustainable human settlement does not mean only the environment but also the other aspects of human being's life through which he can sustain basically. There should be sustainable economy providing employment, sustainable community with social coherence stability and solidarity, sustainable living environment with stable ecosystems and a sustainable democracy through active participation of local people.

Poor localities in urban areas need to be seen not as illegal encroachments or inhuman settlements, but as service guilds. This perception must be based on the real valuable contribution of the poor dwellers, without appropriate, just remuneration. Unhygienic conditions, a major criterion to define a slum is a result of inadequate services, sanitation, infrastructure, health and shelter and related also to poverty. The slum dwellers deserve and should be granted adequate space—physical (land), economic (budget allocation) and political (decision-making) as a policy decision, taking cognisance of their numbers, social and economic needs and contribution. No authority or society can aim to allow perpetuation of slums. However, slum demolitions or clearance should not be an activity recognised as legal and only "slum development" should be part of the metropolitan region development plan.

Slum development can have the multi-components of slum improvement that would include augmentation of basic civic services like water supply, sanitation, access roads, etc., and slum housing implemented through government supported self-help groups of slum dwellers assisted with part subsidiary and part loan at the lowest interest rate depending on their income category. Rehabilitation should occur on the same land and on alternative land only in exceptional situation. The state exchequer must be the main source for mobilising funds. The employer's contributions, employment guarantee scheme, special schemes for SCs/STs should be brought together and used optimally. Financial institutions like the National Housing Bank and HUDCO should be mobilised. Since the infrastructural projects especially highways, flyovers and other road network have benefited the private car owners the most, tax on vehicle owners can be imposed as a source for financing the housing for the poor.

The 74th Constitutional Amendment Act, 1992 brought political decentralization, made the urban local bodies more 'independent', politically weaker and functionally restricted with reduced budgetary allocations and shrunken economic base. There is an immediate need to implement the provisions of the 74th Constitutional Amendment and bring in decentralised people-centric planning, empowering urban local bodies (ULBs) but having stipulated representation of the urban poor on board. In addition all ULBs need to be financially strong to implement urban development schemes such as National Slum Development Programme, Swarna Jayanti Sahakari Rozgar Yojana and Valmiki-Ambedkar Aawaas Yojana (for housing the socially marginalized urban poor). On the lines of gram sabhas, in urban areas, basti sabhas should be the first unit of decision-making, with each developmental plan, policy and scheme being proposed, discussed and approved by it. The processes of centralised planning and decision-making, especially evident in case of development authorities like MMRDA, HUDA, etc., need to be reviewed in the context of their encroaching on the domain of municipalities and municipal corporations. A series of capacity building initiatives should be promoted to enable ULBs to effectively carry out slum development in accordance with National Slum Development Policy. This should include skill development, financial administration and management and human resource development.

Conclusion

Urban areas are henceforth going to witness rapid development. It is, however, important to ensure that the urban poor are not adversely affected in this process. Urban areas in the developing countries have put themselves in the development path but the path has not been people sensitive. In case displacement becomes unavoidable, a detailed resettlement plan with specified time-bound actions and budget are required. Resettlement plans should be built around a development strategy; and cash compensation, resettlement and rehabilitation packages should be designed to generally improve or at least restore social and economic base of those to be relocated. Monetary compensation for land alone may not be adequate. If properly planned, carried out with government's commitment to equity and adequate resource allocation and with the informed involvement of affected in bargaining and in considering options, relocation may ultimately generate certain benefits. Such improvement may result, primarily, from gaining security of ownership over the new house plots, from better housing standards for the relocates and from better environmental and social services. Building sustainable local capacity is the key to tackling the immediate needs of urban displaced and their host communities and to making all our cities better places to live.

There is a need for social mobilisation and the socio-cultural milieu of the local populations should be considered while planning and designing habitats and houses. People should be involved in the designing processes and the house design should be taken into consideration the livelihoods/occupational characteristics of the residents. The planning framework for urban involuntary resettlement and relocation policy guidelines must be included in the overall national policies that ought to govern urban planning and investment allocations in developing countries. Given the trend of urbanisation, the need is for re-prioritising our goals as well as the means employed to achieve them towards its citizens and not be a ploy to maximise or earn profit over the needs of people. The reasons for this are varied and interconnected including the absence of international attention, the marginal status and voiceless of the displaced, and the limited attention devoted to the subject by social scientists. One can only hope that the government of India doesn't lose this historic opportunity and recognizes the pains and suffering of millions of uprooted and disappeared citizens in 60 years of Independence. It would be a worthy initiative if the Government of India brings a legislation which not only takes into account the various critiques and suggestions offered by the stakeholders but also moves ahead in creating a National Slum Policy.

REFERENCES

Banerjee, S.G. (2009). Neoliberalising the urban: new geographies of power and injustice in Indian cities. *Economic and Political Weekly*, May 30: 95-107.

Cernea, Michael M. (1995). Urban Settlements and Forced Population Relocation. In H.M. Mathur and M.M. Cernea (eds.), *Development, Displacement and Resettlement: Focus on Asian Experiences,* New Delhi: Vikas Publishing House.

Dorgan, M. and John, D.K. (eds.) (1988). *The Metropolis Era: A World of Giant Cities*. Vol. 1, Beverly Hills CA: Sage Publication.

Dasgupta, S. and Mohan R. (2005). The 21st century: asia becomes urban. *Economic and Political Weekly,* Jan. 15: 213-23.

Fernandes, W. (1995). Power and powerless: development projects and displacement of tribal. *Social Action*, 41(3), 243-270.

Government of India (2001). *Draft National Slum Policy*, Ministry of Urban Development and Poverty Alleviation, New Delhi.

Census of India (2001). Office of the Registrar General and Census Commissioners, New Delhi: Ministry of Home Affairs.

Kundu, A. (2003). Urbanisation and urban governance. *Economic and Political Weekly*, July 19, 3079-3087.

Mathur, H.M. and Marsden, D. (Eds.) (1998). *Development Projects and Impoverishment Risks*, Delhi: Oxford Press.

Mejia, M.C. (1999). Economic dimensions of urban resettlement: experiences from Latin America, In Michael, M.C. (ed.) *The Economic of Involuntary Resettlement: Questions and Challenges*, Washington DC: The World Bank.

Marsden, D. (1998). Resettlement and rehabilitation in India: some lessons from recent experience, In H.M. Mathur and D. Marsden (eds.), *Development Projects and Impoverishment Risks: Resettling Project-Affected People in India*, New Delhi: Oxford University Press.

Modi, R. (2007). Resettlement and rehabilitation in urban centres. *Economic and Political Weekly*, 44, February 7, 20-22.

Monto, M., Ganesh, L.S. and Varghese, K. (2005). *Sustainability and Human Settlements: Fundamental Issue, Modelling and Simulations*. New Delhi: Sage Publications.

Patkar, M. and Singh, S. (2007). Urban renewal: at whose cost? *Economic and Political Weekly,* March 17, 926-928.

Roy, A. (1999). The greater common good. *Frontline*, 4th June, 4-29.

Sharma, K. (2000). *Rediscovering Dharavi: Stories from Asia's Largest Slum*. New Delhi: Penguin Books, India.

Verma, G.D. (2002). *Slumming India: A Chronicle of Slums and their Saviours*, New Delhi: Penguin Books India.

10
Spatial Patterns of Urbanisation and Development in Odisha, India

Sibabrata Das

Introduction

Urbanization has always been believed as an engine for development. In the context of regional development 'urban places' are regarded as 'growth poles'. Growth poles are the points and locations from which development initiates and percolates to surrounding regions. In this point of view urbanization has been regarded as the most essential precondition for the development of the regions. In geographical literature urbanization has been defined as a spatial process. Urbanization being a dynamic spatial process involves spatio-temporal variations and unequal urbanization results in unequal development across space. The process of urbanization also gets influenced by surrounding areas in terms of mobility of the people, goods and services. The reciprocal and dynamic relationship between city and its region reveals that the city is product of region and the region is served and transformed by its city (Rao, 1983:128). Empirical research findings from different regions of the world including India have also shown that urbanization and development are related. Njoh (2003) observed positive relationship between urbanisation and development in Sub-Saharan Africa, measured in terms of Human Development Index (HDI). Urbanisation has a substantial and systematic poverty reducing effect in surrounding rural areas (Cali and Menon, 2009). It was further observed that this effect is largely attributable to positive spillovers from urbanisation rather than to the movement of the rural poor to urban areas *per se*. Shafiqullah (1999) found significant association of urbanization and development indicators.

Odisha is a socio-economically backward state of India with relatively low level of urbanization. According to NSS data, around 40 per cent of the population were living Below Poverty Line in the year 2004-05. But an important feature of Odisha is intra-regional variations in level of development and urbanization. So it would be interesting to look at the interrelationship between development and urbanization in a state with low level of development and urbanization. In this context, this paper tries to examine whether the commonly accepted relationship between urbanization and development prevails in Odisha. This paper is also aimed at presenting the salient features of Odisha's urbanization paying due attention to inter-district variations. There are six sections in this paper including this introductory section. Section two gives an account on data source and methodology used in the paper. Third section deals with the analysis of the levels and trends of urbanization in Odisha. Section four emphasizes on inter-district patterns of urbanization in the state. An attempt has been made to analyse the link between urbanization and development in Odisha in spatial context in the fifth section. This section also tries to examine the linkages of migration and urbanization in the state by using existing evidences. Last section deals with concluding remarks.

Methods and Materials

The Census has been the main source of data on urban population in India. The Indian Census defines urban areas based on certain criteria. All places with a municipality, corporation, cantonment board, or notified area committee are considered urban centres. These towns are known as statutory towns. Other settlements which do not satisfy the preceding criterion but fulfil the following conditions are also considered urban in the Indian census: (1) having a minimum population of 5,000; (2) having at least 75 per cent or more male

Table 10.1: Urbanisation in Odisha and India, 1901-2001

Year	*Urbanisation level*		*Decennial growth rate of Population*	
	Odisha	*India*	*Odisha*	*India*
1901	2.5	10.84	—	—
1911	2.4	10.29	8.04	0.35
1921	2.5	11.18	2.30	8.27
1931	2.4	11.99	12.70	19.12
1941	3	13.86	30.03	31.97
1951	4.06	17.29	44.01	41.42
1961	6.32	17.97	86.79	26.41
1971	8.41	19.91	66.30	38.23
1981	11.79	23.34	68.54	46.14
1991	13.43	25.72	36.16	36.10
2001	14.99	27.80	30.28	31.74

Source: General Population Tables, Odisha, Census of India, 2001.

working population engaged in non-agricultural activities and (3) having a density of population of at least 400 persons per sq. km. The towns which are classified on the basis of above three criteria are known as census towns. In addition to these, some places that do not satisfy these three criteria are also classified as urban, if they have distinct urban characteristics such as major project colonies, areas of intensive industrial development, railway colonies, university campus, important tourist centres etc. Level of urbanization (proportion of the population living in urban areas), urban population growth rate, urban-rural growth differential etc. have been used as indicators of urbanization. Besides, for different development indicators such as female literacy rate, workers in non-agricultural activity etc. census data have been used. District level estimates of infant and child mortality rate provided in a report by the Population Foundation of India (Rajan *et al.*, 2008) have also been used. The Odisha State Human Development Report, 2004 has been the source of district level Human Development Index values. The Pearson Correlation method has been adopted to see the association between urbanization and development.

Levels and Trends of Urbanisation in Odisha

In demographic point of view, Urbanization refers to the growth in population defined as urban, i.e. population residing in designated urban areas. Increase in the number of urban places and expansion of existing urban areas are the two ways in which urbanization proceeds. Definitional changes adopted by different Indian censuses to define a place as urban, however, should be kept in mind while interpreting the trends from the census data. The major changes in the definition of urban in India took place between 1951 and 1961 as a result of which 810 towns of 1951 were declassified as rural in 1961. From 1961 onwards the definition of an urban place in Indian census has remained more or less stable.

At the beginning of the last century, the level of urbanization (the per cent of population living in urban areas) in Odisha was only 2.5 per cent compared to 10.8 per cent in whole India. There was no substantial change in the percentage of urban population during 1901-1941 in the state (Table 10.1). The level of urbanization in Odisha was about 4 per cent around independence and has increased to 14.99 per cent in 2001. During 1901-2001, the level of urbanization in Odisha continues to be lower than the national figures (Table 10.1). Since 1951, the level of urbanization in Odisha has been increasing which indicates that proportion of rural population is on continuous decline. However, still Odisha is predominantly a rural state where around 85 per cent of the population lives in the villages. As far as urban growth rate is concerned, from 1951 to 1991, Odisha recorded a higher rate of growth of urban population compared to the national average (Table 10.1). During 1991-2001, though rate is lower for Odisha compared to India, the difference is nominal.

Effective urbanization rate, which is computed on the basis of urban population living in only those towns having population above 20,000, is generally used to overcome the problem of data comparability resulting out of declassification and reclassification. Effective urbanization data also reflects low level of urbanization (Table 10.2). It increased from 1.41 per cent in 1901 to 12.53 per cent in 2001. Urban-Rural growth differential in Odisha is on continuous decline from 1971. This is due to the combination of two processes, viz. rapid rate of decline in urban population growth rate and slower rate of decline in rural population growth rate.

Table 10.2: Odisha: Indicators of Urbanisation

Census year	*No. of UA/Towns*	*Urban population*	*Percentage urban population*	*Effective urbanisation**	*Percentage growth of urban population*	*Percentage growth of rural population*	*Urban-rural growth differentials*
1901	14	254,684	2.5	1.41	—	—	—
1911	18	275,159	2.4	1.26	8.04	10.50	−2.46
1921	20	281,498	2.5	1.06	2.30	−2.04	4.34
1931	21	317,254	2.4	1.28	12.70	11.92	0.78
1941	29	412,528	3.0	1.3	30.03	9.71	20.32
1951	39	594,070	4.06	2.06	44.01	5.21	38.8
1961	62	1,109,650	6.32	3.41	86.79	16.99	69.8
1971	81	1,845,395	8.41	5.86	66.30	22.26	44.04
1981	108	3,110,287	11.79	8.94	68.54	15.73	52.81
1991	124	4,234,983	13.43	10.54	36.16	17.91	18.25
2001	138	5,517,238	14.99	12.53	30.28	14.08	16.20

* Calculated from census data.

Source: General Population Tables, Odisha, Census of India, 2001.

Urban Growth and Size of Urban Places

In 1901, Odisha had only 14 urban centres which increased to 39 in 1951 and 138 in 2001 (Table 10.2). Census of India has adopted six-fold classification of urban places. While the first size class comprises cities, i.e. urban places having 100,000 and more population, the last category comprises the tiny towns, i.e. places having less than 5,000 population. In order to provide a meaningful analysis on changes in size-class composition of urban population during the 50 years period since 1951, the last three categories (IV to VI) comprising the urban places having less than 20,000 population are grouped together and termed as small towns. Till 1951, there was no class-I city in Odisha. In 1951, Cuttack emerged as the first class-1 city of the state. According to 2001 census there were nine class-I cities and none of them was million plus cities. Out of these nine cities, only two cities were having population between the range of 500,000 to 999,999 (Table 10.3). In 2001, the number of class-IV towns was 48, followed by class-III (36), class-V (24), class-II (15) and class-VI (6).

During 1951-61, urban population was largely concentrated in intermediate towns (class-II, III) or medium and small towns (Table 10.4). Suddenly, in 1971, the share of class-1 cities rose to 39.6 per cent from 14.4 per cent in 1961. Since then class-1 towns continues to constitute the largest share of urban population. As per 2001 census, the population in class-I towns was around 48 per cent. Data reveals the declining trend of share of urban population in class-III, class-IV and class-V towns. However, the share of class-II towns has been increasing since 1971. The share of class-V towns in total urban population has declined drastically from 30.86 per cent to 3.62 per cent. Class-VI towns constitute very negligible share of total urban population of the state. The stagnation or decline in share of the class-V and VI towns can be attributed to upward mobility due to population growth as well as declassification into rural areas. Another important aspect of Odisha's urbanization is that the share of small towns having less than 20,000 population has been on continuous decline from 49.01 per cent in 1951 to 16.22 per cent in 2001. It is important here to note that during the 1980s and 1990s growth of population in all the categories except class-V and VI towns declined substantially. Class-II towns had a faster rate of population growth than the cities during above period (Table 10.5).

Table 10.3: Distribution of Towns by Size, Odisha

Classification	*1941*	*1951*	*1961*	*1971*	*1981*	*1991*	*2001*
All Classes	29	39	62	81	108	124	138
Class-I	—	1	1	4	6	8	9
Cities with one million and above population	—	—	—	—	—	—	—
Cities with 500,000 to 999,999 population	—	—	—	—	—	—	2
Class-II	1	1	3	2	8	10	15
Class-III	3	5	8	20	26	29	36
Class-IV	8	8	22	23	40	52	48
Class-V	16	23	25	30	25	22	24
Class-VI	1	1	3	2	3	3	6
Class IV+Class V+ Class VI	25	32	50	55	68	77	78

Source: General Population Tables, Odisha, Census of India, 2001.

Table 10.4: Percentage of Population in Each Size Class, Odisha, 1951-2001

Classification	*1951*	*1961*	*1971*	*1981*	*1991*	*2001*
All Classes	100	100	100	100	100	100
Class-I	17.26	14.4	39.6	41.63	44.45	48.56
Class-II	10.49	25.75	3.94	12.77	14.65	17.08
Class-III	23.24	16.77	27.85	21.81	19.74	18.14
Class-IV	18.15	25.86	15.47	17.04	16.96	12.61
Class-V	30.03	16.35	12.69	6.31	3.81	3.16
Class-VI	0.83	0.87	0.45	0.44	0.28	0.45
Class IV + Class V + Class VI	49.01	43.08	28.61	23.79	21.19	16.22

Source: General Population Tables, Odisha, Census of India, 2001.

Table 10.5: Percentage Decadal Urban Growth, Odisha, 1951-2001

Classification	*1951-1961*	*1961-1971*	*1971-1981*	*1981-1991*	*1991-2001*
All Classes	86.79	66.30	68.54	36.16	30.28
Class-I	55.88	357.36	77.19	45.71	42.00
Class-II	358.39	–74.57	446.53	56.21	51.87
Class-III	34.75	176.25	32.00	23.22	19.74
Class-IV	166.19	–0.53	85.63	35.6	–3.18
Class-V	1.71	29.07	–16.24	–17.78	8.09
Class-VI	94.81	–13.44	65.12	–13.00	105.54

Note: The inter censal growth rates of population have been worked out according to the size-class of the town in the 1981 census in respect of the decade 1981-91 and in the 1991 census in respect of the decade 1991-2001.

Source: General Population Tables, Odisha, Census of India, 2001.

Among the major towns Bhubaneswar has the maximum population of 6,47,302 followed by Cuttack (5,35,139 population) and Berhampur (289,724 population). In terms of share of cities population with total urban population of the state Bhubaneswar shares 13.28 per cent followed by Cuttack 10.98 per cent and Berhampur 5.94 per cent.

Spatial Pattern of Urbanisation in Odisha

The spatial pattern of urbanisation can be looked in different zones. Another way of analysing pattern of urbanisation would be in terms of districts. Table 10.6 shows the level of urbanisation at district level. The level of urbanisation varies widely among districts, as shown in table. Among all the districts of Odisha, the Khordha district is most urbanized with 42.93 per cent urban population followed by Jharsuguda (36.40 per cent), Sundargarh (34.38 per cent), Cuttack (27.41 per cent), Sambalpur (27.37 per cent), Ganjam (17.16 per cent), Koraput (16.82 per cent). Levels of urbanization in these districts are more than that of state average (15.97 per cent). The proportion of urban population is the lowest in Nayagarh with 4.29 per cent followed by Jajapur with 4.49 per cent, Boudh (4.82 per cent) and Nuapada (5.66 per cent). Out of the 30 districts in the state, the level of urbanisation is less than 10 per cent in 16 districts. On the other hand the level is more than 30 per cent in only five districts of the state. The figure varies between 10 to 20 per cent in rest of the ten districts.

The pattern of urban growth across districts is significantly different from that of levels of urbanization. There is no statistical significant correlation between level of urbanization and urban growth rate though they have positive relationship (Table 10.10). Comparison of urban growth rate during 1981-91 and 1991-2001 shows that 24 out of 30 districts showed a declining trend (Table 10.7). Districts, namely Ganjam, Jajapur, Mayurbhanj, Nabarangpur, Rayagada, Sambalpur have picked of urban growth in the 1980s compared to 1980s.

The share of total urban population of the state is less than one per cent for eight districts (Table 10.8). 15 out of 30 districts accounts for around 84 percentage of total urban population of the state. The five most urbanized districts comprised around half (47.75 per cent)

Table 10.6: Distribution of Districts by Level of Urbanisation

Level of urbanisation	*Districts*
0-10	Nayagarh (4.29), Jajapur(4.49), Baudh(4.82), Nuapada (5.66), Nabarangpur (5.82), Kandhamal (6.81), Mayurbhanj (6.99), Kalahandi (7.51), Malkanagiri (7.21), Debagarh (7.33), Sonapur (7.41), Bargarh (7.69), Kalahandi (7.51), Kendrapara (5.69), Dhenkanal (8.70), Jagatshingpur (9.87)
10-20	Gajapati (10.18), Bhadrak (10.58), Balesware (10.88), Balangir (11.55), Puri (13.60), Kendujhar (13.64), Angul (13.90), Rayagada (14.02), Koraput (16.81), Ganjam (17.16)
20-30	Sambalpur (27.37), Cuttack (27.41), Sundargarh (34.38), Jharsuguda (36.40), Khorda (42.93).

Source: Provisional Population Totals, Paper-2 of 2001, Rural-Urban Distribution of Population, Census of India-2001.

Table 10.7: Inter-district Differences in Urban Growth Rate in Odisha, 1981-91 and 1991-01

State/districts	*Urban growth*		*State/districts*	*Urban growth*	
	1981-1991	1991-2001		1981-1991	1991-2001
Odisha	36.16	29.78	Kandhamal	41.67	23.64
Angul	76.51	43.83	Kendrapara	29.41	17.23
Balangir	23.44	19.17	Kendujhar	32.07	27.61
Balesswar	58.07	42.84	Khorda	76.2	55.88
Bargarh	35.66	28.5	Koraput	24.76	15.38
Bhadrak	24.24	28.32	Malkanagiri	5.01	0.53
Boudh	22.79	16.42	Mayurbhanj	28.4	33.68
Cuttack	38.38	32.05	Nabarangpur	5.78	40.62
Debagarh	27.2	16.27	Nayagarh	67.83	41.18
Dhenkanal	27.38	18.79	Nuapada	20.43	16.55
Gajapati	12.71	12.84	Puri	32.45	24.67
Ganjam	25.04	27.01	Rayagada	26.28	29.16
Jagatshingpur	40.47	37.1	Sambalpur	19.74	23.76
Jajapur	23.78	36.52	Sonapur	22.72	15.03
Jharsuguda	46.32	16.31	Sundargarh	28.26	19.79
Kalahandi	32.15	28.29			

Source: Provisional Population Totals, Paper-2 of 2001, Rural-Urban Distribution of Population, Census of India-2001.

of the total urban population of the state. Besides, the top three urbanised districts, namely Khorda (14.64 per cent), Cuttack (11.64 per cent), and Sundargarh (11.44 per cent) shares 38 per cent of urban

Table 10.8: District Level Indicators of Urbanisation in Odisha

Level of urbanization (2001)		*Share of urban population (in %)*		*Decadal urban growth rate (1991-01)*		*Urban–rural growth differentials (2001)*	
Khorda	42.93	Khorda	14.64	Khorda	55.88	Khorda	47.37
Jharsuguda	36.4	Cuttack	11.67	Angul	43.83	Nayagarh	31.86
Sundargarh	34.38	Sundargarh	11.44	Baleswar	42.84	Angul	28.55
Cuttack	27.41	Ganjam	9.79	Nayagarh	41.18	Jagatshingpur	26.08
Sambalpur	27.37	Sambalpur	4.63	Nabarangpur	40.62	Baleswar	25.96
Ganjam	17.16	Baleswar	4.01	Jagatshingpur	37.1	Cuttack	23.64
Koraput	16.82	Kendujhar	3.87	Jajapur	36.52	Nabarangpur	21.43
Rayagada	14.02	Puri	3.71	Mayurbhanj	33.68	Jajapur	20.22
Angul	13.9	Koraput	3.6	Cuttack	32.05	Bargarh	18.25
Kendujhar	13.64	Jharsuguda	3.37	Rayagada	29.16	Mayurbhanj	16.83
Puri	13.6	Angul	2.89	Bargarh	28.5	Rayagada	15.87
Balangir	11.51	Mayurbhanj	2.83	Bhadrak	28.32	Ganjam	13.05
Baleswar	10.88	Balangir	2.81	Kalahandi	28.29	Sambalpur	12.83
Bhadrak	10.58	Bhadrak	2.56	Kendujhar	27.61	Kendujhar	12.36
Gajapati	10.16	Rayagada	2.1	Ganjam	27.01	Balangir	11.91
Jagatshingpur	9.87	Jagatshingpur	1.9	Puri	24.67	Puri	11.28
Dhenkanal	8.7	Bargarh	1.88	Sambalpur	23.76	Kalahandi	11.06
Debagarh	7.83	Kalahandi	1.82	Kandhamal	23.64	Bhadrak	8.71
Bargarh	7.69	Dhenkanal	1.69	Sundargarh	19.79	Dhenkanal	6.9
Kalahandi	7.51	Kendrapara	1.35	Balangir	19.17	Kandhamal	5.39
Sonapur	7.41	Jajapur	1.33	Dhenkanal	18.79	Sundargarh	5.3
Malkanagiri	7.21	Nabarangpur	1.08	Kendrapara	17.23	Kendrapara	4.21
Mayurbhanj	6.99	Gajapati	0.96	Nuapada	16.55	Nuapada	3.75
Kandhamal	6.81	Kandhamal	0.8	Boudh	16.42	Jharsuguda	1.84
Nabarangpur	5.82	Sonapur	0.73	Jharsuguda	16.31	Sonapur	1.77
Kendrapara	5.69	Nayagarh	0.67	Debagarh	16.27	Koraput	1.16
Nuapada	5.66	Malkanagiri	0.63	Koraput	15.38	Debagarh	–0.81
Boudh	4.82	Nuapada	0.55	Sonapur	15.03	Boudh	–1.08
Jajapur	4.49	Debagarh	0.37	Gajapati	12.84	Gajapati	–1.31
Nayagarh	4.29	Boudh	0.33	Malkanagiri	0.53	Malkanagiri	–14.35
Coefficient of Variations	75.88		110.17		44.35		

Source: Provisional Population Totals, Paper-2 of 2001, Rural-Urban Distribution of Population, Census of India-2001.

population of the state. Higher proportion of urban population in these districts is simply due to the presence of administrative (Bhubaneshwar), business (Cuttuck) and industrial centres. In Sundargarh, higher level of urbanization is due to the presence of Rourkela city, which is the biggest industrial centre of the state. Tribal dominated western and central districts have recorded lower level of urbanization. Sambalpur, Sundargarh, Mayurbhanj, Baleswar, Cuttack, Khorda, Puri and Ganjam have one class-1 city each. Southern part of Odisha does not have a single class-I city. One could find very low level of urbanization in the relatively more urbanized districts of the state after subtracting the population of the corresponding district headquarters.

Broad regional patterns reflects that eastern and northern parts of Odisha fare relatively better and southern and western parts of the state are lagging behind as far as urbanisation indicators are considered. Of course, there are exceptions; Kendrapara and Bhadrak being coastal districts are less urbanized and fare poorly in terms of urbanization indicators. One thing should be kept in mind when making a regional analysis of urbanization in Odisha, is that level of urbanization is very low in the newly created districts even in the eastern Odisha such as Kendrapara, Bhadrak, Jajapur, Jagatshingpur etc. because of their short history and of course, weak pull factors.

The geographical pattern of urbanisation reflects large regional imbalances as coefficient of variation figures are high for level of urbanization, share of districts to the total urban population of the state (Table 10.8). Urbanization has not extended much to the relatively backward regions of the state. It has been observed that urban clusters have been formed around big cities and in the industrial belts. Besides, relatively higher level of urbanisation has also been observed in the agriculturally progressive districts of the state. It has also been noticed that urban population in the most urbanised districts are concentrated in and around big cities. This shows that major parts of the most urbanized districts are least urbanized. However, the presence of these major urban centres may have trickle down effect on the overall development of the districts. So in this paper an attempt has also been made to look at the linkages between urbanization and development in the state.

Explaining the Low Level of Urbanization in Odisha

From the above discussion it is clear that the level of urbanization in Odisha is very low. Why is it so? Is it due to low level of industrialization? Is it that people of Odisha are less mobile in the sense rural-urban migration is not significant? At the demographic point of view the slower rate of urbanization in Odisha can be explained in terms of a relatively higher growth rate for rural population compared to the urban growth rate in some of the districts, indicating in turn a lower rate of rural-urban migration. Table 10.8 shows that in four districts such as Debagarh, Baudh, Gajapati and Malkanagiri rural population growth rate was higher than urban growth rate. In five districts urban-rural growth differentials are less than five per cent. The slower rate of urbanization in the state can be partly attributed to the relatively lower level of industrialization. The share of Odisha in all-India value added has practically remained unchanged at about 2 per cent between 1970-71 and 1997-98 (GoO-2004) The process of industrialization in Odisha has been slow rendering the economy to depend heavily on primary sector in general and Agriculture and allied activities in particular, hence low level of urbanisation. Another aspect of industrialization in Odisha is lesser degree of diversification in the industrial structure.

Besides, urbanization in Odisha is its high degree of spatial concentration. One could argue that overall low level of urbanization in the state is the result of very low level of urban growth in some districts of the state. In 16 out of 30 districts the level of urbanization was found to be less than 10 per cent. Most of these districts are located in southern and western part of Odisha. In these districts number of towns is also comparatively lower. Low level of urbanization in this region can be attributed to the rough physiography, slower pace of infrastructural development compared to eastern region, higher concentration of tribal population etc. In Eastern Odisha, except few districts such as Jajapur, Bhadrak, Jagatshingpur, all other districts recorded urbanization more than 10 per cent. In north-western Odisha relative higher level of urbanization has been an outcome of presence of industrial clusters. In the western and southern region, existing towns are not sufficient to cater the needs of the regions. More towns are necessary to serve as focal points and centres of trade, business and other services.

Coastal Odisha has strong agricultural economy. It also has some of the big towns of the state. Examination of data also reveals that high level of urbanization in some of the coastal districts is due to the fact that presence of few big towns and cities. For example, the urban population in Khurda district is substantially high because of the presence of the capital city of Bhubaneswar. Bhubaneswar alone accounts for around 60 per cent of the total urban population of the district. It indicates that even in the districts of eastern coastal Odisha the share of small and medium towns to the total urban population is not significant suggesting the need of more towns of these categories. Why small and medium towns have not been able to come up in Odisha is an important issue which needs further research and this does not come under the purview of this paper?

Urbanisation and Development in Odisha: A District Level Analysis

Development in a broader sense can be defined as the improvement in the standard of living of the population of a region. Traditionally, development refers to the economic growth. But it includes material as well as social and cultural well-being of the people in a region. The development of the region can be identified by many indicators like per capita income, proportion of the people engaged in non-agricultural sectors, literacy rate, infant mortality etc. Human Development Index which is a composite measure of development has been widely used after 1990, the year in which the first Human Development Report was published by the United Nations Development Programme. District level HDI values are shown in Table 10.9. On the whole, inter-district disparity in HDI values is rather low; Coefficient of Variation (CV) is 16.95. However, data reflects relatively lower values of HDI southern districts.

The level of urbanization of a region is classically regarded as a meaningful indicator of development. In this section an attempt has been made to see the correlation between urbanization and other development indicators. The correlation co-efficients of the indicators presented in Table 10.10 show that the level of urbanization is significantly positively associated with Human Development Index (HDI) value, female literacy rate and workers in non-agricultural

Table 10.9: Distribution of Districts by Range of HDI Values

HDI Values	*Districts/state*
≤ 0.5	Malkanagir (0.37), Kandhamal (0.389), Gajapati (0.431), Koraput (0.431), Nabarangpur (0.436), Rayagada (0.443)
>0.5 and < 0.6	Kendujhar (0.53), Baudh (0.536), Jajapur (0.54), Balangir (0.546), Ganjam (0.551), Jagatshingpur (0.557), Balesware (0.559), Bargarh (0.565), Sonepur (0.566), Nayagarh (0.571), Odisha (0.579), Nuapada (0.581), Sambalpur (0.589), Dhenkanal (0.591)
≥0.6	Kalahandi (0.606), Kendrapada (0.626), Mayurbhanj (0.639), Bhadrak (0.646), Puri (0.657), Angul (0.663), Deogarh (0.669), Sundargarh (0.683), Cuttack (0.695), Jharsuguda (0.722), Khurda (0.736)

Source: Government of Odisha, Human Development Report 2004.

sectors and negatively related with child mortality rate. The deccenial urban population growth rate has statistically significant positive relationship with female literacy rate and non-agricultural workforce. Infant Mortality is negatively related with both indicators of urbanization, although correlation co-efficient are low and insignificant. The correlation co-efficients show that the female literacy rate and proportion of non-agricultural workforce are relatively higher in the more urbanized districts. It is important here to mention that the correlation coefficients between level of urbanization and Human Development Index (HDI) value is statistically significant at the 1 per cent level which indicates that districts with higher level of urbanization are relatively developed. These findings reaffirm the earlier research findings from different parts of the world that urbanization and development are interrelated.

Migration and Urbanization in Odisha

Urbanization can result from: (a) rural-urban differentials in the rate of natural increase; (b) net migration from rural to urban areas; and (c) reclassification of villages as towns as a result or acquisition of urban characteristics and/or the spillover of urban population growth beyond designated boundaries. Table 10.11 shows the contributions of various factors in urban population growth. Share of net migration in the decadal urban growth declined from 48.95 per cent during

Table 10.10: Correlation Matrix of Different Indicators of Urbanisation and Socio-Economic Variables, India, (N = 30)

	% urban	*Urban growth (%)*	*Female literacy rate*	*Infant Mortality Rate*	*Child mortality rate*	*% Workers in non–agricultural activity*	*HDI Value*
% urban	1.0						
Urban growth (%)	0.192	1.0					
Female literacy rate	0.366*	0.422*	1.0				
Infant Mortality rate	–0.327	–0.317	–0.769**	1.0			
Child mortality rate	–0.367*	–0.297	0.763**	0.963**	1.0		
% Workers in non-agricultural activity	0.737**	0.515**	0.770**	–0.607**	–0.618**	1.0	
HDI Value	0.519**	0.317	0.714**	–0.750**	–0.762**	0.770**	1.0

* Significant at the level of 5 per cent.

**Significant at the level of 1 per cent.

Table 10.11: Estimates of Relative Share of Natural Increase, Net Migration and Reclassification in the Decadal Urban Growth of India, 1961-1991

Year	*Natural Increase (N!)*	*Net Migratio (NM)*	*Areal Reclassification (AR)*
1961–1971	30.97	39.92	29.11
1971–1981	29.18	48.95	21.87
1981–1991	60.70	24.80	14.50

Notes: Figures pertaining to natural increase refers to the residual, i.e. growth-number of net intercensal migrants and the population added due to reclassification of towns as new and declassified towns.

Figures on net migration have been derived by subtracting estimated number of persons added on account of Natural Increase and Reclassification.

Source: Bose (1996).

1971-1981 to 24.80 in 1981-1991. On the other hand, the share of natural increases substantially from 29.18 per cent to 60.70 per cent during the same period. This shows the increasing nativity in the urban areas confirming the role of increasing role of natural increase in Odisha's urbanization.

According to the Census 2001 figures, 11 million out of the state's population of 37 million were migrants out of which 7.9 million were intra-district migrants, 2.4 million were inter-district migrants, 0.66 million were inter-state migrants and the remaining 46,000 people being international migrants. As per 2001 census, around 229,687 people migrated to Odisha from other states while the number of out-migrants were 440,893. The number of in-migrants from other countries was recorded to be 3931. So the net migrants in 2001 was around—207,275. This implies that out-migrants from Odisha out—numbered in-migrants. It has also been observed that Odisha registered a loss in proportion of the net migrants to gross migrants by 33.13 per cent as far as rural-urban migration is concerned (Census of Atlas of India, 2001). This indicates towards the weak pull factors of particularly urban places in the state. One can also infer that people of the state prefer cities outside the state to the cities inside the state in the context of migration. This is reaffirmed by the fact that Intra-state rural-urban migration is low and rural-rural migration predominates in Odisha which comprises 67.5 per cent total intra-state migrants by last residence (0-9 years) in the state.

Out-migration from Odisha to different parts of the country has been noticed. Out migrants from Odisha preferred Gujarat and Maharashtra as the destination. Out-migration to these states made up to 34 per cent of total out-migrants from Odisha (Jagori, 2001). Industry in developed states such as Gujarat, Maharashtra, and Punjab is drawing labourers from Odisha particularly urban Odisha. Western Odisha has long been a major source area for migrants because of its highly unequal land distribution, high levels of poverty among landless and marginal farmers and low levels of human capital, industrialization, urbanization and diversification into non-farm occupations combined with poor governance. A study by Jagori, an NGO in Delhi put the number of Oriya power loom workers in Surat at roughly 100,000 in 2001 (Jagori, 2001). Migration is more permanent among people from eastern districts like Khorda, Nayagarh and Ganjam. These migrants are generally better off economically, with some education. They migrate to other states mainly in search of better employment opportunities, not so much because of a vulnerability to drought or famine, like people in western Odisha (Narasimham, 2004).

Conclusion

Odisha has been charecterised by very low level of urbanization with wide regional variations. It is found that the urban population is highly concentrated in some of the districts with cities and industrial towns. South and western part of the state is found to be least urbanized and also least developed. It was believed that the process of economic liberalization would accelerate rural to urban migration and boost urban growth in post liberalized period (in 1990s and the following decade). But in case of Odisha it is observed that the pace of urbanization has slowed down in 1990s compared to 1980s. Estimates of Relative share of Natural increase, Net Migration and Reclassification in the decadal urban growth of India, 1961-1991, reflects a declining role of migration and increasing role of urbanization. It has also been observed that out migrants outnumber the in-migrants in Odisha reflecting weak pull factors of urban places of the state. However, findings of district level analysis indicate a very close association between urbanization and development. Small and medium size towns need to be developed for accelerating the process of development in the state.

REFERENCES

Njoh, A.J. (2003). Urbanization and development in sub-Saharan Africa, *Cities,* 20(3), 167-174.

Bose, A. (1996). *India's Basic Demographic Statistics*, Delhi: B.R. Publishing Corporation.

Cali, M. and Menon, C. (2009). Does urbanisation affect rural poverty? Evidence from Indian districts, *SERC Discussion Paper*, 14 London: LSE.

Registrar General of India (2009). General Population Tables, *Odisha*, New Delhi: Office of the Registrar General and Census Commissioners, India.

Census of India (2001). Rural-Urban Distribution of Population, Provisional Population Totals, *Odisha*, Paper-2 of 2001, New Delhi: Office of the Registrar General and Census Commissioners.

Census of India (2008). *Census Atlas of India, 2001*, New Delhi: Registrar General and Census Commissioners.

Government of Odisha (2004). *Human Development Report 2004*, Bhubaneswar: Planning and Coordination Department.

Jagori (2001). Migration, trafficking and sites of work: vulnerabilities and rights, a pilot study. *Initial findings from Rajasthan and Gujarat*, New Delhi: Jagori.

Narasimham, B.V. (2004). Money talks and walks, *India Together*, March 2004.

Rajan, S.I., Nair, P.M., Sheela, K.L., Jagatdeb, L. and Mishra, N.R. (2008). *Infant and Child Mortality in India: District Level Estimates*. New Delhi: Population Foundation of India.

Rao, P.V.L.S. (1983). *Urbanisation in India: Spatial Dimension*, New Delhi: Concept Publishing Company.

Shafiqullah (1999). Urbanisation and levels of development in Uttar Pradesh: a regional analysis, *National Geographical Journal of India*, 45(1-4), 71-79.

UNDP (2007). *Journey to Safety Information and Communication Technologies (ICT) for Reducing HIV and other Vulnerabilities of Migrants, their Families and Communities*, New Delhi: United Nations Development Programme.

Part III
Socio-Economic Development

11
Development of Eastern Uttar Pradesh: Challenges and Opportunities

Ram Bali Singh

Introduction

Eastern U.P., as defined for the present study, covers an area of 84,065 sq.km. and had a population of 65,176,000 persons (2001). Bahraich, Gonda, Faizabad, Sultanpur, Pratapgarh and Allahabad districts are located on the western margin while the state boundary delimits the region on the north, east and south. Thus the region covers 34.9 per cent area and contains 39.3 per cent population of the state of Uttar Pradesh. It is a typically depressed region with a lower level of development and a still lower rate of economic performance *vis-à-vis* the state or the nation. It is groaning under the acute pressure of poverty, deprivation, polarization and an unholy nexus of vested interest groups consisting of politicians, officials and criminals. One may be tempted to picturize the situation in vivid detail through a *Bhojpuri lyric.* It says *Laga jhulaniyan Ka dhakka balam Calcutta neekal gayo. Jhulani* (nose-ring) is a very low-priced gold ornament yet it carries much more significance and importance for a married Hindu woman. It is the mark of her being married and her husband being alive. The wife asks her husband to purchase a nose-ring for her. The husband could not afford to fulfil the wish of his wife and in utter disgust he leaves the village and goes to a distant place like Kolkata, because there were no intervening opportunities, in order to earn some money to fulfil the small wish of his wife. So has been the situation in eastern Uttar Pradesh during the British and it has not improved since then. The downward spiral has continued unabated and the deprivation trap has become rather much more entrenched for a large segment of the regional population.

An attempt has been made in this paper to bring out the level of development in eastern U.P. in relation to the state and the country in terms of 13 variables and the cumulative development index and inter-district variation in performance level with respect to 11 development indicators as well as the composite index of development worked out by using 36 indicators. Thereafter, the challenges and opportunities have been briefly summarized. Lastly, some suggestions have been put forth to ameliorate the existing situation and to accelerate the process of development and transformation in the region.

Eastern U.P. in Relation to India and Uttar Pradesh

A comparison of the first and the last ranking districts of eastern U.P. to the first ranking district in India and in the state of Uttar Pradesh in terms of 14 indicators provide a glimpse of very depressing conditions prevailing in eastern U.P. (Table 11.1). The 1st ranking districts of Uttar Pradesh, the largest state in the country in term of population, are placed within one hundred ranks with respect to only 3 indicators—households using safe drinking water (5), households having toilet facility (68) and births below age 20 years (80) whereas the districts ranking first in eastern U.P. with respect to the same indicators occupy 9th, 180th and 140th position respectively in the country. The districts ranking first in Uttar Pradesh occupy between 101 and 150 position in India with respect to another 4 indicators—electrified households (112), women receiving at least two T.T. injections (120), immunization drop out (126) and population below age 6 years (140) whereas districts ranking first in eastern U.P. occupy 246th, 120th, 200th and 378th position respectively. The 1st ranking districts of Uttar Pradesh occupy between 151 and 200 position in India with respect to 2 more indicators—female literacy (184) and under 5 mortality rate (198) whereas districts occupying first rank in

Table 11.1: Eastern Uttar Pradesh *vis-à-vis* India and U.P.

Variables (indicators)	*India: district ranking first*		*U.P.: district ranking first*		*Eastern U.P.: districts ranking first & last*			
Population below age 6 years	Leh (J & K)	6.2	140 12.8	Kanpur urban	378 17.2	Allahabad	566 20.5	Siddharthnagar
Birth order 3 and above	Alappuzha (Kerala)	12.3	299 48.4	Jhansi	453 57.5	Varanasi	527 66.9	Balrampur
Births below age 20 years	Una (Himachal Pradesh)	0.8	80 2.9	Unnao	140 3.7	Mau	457 7.5	SRN Bhadohi
Under 5 mortality rate	South Goa (Goa)	43.4	198 89.0	Ghaziabad	231 93.0	Mau	557 169	Balrampur
Female literacy	Aizawl (Mizoram)	97.3	184 61.2	Lucknow	324 50.9	Mau	590 18.8	Shravasti
Households using safe drinking water	Chandigarh	99.8	5 99.2	J.P. Nagar	9	Deoria	431 62.3	Mirzapur
Households having toilet facility	Aizawl (Mizoram)	97.3	68 72.4	Ghaziabad	180 46.6	Varanasi	532 12.3	Kushinagar
Electrified households	Lakshadweep	99.7	112 70.6	Ghaziabad	246 65.8	Varanasi	563 9.0	Shravasti
Women receiving 3 or more ANC visits	Yanam (Pondicherry)	100.0	315 46.5	Lucknow	324 45.5	Ballia	590 9.0	Shravasti
Women receiving at least 2TT infections	Erode (Tamil Nadu)	99.3	120 85.2	Deoria			528 49.4	Mirzapur
Full immunization coverage	Taruvannamalai (Tamil Nadu)	98.8	126 55.3	Deoria			516 16.3	Balrampur
Immunization drop-out	Tiruvannamalai (Tamil Nadu)	1.2	126 2.6	Ferozabad	200 28.8	Allahabad	446 47.8	Basti
Contraceptive prevalence rate	Leh (J & K)	90.7	265 55.1	Jhansi	308 52.8	Varanasi	502 17.9	Balrampur
Cumulative development index	Chennai (Tamil Nadu)	0.86750	212 0.61554	Kanpur urban	313 0.52723	Varanasi	579 0.32299	Balrampur

1. Figures (except rank and CDI) are in percentage. 2. Data available for 591 to 593 districts for different indicators.

Source: Ram and Shekhar, 2006.

eastern U.P. occupy 324th and 231st ranks respectively in India. The districts ranking 1st in Uttar Pradesh occupy between 201 and 300 positions in India with respect to four more indicators—cumulative development index (212), full immunization coverage (226), contraceptive prevalence rate (265) and birth order 3 and above (299) whereas the first ranking districts of eastern U.P. occupy 313th, 226th, 300th and 453rd ranks respectively in India. Lucknow, the first ranking district of Uttar Pradesh with respect to women receiving 3 or more ANC visits occupies 315th rank in India and Ballia, the first ranking district of eastern U.P. occupies 324th rank in India. The last ranking districts of eastern U.P. occupy ranks varying from 431st to 590th ranks in India with respect to different indicators. It is pertinent to mention here that Deoria district of eastern U.P. ranks first in Uttar Pradesh with respect to two more indicators—women receiving at least two T.T. injections and full immunization coverage. The last ranking districts of eastern U.P. with respect to these two indicators occupy 528th (Mirzapur) and 516th (Balrampur) ranks respectively implying thereby large inter-district variations in performance level. The last ranking districts of eastern U.P. occupy ranks beyond 500 in as many as 11 indicators whereas the total number of districts included in the analysis was 591 to 593 in case of different indicators.

Regional Economy: Sectoral Characteristics

The status of agriculture, the mainstay of the regional economy, is not very encouraging. The cumulative development index (CDI) of land use is 83.79 for the state of Uttar Pradesh. As many as nine districts of eastern U.P. (Allahabad, S.R.N. Bhadohi, Mirzapur, Sonbhadra, Varanasi, Jaunpur, Pratapgarh, Sultanpur and Faizabad) have a CDI value lower than the state average and these form a compact area occupying the south-western portion of the region (Table 11.2). The percentage of net irrigated area to net sown area is 78.61 for the state of Uttar Pradesh. As many as 11 districts of eastern U.P. (S.R.N. Bhadohi, Mirzapur, Sonbhadra, Ballia, Kushinagar, Siddharthnagar, Basti, Gonda, Balrampur, Shravasti and Bahraich) have a lower proportion of irrigated area than the state average and 8 of these districts have a peripheral location adjoining the international boundary in the north and the state boundary in the east and the south.

The intensity of cropping is 152.15 for the state of Uttar Pradesh. Eight districts of eastern U.P., namely Pratapgarh, S.R.N. Bhadohi, Mirzapur, Sonbhadra, Basti, Gonda, Balrampur and Shravasti lie below the state average. The low value of cumulative development index of land use and a somewhat higher proportion of irrigated area and cropping intensity are clearly reflected in a low level of productivity in agriculture. The situation becomes much more depressing when one considers the level of productivity.

The average yield is 2050 kg per hectare for the state of Uttar Pradesh. As many as 21 districts of eastern U.P. record a level of productivity lower than the state. Ambedkarnagar (25.8), Faizabad (23.9) and Sultanpur (21.6) in the middle and Kushinagar (22.6) and Siddharthnagar (20.9) adjoining the eastern and northern boundary respectively report a level of productivity in agriculture higher than the state average (Fig. 11.1A). It is as low as 780 kg/ha in Sonbhadra a district located in southern plateau region. The advantages of a somewhat higher proportion of net irrigated area to net sown area and a relatively higher value of cropping intensity in the districts of eastern U.P. are negated by the lower level of yield per hectare in the region.

The scenario in urban industrial sector is in no way better. Though a little over one-fifth (20.8 per cent), of the total population of the state of Uttar Pradesh lives in settlements defined as urban, however, only two districts of eastern U.P. (Varanasi 40.2 per cent and Allahabad 24.4 per cent) claim a level of urbanization higher than the state. In districts of Shravasti (2.8 per cent), Siddharthnagar (3.8 per cent), Sultanpur (4.7 per cent) and Kushinagar (4.6 per cent) as low as less than 5 per cent of the total population lives in urban settlements. The level of urbanization is less than half of the state average in as many as 12 other districts—Pratapgarh (5.3), Maharajganj (5.1), Sant Kabirnagar (7.1), Basti (5.6), Azamgarh (7.5), Ghazipur (7.7), Gonda (7.0), Balrampur (8.1), Ambedkarnagar (8.9), Ballia (9.8), Deoria (9.9) and Bahraich (10). The contribution of the industrial sector to the net domestic product of the region is a good indicator of the level of industrialization. The industrial sector's contribution (2005-06) to the net domestic product of the state of Uttar Pradesh at current prices has been 11.03 per cent. However, it has been less than five per cent in 12 districts of eastern U.P.—Pratapgarh (1.42), Siddharthnagar

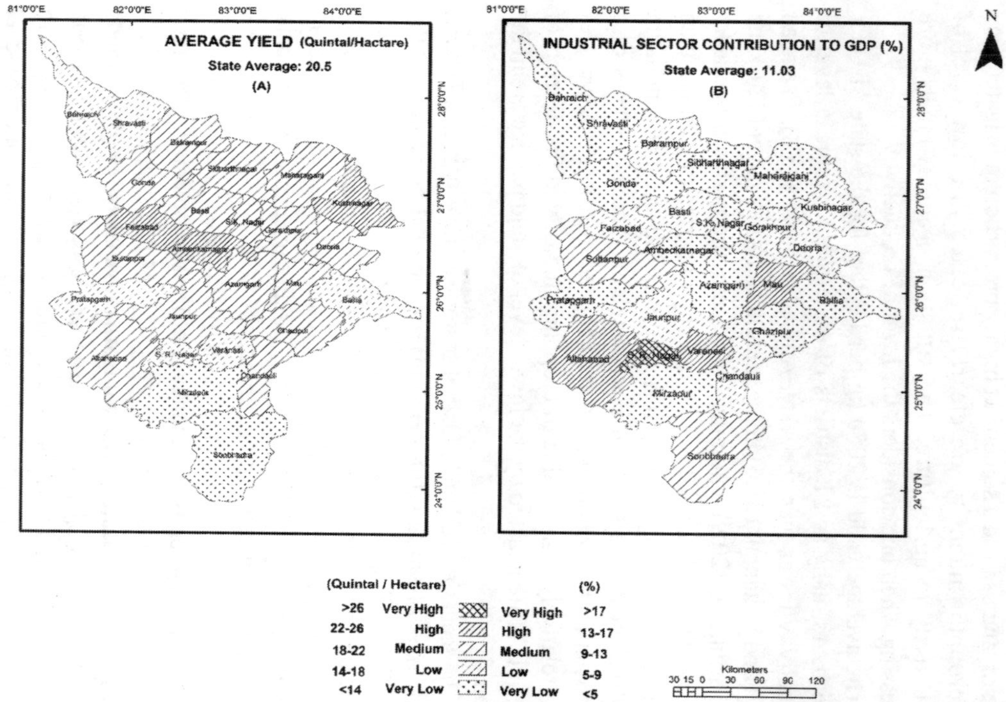

Fig. 11.1: Eastern Uttar Pradesh

Table 11.2: Eastern Uttar Pradesh: Development Indicators

Districts	Urban (%)	Income/person (Rs.) Current Prices	Credit/deposit ratio (%)	Electricity consumption per person KWH	Net irrigated/Net sown area (%)	Beds /Lakh population	Yield quintals /ha	Intensity of cropping	Industrial worker/ lakh population	Industrial sector to DDP (%)	Composite index of development
1. Allahabad	24.4	12737	26.98	303.6	78.65	63.33	18.1	153.62	202	14.72	103.97
2. S.R.N. Bhadohi	12.8	10822	47.59	90.6	74.69	15.27	17.3	141.76	252	25.76	94.56
3. Mirzapur	13.5	9192	39.51	125.8	60.22	17.96	13.8	136.47	46	4.87	73.17
4. Sonbhadra	18.8	21359	29.55	134.9	19.17	14.66	7.8	122.78	801	10.73	98.23
5. Chandauli	10.6	10900	36.14	150.4	90.40	17.82	18.4	179.45	138	7.44	81.67
6. Varanasi	40.2	12032	34.33	274.5	86.97	108.28	17.8	164.30	116	16.58	111.12
7. Jaunpur	7.4	7652	20.59	95.7	86.81	22.49	20.2	162.27	65	7.32	79.10
8. Ghazipur	7.7	8531	21.91	110.1	84.26	23.75	19.4	163.25	50	3.59	76.96
9. Ballia	9.8	8634	21.97	73.5	77.27	27.74	16.4	156.80	15	4.32	79.86
10 Mau	19.4	10243	21.86	181.3	92.16	9.22	20.3	168.34	60	14.01	90.01
11. Azamgarh	7.5	8249	22.23	82.5	86.31	27.07	20.2	170.09	19	4.99	75.57
12. Deoria	9.9	7566	26.63	59.9	86.55	37.30	19.6	161.32	56	5.05	83.37
13. Kushinagar	4.6	8024	37.88	36.0	76.28	14.47	22.6	152.30	112	6.90	76.93
14. Gorakhpur	19.6	10606	33.00	179.5	81.72	80.44	19.6	152.18	175	5.32	98.97
15. Maharajganj	5.1	7998	41.63	41.1	82.03	17.20	18.5	180.48	43	4.03	76.97
16. Siddharthnagar	3.8	8096	33.88	41.5	77.04	20.39	20.9	162.58	–	3.00	76.14
17. Basti	5.6	8165	34.09	72.3	63.21	47.43	19.1	142.60	58	6.24	93.80
18. S.Kabirnagar	7.1	7962	28.97	63.5	83.14	12.52	19.3	172.29	27	4.03	73.40
19. Gonda	7.0	8964	37.22	48.4	66.67	16.56	19.2	149.13	34	3.58	84.19
20. Balrampur	8.1	9312	32.15	41.9	33.91	20.15	18.2	145.28	130	6.05	76.45
21. Shrawasti	2.8	5843	60.82	–	41.02	10.91	16.5	140.05	–	3.31	72.74
22. Bahraich	10.0	8114	56.89	59.8	49.81	18.46	17.7	152.80	87	4.29	77.80
23. Faizabad	13.5	10081	36.83	116.9	89.41	27.82	23.9	157.13	70	6.65	100.37
24. Ambedkarnagar	8.9	8814	33.37	103.9	94.35	17.14	25.8	168.04	63	3.45	88.59
25. Pratapgarh	5.3	7137	23.76	68.2	85.07	16.62	17.7	139.80	6	1.42	76.53
26. Sultanpur	4.7	9769	32.20	91.3	82.90	33.80	21.6	152.61	113	10.60	99.38
Uttar Pradesh	20.8	13262	45.14	180.4	78.61	35.82	20.5	152.15	331	11.03	100.00
Highest in U.P.	67.0	47220	86.43	1381.1	99.36	112.89	33.2	187.55	8876	48.50	413.15

Source: Govt. of U.P., Lucknow (2007). Planning Atlas, 2000.

(3.00), Shravasti (3.31), Ambedkarnagar (3.45), Ghazipur (3.59), Gonda (3.58), Maharajganj (4.03), Sant Kabirnagar (4.3), Bahraich (4.29), Ballia (4.32), Mirzapur (4.87) and Azamgarh (4.99) (Fig. 1B), Only 4 districts—S.R.N. Bhadohi (25.76), Varanasi (16.58), Allahabad (14.75), and Mau (14.01) claim a higher contribution of the industrial sector to the net domestic product of the district than the state average.

Only 3 districts—Shravasti (60.82), Bahraich (56.89) and S.R.N. Bhadohi (47.59) of the study area record a credit-deposit ratio above the state average (45.14). The first two districts are located in the north adjoining the Nepal border while the third one, located in southern part, is world famous for carpet manufacture (Fig. 11.2B). In five districts—Jaunpur, Azamgarh, Mau, Ghazipur and Ballia—located in the middle of the region, the C/D ratio is less than half of the state average. The C/D ratio is between half and three-fourths of the state average in nine more districts—Allahabad, Sonbhadra, Deoria, Gorakhpur, Sant Kabirnagar, Balrampur, Ambedkarnagar, Pratapgarh and Sultanpur scattered throughout the study area. A low C/D ratio enables to infer that banks are, for one reason or the other; siphoning off the savings of the region to other parts of the country. The per capita consumption of electricity is an important indicator of development and of performance of the non-farm sector. It is as low as 36 kwh/person (less than 20 per cent of the state average) in Kushinagar (Fig. 11.2A). The electricity consumption is less than half of the state average (180.4 kwh/person) in 11 more districts of eastern U.P. It is more than the state average in only 3 districts of the region—Allahabad (303.6), Varanasi (274.8) and Mau (181.3). The districts of eastern U.P. also suffer from daily load-shedding for quite a few hours.

Though per capita income is a disputed and insufficient measure to identify the level of overall socio-economic development, however, it is the most often used indicator to define level of welfare, of economic progress and performance of the economy. In case one considers per capita income (2005-06) at current prices; it is Rs. 13,262 for the state of Uttar Pradesh. It is the highest in Gautam Buddhanagar (Rs. 47,220) and the lowest in Shravasti (Rs. 5,843) district. The per capita income is higher than the state average in only one district Sonbhadra (Rs. 21,359) of eastern U.P. (Fig. 11.3A).

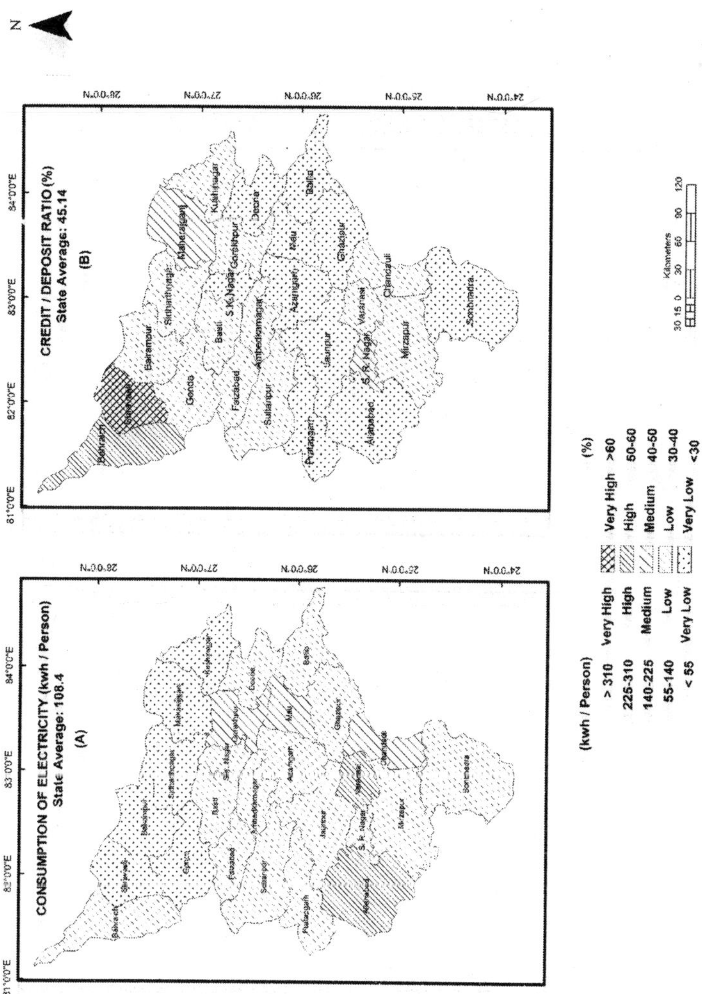

Fig. 11.2: Eastern Uttar Pradesh

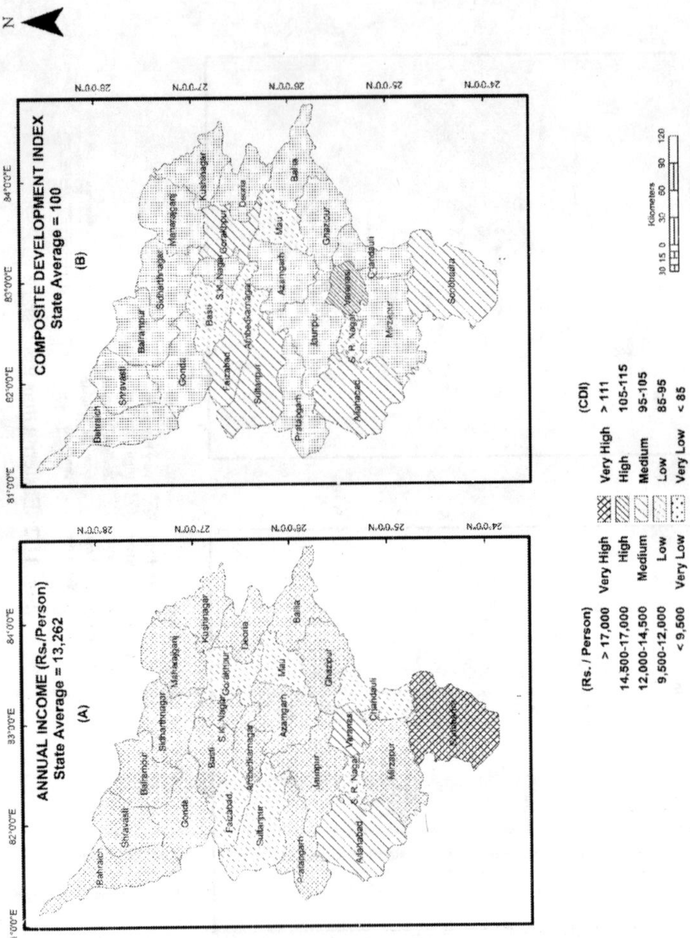

Fig. 11.3: Eastern Uttar Pradesh

Sonbhadra and the surrounding region is fast emerging as the largest thermal energy complex of the country. The per capita income is less than two-thirds of the state average in 14 districts of eastern U.P. (Pratapgarh, Jaunpur, Azamgarh, Ghazipur, Ballia, Deoria, Kushinagar, Maharajganj, Siddharthnagar, Shravasti, Bahraich, Basti, Sant Kabirnagar and Ambedkarnagar). The districts may be defined as economically backward and pose a problem to the state polity. It is quite pertinent to mention here that Chitrakoot is the only other district in the state of Uttar Pradesh to be defined as economically backward on this ground.

Levels of Development

Ram and Chandra Shekhar (2006) of International Institute for Population Sciences (Mumbai) worked out a composite index of development on the basis of 13 socio-economic indicators as listed in Table 11.1. They ranked all the districts of the country and classified 148 districts as 'More Developed', 293 districts as 'Developing' and 150 districts as the 'Least Developed'. As many as 16 districts of eastern U.P. were identified as the least developed and 10 districts as 'developing' on the basis of composite index of development. In Saryupar plain 9 districts out of total of 12 districts were classified as the least developed. It is relevant to mention here that none of the districts of Uttar Pradesh was classified as 'more developed'. The composite index of development worked out by the authors laid much more emphasis on health and family welfare related aspects.

The State Planning Institute's Area Planning Division of Planning Department, Government of Uttar Pradesh worked out a composite index of development (CID) for all the districts of the state (Planning Atlas, 2007). The Division used as many as 36 important indicators—agricultural development (10), industrial development (07), economic infrastructure (10) and social infrastructure (09) (Table 11.3).

As expected the index was the highest for Gautam Buddhanagar (413.15) located in the west adjoining Delhi and the lowest for Shravasti (72.74) located in the north adjoining the Nepal border. The Division identified 5 development categories: very high (125 and above), high (110-125), medium (95-110), low (80-95) and very low (80 and below). The level of development was categorized, as

Table 11.3: Development Indicators used by Area Planning Division, Government of Uttar Pradesh

A. Agriculture	
1.	Gross value of agricultural product per ha of gross area sown (Rs.)
2.	Gross value of agricultural product per rural person (Rs.)
3.	Area under commercial crops to gross sown area (%)
4.	Net irrigated area to net area sown (%)
5.	Fertilizer distribution/ha of gross area sown (Kg.)
6.	Production of food grains/person (Kg.)
7.	Pump sets thousand ha of gross cropped area.
8.	Area irrigated by tube wells to net irrigated area (%)
9.	Milk production/person (Kg.)
10.	Intensity of cropping
B. Industrial Infrastructure	
11.	Workers engaged in household industry to total workers (%)
12.	Workers engaged in registered factories/lakh population
13.	Industrial sector contribution to N.D P. of district (%)
14.	Gross value by industrial produce/person (Rs.)
15.	Invested capital/industrial worker ('000 Rs.)
16.	Registered working factories/lakh of population
17.	Consumption of electricity in industries/total consumption (%)
C. Economic Infrastructure	
18.	Pucca roads under PWD/lakh population (Km)
19.	Pucca roads under PWD/thousand sq.km. of area (Km)
20.	Electricity consumption/person (Kwh)
21.	Electrified villages to total villages (%)
22.	Telephone connections (BSNL)/lakh population
23.	Post offices/lakh population
24.	Scheduled Commercial Banks/lakh population
25.	Credit–Deposit ratio (%)
26.	Registered motor vehicles/lakh population
27.	Income/person (Rs.)
D. Social Infrastructure	
28.	Literacy rate (total)
29.	Gender gap in literacy
30.	Sex ratio (0–6 years)
31.	Primary schools/lakh population
32.	Upper primary schools/lakh population
33.	Polytechniques/lakh population
34.	I.T.Is. /lakh population
35.	Allopathic hospitals dispensaries/lakh population (including PHCs)
36.	Beds in allopathic hospitals–dispensaries/lakh population (including PHCs)

Note: Rs. figures have been taken at current prices. Agriculture data is mostly for the year 2005-06. Most of the data pertaining to industrial infrastructure is for the year 2004-05. Economic and social infrastructure data is mostly for the year 2006-07. Census based data is for the year 2001.

Source: Planning Atlas, State Planning Department, Govt. of U.P., Lucknow, 2007.

'very high' in 6 districts (Gautam Buddhanagar, Ghaziabad, Meerut, Agra, Kanpur and Lucknow) and none of these districts is located in eastern U.P. The level of development was identified as 'high' in seven districts (J.P. Nagar, Muzaffarnagar, Mathura, Saharanpur, Bulandshahar, Mahamayanagar and Varanasi) and only the last one, with the lowest index of development in this category is located in eastern U.P. The level of development has been categorized as 'medium' in five districts and as 'low' in seven districts of eastern U.P. The level of development has been classified as 'very low' in 16 districts of Uttar Pradesh—Bahraich, Shravasti, Balrampur, Siddharth-nagar, Maharajganj, Kushinagar, Sant Kabirnagar, Ballia, Ghazipur, Azamgarh, Jaunpur, Pratapgarh, Mirzapur, Kaushambi, Chitrakoot and Banda and as many as 13 districts (except the last three districts) are located in eastern U.P. It is interesting to note that 9 districts of eastern U.P. with 'very low' level of development have peripheral location adjoining the boundary of the state of Uttar Pradesh. The same five development levels have been identified in eastern U.P. with some minor changes in CDI values (Fig. 11.3b). The map shows one district as High, five districts as Medium, four districts as Low and as many as 16 districts as Very Low. As such 12 districts with very low level of development have peripheral location in the region.

Development Challenges

In a depressed region like eastern U.P. the greatest challenge to progress and social reform is to transform the downward spiral of Growth process into upward spiral by putting in place all the growth inducing factors and facilities. With a level of urbanization as low as less than 10 per cent in 17 districts and an industrial sector contributing less than 10 per cent towards the domestic product of the district in as many as 20 districts and a credit/deposit ratio of less than 35 per cent in 17 districts signifying a lower propensity of local entrepreneurs to take investment risks, it is the agriculture and allied sectors along with the social sector that must form the engine of growth and progress of the regional economy during the initial phase. As discussed in the sequel, the challenges to economic progress in eastern U.P. seem to be mostly of socio-political in nature rather than economic. A survey conducted on the occasion of Hindustan Conclave 'Uttar Pradesh

2020: Retrospect and Prospect' organized in October 2009 put forth corruption as by far the most formidable challenge to progress in the state. It was cited by 35.57 per cent of the respondents. Problems related to law and order were put forth as the second (20.19 per cent) closely followed by lack of physical and social infrastructures (19.30 per cent). The most severe economic problems of poverty and unemployment were cited by 15.01 per cent of the respondents as the most formidable challenge to economic progress. The survey findings are equally, if not more, valid for eastern U.P.

The region suffers from an unholy nexus of politicians, bureaucrats and criminals. An increasing concentration of political, muscle and money power in one and the same individual and a tendency of bureaucrats and administrators to seek political patronage of such individuals has put the process of development out of gear and has jeopardized the development administration and programme implementation leading to rampant corruption and malpractices. The challenge is to ensure the rule of law and to establish a free and fair but responsive and responsible system of governance and administration in eastern U.P. It is essential in order to mobilize local investment and to attract outside entrepreneurs to invest in the region in farm and non-farm enterprises.

Improvement in infrastructures is another challenge. Services like electric supply, transportation and communication, education, health, local governance and banking and finance are not playing their intended role in promoting growth and facilitating politico-social transformation in the region. Electricity supply in eastern U.P. is very erratic. The rural areas suffer from about 14 hours of compulsory load shedding. Even urban areas have to undergo 3 to 6 hours of load shedding and local break downs. Majority of towns are small and have been clubbed together with rural areas for the propose of electricity supply. In order to ensure regular and reliable electricity supply the transmission system needs to be improved. Transmission and distribution losses need to be minimized and supply needs to be augmented. The number of educational institutions has increased significantly, however, the elementary education is in bad shape, the higher education lacks in quality and the academic curriculum is not in tune with needs of the society and the local economy. Development of transport has been lopsided. Health facilities are deficient and suffer from shortage of supplies and skilled man power. Banking and financial institutions do not cater to the needs of rural poor because of a cumbersome procedure and lack of awareness and entrepreneurship in local population.

Proper development and management of regional resources like land, water and human is another challenge. The study area is rich in surface and ground water and its proper development and management will facilitate increase in cropping and irrigation intensity, introduction of more remunerative new crops and diversification of agriculture and thus leading to significant improvement in productivity in agriculture sector. It will foster better access to safe water and improved sanitation thus resulting in improved living condition and better quality of life. Proper management of surface water will augment total water supply and control floods and prevent loss of life and property. Though human resource is the most important resource of the region however its quality and employability is very low. A proper management of land resources by improving irrigation practices, by introducing a proper crop rotation system, by providing modern agricultural inputs and farm supporting services and by putting agro-based enterprises, is another major challenge to economic progress in the region. Level of urbanization and industrialization is very low. Majority of the urban centres are small. Urban amenities in them are very limited and in many cases non-existent. Their role in promoting growth and social transformation is almost negligible. The distribution of services has been very much diffused without any locational considerations. The challenge is to create a well articulated system of towns and service centres equipped with urban functions and services necessary for agricultural diversification, for growth of non-farm industrial activities, for social transformation and rural sector regeneration and for strengthening mutually beneficial functional interaction between rural-agricultural and urban-industrial sectors.

Development Opportunities

Challenges and opportunities seem to be interwoven. A low performance level of regional economy shows much more scope for improvement provided material and technological inputs are managed by the society in a better way. A small increase in absolute terms may exhibit a relatively larger proportionate increase. A lower level of productivity in agriculture signifies much scope for improvement provided management practices are transformed. In case 22 districts of eastern U.P. with a productivity level lower than the state average can achieve a productivity level equal to the state average it may change the entire scenario of the rural-agricultural sector. The area is almost a level plain and has fertile soil. It receives abut 100 cm of rainfall and the crop growing season extends throughout the year. Eastern U.P. is having large water resource potentials both—surface and ground water. A conjunctive use and proper management of surface water will solve the problem of drought and flood, will augment the irrigation potential and will ensure safe water and better sanitation to a larger population. There exists ample opportunity to introduce new crops, rationalize crop rotation system, expand area under multiple cropping, diversify agriculture and make provisions for agriculture supporting services and functions.

A large number of rural service centres do exist in the study area which can play an important role in the process of agricultural development and rural transformation provided these centres are equipped with growth promoting services and functions. District centres like Varanasi, Allahabad, Gorakhpur, Faizabad and Mau have the potentials to develop as regional growth centres performing important function of regional articulation and of interregional functional linkage. Growth activities, development functions and services of socio-economic nature may be established giving due considerations to the existing system of service centres and in accordance with guidelines of the envisaged locational policy. A functionally well articulated and economically viable system of urban centres and other central places will generate growth, will promote economic transformation and will accelerate the pace of socio-cultural change and modernization in this depressed region.

The study area has a large population in active age group and has shown significant gains in adult literacy. There are quite a large number of educational institutions and their number has been increasing. Recently there has been a tendency to establish vocational institutions. However, the quality of education and training provided by these institutions is very poor and the level of employability of the candidates is quite low. Though an elaborate system of local governance from the village level to the district level is in place yet it suffers from favouritism, corruption, nepotism and proxy workers. The government has put in place health facilities and has appointed medical and para-medical personnel, however, essential supplies are inadequate and men are not responsive to the needs of the local population. The system has become inefficient. A proper management and effective monitoring is needed to improve the situation. A responsive and responsible system of governance and administration, if in place, can usher in a new era of progress and prosperity. Education and awareness will definitely play in important role in the entire process.

Tourism may provide an important opportunity for the regeneration of the regional economy. Eastern U.P. has great potentials for the development of religio-cultural tourism. Places like Varanasi, Sarnath, Kushinagar, Shravasti, Vindhyachal, Ayodhya, Allahabad, Pawanagar and many more need to be developed as centres of tourism and pilgrimage for domestic as well as international tourists. 'Discover your Roots' programme should be encouraged. Carpets of Bhadohi, silk sarees and other silk products of Varanasi and Mubarakpur need to be promoted in India and abroad. There is need to streamline the tourism and pilgrimage related facilities and services with an objective to generate additional income and employment.

The challenges to economic progress in a depressed region like eastern U.P. are great. The politician-bureaucrat-criminal nexus resulting in concentration of muscle, money and decision making power at one place, the deprivation trap leading to alienation and marginalization of a large section of population, deteriorating law and order situation hampering investment in the region and an inefficient and corrupt system of administration make the situation even worse. However, firm political commitment, local awareness, an effective and responsive system of administration, a sincere effort towards implementation and an efficient and impartial system of monitoring and evaluation coupled with the suggestions put forth below will definitely improve the situation.

Development Suggestions

A responsive system of governance and an equally responsible system of administration is a must to establish the rule of law in the study region. It should facilitate the dismantling of the politician, bureaucrat and criminal nexus, assist in controlling the increasing concentration of money and muscle power in the nexus and be helpful in neutralizing the deprivation trap and the divisive forces active in the region. It will not be too difficult to realize. About five years before the conditions were equally bad, if not worse, in the neighbouring state of Bihar. The level of development was one of the lowest among the states. The rate of performance of the economy was miserably low. The law and order situation was equally bad. Investment was not forthcoming. Entrepreneurs were fleeing the state. Education system had become irrelevant. Sincerity to purpose, commitment to action and determination to improve shown by the state government have transformed Bihar into a performing state. It has been the second fastest growing state in the country after Gujarat. A rule of law has been established. The society has been transformed with some purpose and so has been the case of bureaucracy in the state.

A special financial package be provided by the national government to finance identified projects and designated programmes. The state government should provide a matching share to finance the development of project sites, transportation, irrigation, flood control, storage and marketing, electricity transmission and distribution as the private sector is always hesitant to enter these sectors more so in an economically backward region having law and order problems. Poorvanchal Development and Planning Board (PDPB) be vested with the authority to coordinate the development activities and to monitor their progress.

Eastern U.P. be declared a tax-free zone for at least 10 years (2011-2021) with an objective to attract investment in the region in industrial sector and agriculture based non-farm sector. Banks and other financial institutions may be persuaded to streamline their lending procedure and take necessary measures to improve their credit/deposit ratio. Banking institutions and cooperatives can extend short term credit facilities to finance the process of diversification and modernization of agriculture and establishment of small scale enterprises and village crafts.

Quite a few development programme, projects and schemes have been launched by the national and the state governments; however, their implementation is wrought with corruption and favouritism and suffers from inefficiency and indifference of the administration. A time-bound and proper implementation as well as effective monitoring and evaluation of various rural development progrmmes and schemes will definitely ensure achievement of their desired objectives and thus will benefit the local population in general and the weaker sections in particular. Facilities and functions need to be made more accessible to the general public. There is need to create general awareness and to educate population to make proper use of Right to Information Act.

Education and health need to be linked to the demand of the regional economy and the aspirations of the local population. Education needs to be linked to industry and agriculture and the three be linked to health, nutrition and empowerment of people in general and vulnerable sections in particular. Development means enlargement of people's choice and their freedom. Education enables them to make right choice and empowers them to enjoy increased freedom. Education, family welfare, health should become a political agenda. Education facilities existing in the region need to be strengthened and consolidated. The telecommunication system is rather developed, however, the use of information technology in education, health, development administration and social and economic transformation needs to be encouraged.

There have been now and then demands raised to subdivide the state into four smaller states, namely Purvanchal (eastern U.P.), Harit Pradesh (western U.P.), Bundelkhand (south-western plateau region) and Madhyanchal (central U.P.). The smaller states are in general easy to administer and have performed better on economic, social and administration fronts *vis-à-vis* large states. Pending statehood an autonomous organization with wide ranging decision making powers of trans-sectoral nature cutting across the conventional administrative compartmentalization be created. It may consist of whole time planning experts, development analysts and programme administrators as well as *ex-officio* members. It will formulate, implement and monitor development projects and programmes in the region. The national and the state governments be required to make budget allocation to the organization proportionate to the population and the index of poverty and backwardness of the region. The organization may be given the name of Purvanchal Development and Planning Board (PDPB).

REFERENCES

Govt. of Uttar Pradesh (2007). *Planning Atlas 2000*. Uttar Pradesh Area Planning Division, State Planning Institute, Lucknow: Planning Department.

Ram, F. and Shekhar, C. (2006). *Ranking and Mapping of Districts Based On Socio-Economic and Demographic Indicators*. Mumbai: International Institute of Population Sciences.

12

Progress of Poverty and Health Related Goals of MDG: Special Reference to Eastern Uttar Pradesh

Rajesh Kumar Chauhan

Introduction

State of Uttar Pradesh (UP) with over 198 million population in 2009 ranks as fifth largest entity (first four being China, India, USA and Indonesial, Wikipedia, 2009). In other words, every thirty-fourth person on the globe belongs to state of Uttar Pradesh of India. The state has been infamously known for its laggard demographic and development performance and affected overall demographic and economic performance of India to a great extent. UP has also been subject of concern for planners due to plausible impacts it can bring upon national performance. Millennium Development Goals have been devised for nations to monitor and take appropriate corrective actions for achieving them. India being largest democracy and signatory to the millennium declaration, it is important to track and monitor the status of MDGs, specially those related to poverty and health. Poverty and health are considerably interlinked and improvement in one set is likely to bring improvement in other set. The goals 1, 4 and 5 respectively are on poverty, child mortality and maternal mortality. This paper aims to assess the current status along with the progress on these three goals of MDG. The indicators in the selected goals will be dealt are listed as:

Goal 1: Eradicate Extreme Poverty and Hunger

1. Poverty headcount ratio (% of population below poverty line).
2. Poverty gap ratio (incidence times depth of poverty).
3. Squared Poverty gap ratio.

Goal 4: Reduce Child Mortality

4. Underfive mortality rate.
5. Infant mortality rate.
6. Proportion of 1 year old children immunized against measles.

Goal 5: Improve Maternal Health

7. Maternal mortality ratio.
8. Proportion of births attended by skilled health personnel.

More often single estimate of indicators being available for the state, State of UP is placed in a situation where geographic and cultural diversities are ironed out and their impact on related phenomenon are not well known. This calls for the lower (substate) level estimates to be generated for proper monitoring and evaluation of programmes and schemes aimed at improving MDGs in the state. Specific objectives of this paper are listed as:

Objectives

1. To assess the current status of indicators of poverty and health related goals in regions of Uttar Pradesh.
2. To track the progress of these indicators.

Source of Data

1. Planning Commission, GoI.
2. Census Reports, RGI, GoI.
3. DLHS3 District and State Level Fact sheets.
4. Various research papers/reports providing estimates of selected indicators of regions in UP.

Methodology

Direct or indirect estimation of these indices would be done where ready estimates would not be available. The methods suggested in Progress and Prospects of Millennium Development Goals in India, IIPS (2009) has been adopted here. Here actual annual rate of progress (ARP) has been estimated as

$$ARP_i = [X_C - X_B]/[X_B * (t_C - t_B)] \quad \text{... (1a)}$$

$$ARP_i = [X_C - X_B]/[(100 - X_B) * (t_C - t_B)] \quad \text{... (1b)}$$

Where

t_C and t_B are current and base years

X_C and X_B are the values of indicators at current and base years

Equations (1a) to be used where anticipated value of indicator is to reach to 0 while (1b) is to be used for the scenario where indicator needs to attain the value of 100. Likewise the required rate of annual progress (RRP) is to be estimated as dictated by the need for reaching the level of indicator according to commitment.

$$RRP_i = \alpha / (t_{mdg} - t_B) \quad \text{... (2)}$$

Where

t_{mdg} and t_B are 2015 and base year

α is the targeted value of indicators at year 2015

Estimation Related to Goal 1: Eradicate Extreme Poverty and Hunger

All three measures are estimated as per Planning Commission methodology, i.e. based on consumer expenditure data gathered by NSSO were used to identify poor households (persons) against the set Poverty Lines in money terms.

Estimation Related to Goal 4: Reduce Child Mortality

The available census based estimates of district level CMRs and IMRs has been taken from PFI report and regional estimates have been generated using "births last year" and "population aged 6 years" as weights. Another indicator 'proportion children 12-23 immunized against measles' has been computed from raw datasets of NFHS round-1 and DLHS round-3.

Estimation Related to Goal 5: Improve Maternal Health

Maternal mortality ratio; using the indirect method [Ln (MMR)= 8.845–0.831 Ln (safe delivery)] as described in Chauhan, 2010 have been applied here. For this input dataset of 'proportion of births attended by skilled health personnel' have been computed from raw datasets of NFHS round-1 and DLHS round-3.

Thus Details of Base Year and Current Datasets are on 8 of the Selected Indicators are provided as under in Table 12.1.

Targets of MGDs for Year 2015 [Base Year 1990]

As decided base year ought to be year 1990 so a year close to 1990 has been chosen as base year for which data are available at the desired level of geographic approximation.

Goal 1: Eradicate extreme poverty and hunger (Half)
HCR, Depth, Severity (Half)
Goal 4: Reduce child mortality (Reduce 2/3)
CMR and IMR (reduce 2/3)
Proportion children 12-23 immunized against measles (3 times)
Goal 5: Improve maternal health (Reduce 3/4)
MMR (Reduce 3/4)

Poverty HCR

The poverty head count ratio in UP in base year, i.e. 1993-94 was 40.9 and got reduced to 25.7 per cent in recent (comparison) year, i.e. 2004-05. Table 12.2 presents the base year and current year estimates of three indicators of poverty for four regions of Uttar

Table 12.1: Base and Current Year Datasets

Goal/Indicator	*Base*		*Current*	
	Year	*Data source*	*Year*	*Data source*
Goal 1: Eradicate extreme poverty and hunger				
1. Poverty headcount ratio (% of population below poverty line)	1993-94 (NSS 50 Round)	Prabha *et al.*	2004-05 (NSS 61 Round)	Chauhan, 2007
2. Poverty gap ratio (incidence times depth of poverty)	1993-94 (NSS 50 Round)	Prabha *et al.*	2004-05 (NSS 61 Round)	Chauhan, 2007
3. Squared Poverty gap ratio	1993-94 (NSS 50 Round)	Prabha *et al.*	2004-05 (NSS 61 Round)	Chauhan, 2007
Goal 4: Reduce child mortality				
4. Under-five mortality rate.	1991 (Census 1991)	Calculated using PFI District level estimates	2001 (Census 2001)	PFI, 2008
5. Infant mortality rate.	1991 (Census 1991)	PFI	2001 (Census 2001)	PFI, 2008
6. Proportion of 1 year old children immunized against measles	1992-93 (NFHS-1)	Estimated	2007-08 (DLHS-3)	Estimated
Goal 5: Improve maternal health				
7. Maternal mortality ratio	1992-93 (NFHS-1)	Estimated	2007-08 (DLHS-3)	Estimated
8. Proportion of births attended by skilled health personnel	1992-93 (NFHS-1)	Estimated	2007-08 (DLHS-3)	Estimated

Source: Based on personal computations.

Table 12.2: Regional estimates of poverty in UP in base year and current year

Poverty indicator/ Region	*1993-94*			*2004-05*			*Improvement during 1994-2005*		
	Urban	*Rural*	*Overall*	*Urban*	*Rural*	*Overall*	*Urban*	*Rural*	*Overall*
Poverty HCR									
Western	31.1	29.3	29.8	26.4	18.4	20.4	1.38	3.38	2.86
Central	33.9	50.2	46.7	20.1	24.1	23.1	3.70	4.73	4.58
Eastern	38.6	48.8	47.5	36.0	31.9	32.3	0.61	3.14	2.90
Southern	74.4	67.4	68.9	31.8	19.5	22.0	5.21	6.46	6.19
UP	35.1	42.3	40.9	27.3	25.4	25.7	2.02	3.63	3.37
Poverty Depth									
Western	7.7	6.1	6.9	5.0	2.4	3.9	3.2	5.5	3.9
Central	9.5	13.2	11.7	4.6	4.2	4.4	4.7	6.2	5.7
Eastern	9.4	12.0	10.9	6.8	5.9	6.4	2.5	4.6	3.7
Southern	21.1	20.2	20.7	9.7	5.5	8.1	4.9	6.6	5.5
UP	9.0	10.4	9.8	5.5	4.4	5.0	3.5	5.2	4.5
Poverty Severity									
Western	2.8	1.9	2.4	1.3	0.5	1.0	4.9	6.7	5.4
Central	3.6	5.0	4.4	1.7	1.1	1.4	4.8	7.1	6.3
Eastern	3.2	4.0	3.6	1.9	1.7	1.8	3.7	5.2	4.6
Southern	8.2	8.1	8.2	3.4	1.7	2.8	5.3	7.2	6.0
UP	3.3	3.5	3.4	1.7	1.1	1.4	4.4	6.2	5.3

Sources: For poverty indices of 1993-94, Goswami and Chatterji (2009) and 2004-05 Chauhan (2007).

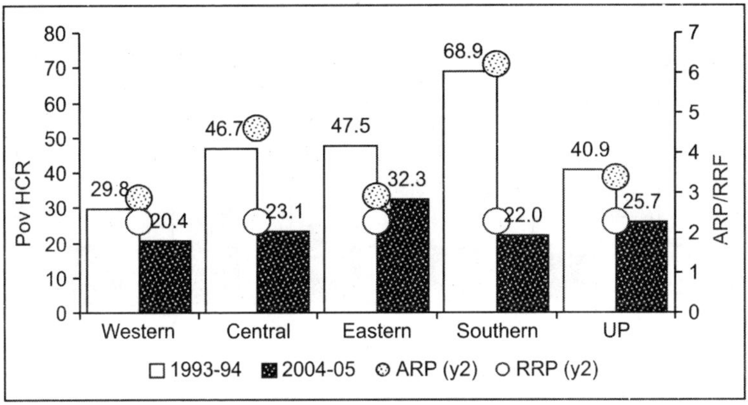

Fig. 12.1: Actual and Required Rates of Progress (Poverty HCR)

Pradesh. In order to achieve this goal the reduction needed by 2004-05 was to be of 2.27 per cent and it actually got reduced by a rate of 3.37 per cent per year. The progress in this indicator has been recorded at the highest annual rate of 6.19 per cent for Bundelkhand Region. All other regions also demonstrated performance, which was far above to that of required progress. Likewise similar declines of higher rate were observed for poverty depth and poverty severity (Fig. 12.1).

Poverty Gap

The poverty gap ratio in UP in base year, i.e. 1993-94 was 9.8 and got reduced to 5.0 per cent in recent (comparison) year, i.e. 2004-05. In order to achieve this goal the reduction needed by 2004-05 was to be of 2.27 per cent and it actually got reduced by a rate of 4.46 per cent per year. The progress in this indicator has been recorded at the highest annual rate of 5.69 per cent for Eastern Region which in fact had second highest rate of inequality in base year. All other regions also noted decline which was far above to that of required progress (Fig. 12.2).

Squared Poverty Gap

The squared poverty gap ratio in UP in base year, i.e. 1993-94 was 3.4 and got reduced to 1.4 per cent in recent (comparison) year, i.e.

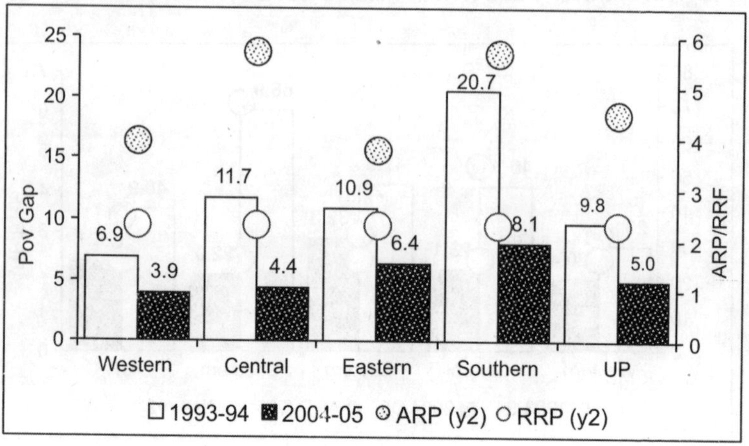

Fig. 12.2: Actual and Required Rates of Progress (Poverty Gap)

2004-05. In order to achieve this goal the reduction needed by 2004-05 was to be of 2.27 per cent and it actually got reduced by a rate of 5.33 per cent per year. The progress in this indicator has been recorded at the highest annual rate of 6.28 per cent for Central Region which in fact had highest rate of inequality in base year. All other regions also noted decline which was far above to that of required progress (Fig. 12.3).

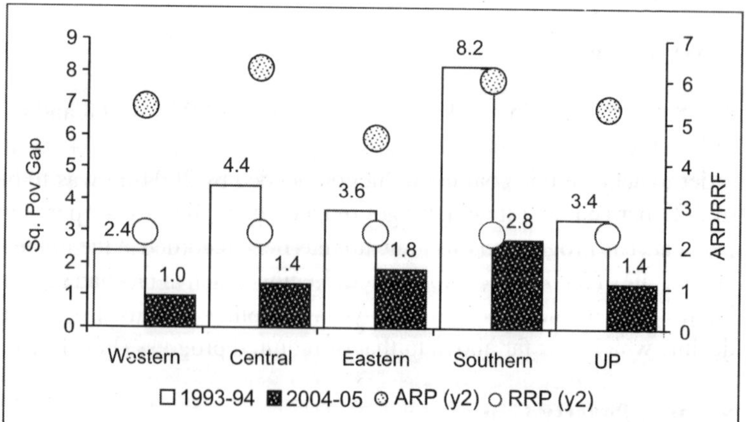

Fig. 12.3: Actual and Required Rates of Progress (Sq. Poverty Gap)

Infant Mortality Rate

IMR in UP was estimated to be 89 in 1991 and in a decade's time period it just came down to 84 per thousand live births. In order to achieve the reduction of three-fourth it is to reach 30 in 2015. Thus an annual decline of 2.78 per cent is needed for the entire period while as in actual it came down by a rate of 0.56 per cent per annum. Southern region registered almost no decline in 10 years period. Performance of all regions on this indicator has been far below than that of desired level (Fig. 12.4).

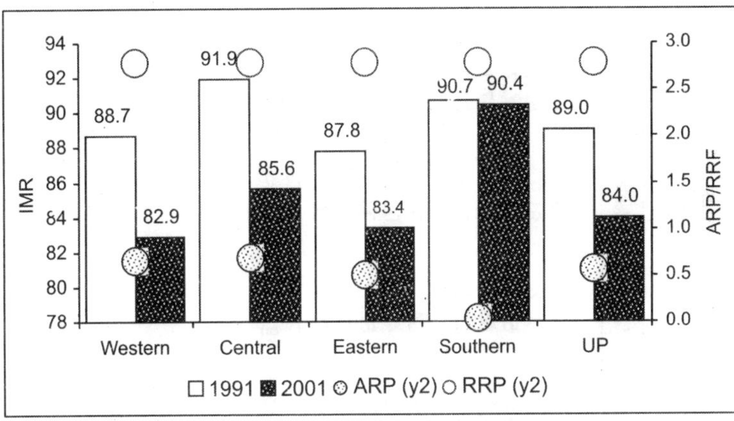

Fig. 12.4: Actual and Required Rates of Progress (IMR)

Child Mortality Rate

CMR in UP was estimated to be 120 in 1991 and in a decade's time period it just came down to 94. In order to achieve the reduction of three-fourth it is to reach 40 in 2015. Thus an annual decline of 2.78 per cent is needed for the entire period while as in actual it came down by a rate of 2.17 per cent per annum. Western region registered highest decline of 2.32 per cent in this indicator in 10 years period, which is slightly lower than that of required. Likewise performance of all regions on this indicator has been far below than that of desired level (Fig. 12.5).

Table 12.3: Regional estimates of infant mortality and child mortality rates in UP in base year and current year

Indicators/Regions	*1991*			*2001*			*Improvement during 1991-2001*		
	Urban	*Rural*	*Overall*	*Urban*	*Rural*	*Overall*	*Urban*	*Rural*	*Overall*
IMR									
Western	62.4	98.4	90.9	57.3	86.7	80.2	0.82	1.19	1.18
Central	46.2	103.0	94.2	40.2	91.9	82.8	1.30	1.08	1.21
Eastern	65.7	91.9	89.9	57.9	82.8	80.7	1.19	0.99	1.02
Southern	59.1	99.4	93.0	55.0	93.9	87.4	0.69	0.55	0.60
UP	58.0	94.0	89.0	52.0	88.0	84.0	1.03	0.64	0.56
CMR									
Western	85.2	134.2	126.9	64.1	99.4	90.9	2.48	2.59	2.84
Central	76.1	133.0	130.2	46.6	108.2	95.0	3.88	1.86	2.70
Eastern	88.7	122.4	122.3	62.4	93.7	90.5	2.97	2.34	2.60
Southern	66.6	146.8	137.0	65.7	111.2	104.4	0.14	2.43	2.38
UP	73.0	132.0	120.0	69.0	100.0	94.0	0.55	2.42	2.17

Source: Authors' calculations for regional estimates using district level estimates of IMR and CMR published by Rajan *et al.* (2008).

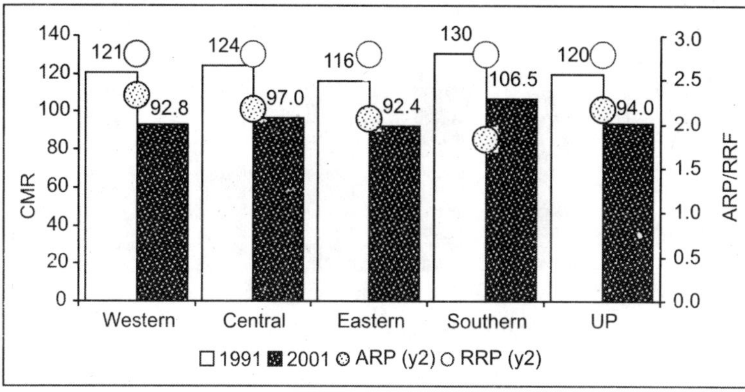

Fig. 12.5: Actual and Required Rates of Progress (IMR)

Measles Immunisation

Proportion of children age one year (age 12-23 months) who have been given Measles vaccine is one of the indicators of study. The base year for this indicator is NFHS-1 (1992-93) and current year is chosen to be 2007-08 (DLHS-3). This is a positive indicator and growth is needed here and ultimately it may assume to be cent per cent. Thus in 25 year (i.e. from 1991 to 2015) this indicator is to be tripled. This sets a goal of increase of 3.09 per cent per annum. The actual rate of progress has been 2.49 per cent per annum which is slightly lower than that of required annual progress. Southern region has displayed slightly higher proportion on this indicator but still actual rate of annual progress have remained well below than that of required rate of progress (Fig. 12.6).

Safe Deliveries

Proportion of safe deliveries is also a positive indicator, this also needs to reach cent-per-cent. The base year for this indicator is NFHS-1 (1992-93) and current year is chosen to be 2007-08 (DLHS-3). This indicator in 23 years (i.e. from 1993 to 2015) is need to grow at a rate of 3.61 per cent per annum to reach the desired level. Table 12.5 provides the estimates at both the time periods and it may be seen that actual progress rates have been far below than that of required rates for UP and its constituent regions. The actual progress have been nearly one-third of what is needed (Fig. 12.7).

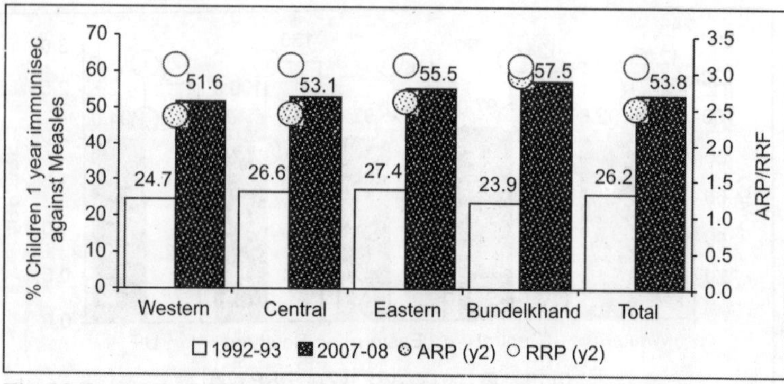

Fig. 12.6: Actual and Required Rates of Progress (Immunisation against Measles)

Table 12.4: Regional estimates of measles immunisation rates in UP in base year and current year (children aged one year)

Regions	*1992-93*			*2007-08*			*Improvement during 1992-2008*		
	Urban	*Rural*	*Overall*	*Urban*	*Rural*	*Overall*	*Urban*	*Rural*	*Overall*
Western	24.7	18.8	45.4	51.6	50.3	55.9	2.38	2.59	1.29
Central	26.6	21.9	39.0	53.1	50.1	68.1	2.40	2.41	3.18
Eastern	27.4	26.6	37.7	55.5	54.9	62.3	2.58	2.57	2.64
Southern	23.9	23.9	23.8	57.5	54.7	74.6	2.95	2.70	4.44
UP	26.2	23.3	41.5	53.8	52.6	60.3	2.49	2.55	2.15

Source: Authors' calculations using NFHS-1 and DLHS-3 raw datasets.

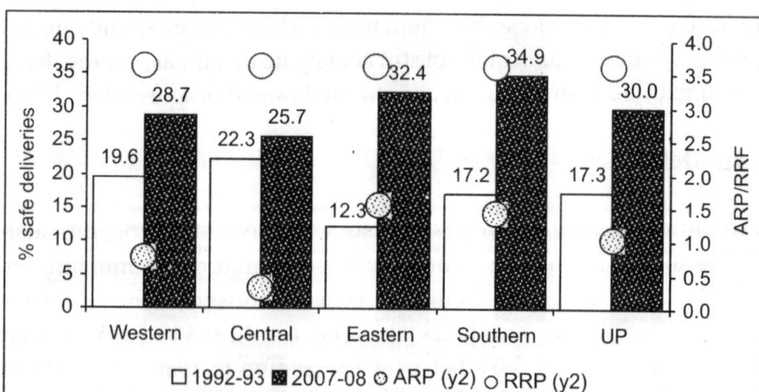

Fig. 12.7: Actual and Required Rates of Progress (Safe Deliveries)

Table 12.5: Regional estimates of safe delivery rates and MMR in UP in base year and current year

Indicators/ Regions	*1992-93*			*2007-08*			*Improvement during 1992-2008*		
	Rural	*Urban*	*Overall*	*Rural*	*Urban*	*Overall*	*Rural*	*Urban*	*Overall*
Safe Deliveries									
Western	12.7	42.2	19.6	24.4	44.0	28.7	0.89	0.21	0.76
Central	9.4	50.0	22.3	21.8	48.9	25.7	0.91	0.15	0.29
Eastern	10.3	39.9	12.3	30.1	58.2	32.4	1.47	2.03	1.53
Southern	16.6	27.4	17.2	30.5	57.9	34.9	1.11	2.80	1.42
UP	11.7	44.7	17.3	26.9	47.2	30.0	1.15	0.30	1.02
MMR									
Western	840	310	586	489	299	426	2.79	0.24	1.82
Central	1074	269	525	536	274	468	3.34	0.12	0.72
Eastern	998	324	864	410	237	385	3.93	1.79	3.70
Southern	672	444	652	405	238	363	2.65	3.09	2.96
UP	899	295	648	450	282	411	3.33	0.29	2.44

Source: Authors' calculation using indirect technique using Chauhan (2010).

Maternal Mortality Ratio

This is a negative indicator and a reduction of three-fourth is set as a goal for the state. The base year for this indicator is also NFHS-1 (1992-93) and current year is chosen to be 2007-08 (DLHS-3). This indicator in 23 years (i.e. from 1993 to 2015) is need to reduce at a rate of 3.26 per cent per annum to reach the desired level. Table 12.6 provides the estimates at both the time periods and it may be seen that actual progress rates have been slightly below than that of required rates for UP. Encouraging to note that Eastern region has had highest MMR of 864 in 1992-93 which came down to lowest of 385 in 2007-08 amongst all four regions of UP. The rate of decline has been 3.7 per cent per annum. Southern region has also displayed a satisfactory level of nearly 3 per cent decline. Western and Central regions have advantage at the base year but could not maintain their positions at the current year.

Status of Selected MDG Goals in UP

Table 12.6 presents the actual progress of the eight selected indicators of MDG in UP. In table the shaded cells are indicative of progress lower to the desired. Thus it may be concluded that at state level the

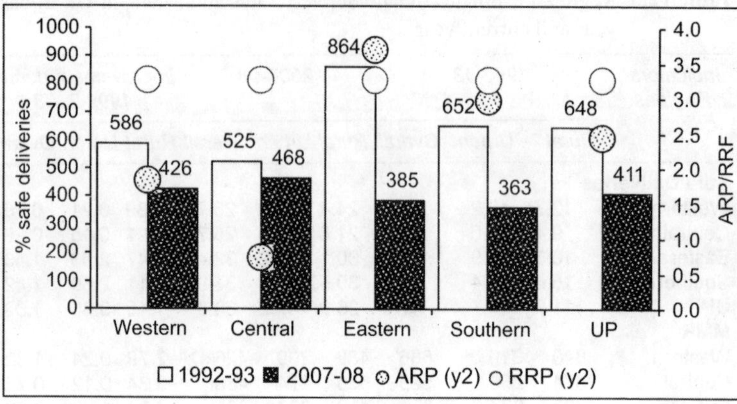

Fig. 12.8: Actual and Required Rates of Progress (MMR)

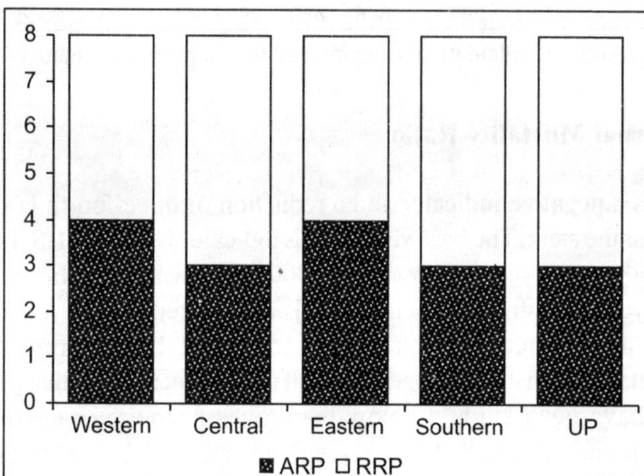

Fig. 12.9: Indicators on Right path

progress has been satisfactory for three of the eight selected indicators. Western and Eastern regions have displayed satisfactory progress for half of the indicators under consideration. Indicators related to poverty are by and large performed well during the intervening period. Same is the case with these indicators in rural areas of state and four regions. While as in urban areas are concerned the state had been slow in realising the decline in poverty because largest regions, i.e. Eastern and Western did not perform adequately. Urban poverty is becoming newer reality and need to be dealt with. All the indicators (five) on

Table 12.6: Actual rate of progress of eight selected indicators in UP during respective base years and current years

Regions	*Poverty*			*Child health*			*Maternal Health*		*Number on indicators on right path*
	HCR	*P1*	*P2*	*CMR*	*IMR*	*Measles*	*Safe Del.*	*MMR*	
Overall UP									
Western	2.86	3.93	5.35	2.84	1.18	2.38	0.76	1.82	4
Central	4.58	5.69	6.28	2.70	1.21	2.40	0.29	0.72	3
Eastern	2.90	3.75	4.59	2.60	1.02	2.58	1.53	3.70	4
Southern	6.19	5.53	6.02	2.38	0.60	2.95	1.42	2.96	3
UP	3.37	4.46	5.33	2.17	0.56	2.49	1.02	2.44	3
Rural									
Western	3.38	5.51	6.70	2.59	1.19	2.59	0.89	2.79	3
Central	4.73	6.20	7.09	1.86	1.08	2.41	0.91	3.34	4
Eastern	3.14	4.62	5.23	2.34	0.99	2.57	1.47	3.93	4
Southern	6.46	6.62	7.18	2.43	0.55	2.70	1.11	2.65	3
UP	3.63	5.24	6.23	2.42	0.64	2.55	1.15	3.33	4
Urban									
Western	1.38	3.19	4.87	2.48	0.82	1.29	0.21	0.24	2
Central	3.70	4.69	4.80	3.88	1.30	3.18	-0.15	-0.12	5
Eastern	0.61	2.51	3.69	2.97	1.19	2.64	2.03	1.79	3
Southern	5.21	4.91	5.32	0.14	0.69	4.44	2.80	3.09	4
UP	2.02	3.54	4.41	0.55	1.03	2.15	0.30	0.29	2
Required rate of annual progress	**2.27**	**2.27**	**2.27**	**2.78**	**2.78**	**3.09**	**3.61**	**3.26**	

Source: Based on personal computations.

child and maternal health has not been up to the mark. If we see rural areas at least MMR has been declining at its desired pace, which is not encouraging but essential to be kept like this. In urban areas all the 5 maternal and child health indicators did not do well (Fig. 12.9).

Conclusion

Thus it may be concluded that a wider set of indicators need to be studied to understand that how many more of these indicators are not doing well. As far as health and poverty related goals of MDG are concerned state has tough task ahead. The status of regions vary according to the performance. Eastern region has a achievement that it has displayed a satisfactory performance on rural MMR decline. Experiences from this region may need to be replicate elsewhere to arrest the high MMR in the state.

REFERENCES

Chauhan, R.K. (2007). Gains and challenges in pooling state and central sample NSS data: case of Uttar Pradesh for 61st (200405) round. Paper presented in 'National Seminar on the Survey Results of NSS 61st Round' organized at New Delhi during October 29-30, 2007.

Chauhan, R.K. (2010). Maternal mortality in districts of Uttar Pradesh: an empirical estimation, In U.V. Somayajulu, K.K. Singh, K.V.R. Subrhamanyam and A. Pandey (eds.), *Population and Reproductive Health: Perspectives and Issues,* New Delhi: Hindustan Publishing Corporation.

Ram, F., Mohanty, S.K. and Ram, U. (2009). *Progress and prospects of millennium development goals in India*, Mumbai: IIPS, India.

International Institute for Population Sciences (2006). *DLHSRCH2, 2002-04,* Mumbai: IIPS.

International Institute for Population Sciences (IIPS) and Macro International (2007). *National Family Health Survey (NFHS3), 2005-06,* Mumbai: IIPS.

Population Foundation of India (PFI) (2008). *Socio Demographic Development Index Uttar Pradesh-2007*, New Delhi: PFI, India.

Goswami, P.K. and Chatterji, B. (2009). Poverty in Uttar Pradesh: an interregional study. *Journal of Rural Development*, 28 (1).

Rajan, S.I., Nair, P.M., Sheela, K.L., Jagatdeb, L. and Mishra, N.R. (2008). *Infant and Child Mortality in India: District Level Estimates*. New Delhi: Population Foundation of India.

Registrar General of India (2009). *Special Bulletin on Maternal Mortality in India: 2004-06,* New Delhi: Sample Registration System (SRS).

Ministry of Health and Family Welfare, Govt. of India and International Institute for Population Sciences (2009). *Fact Sheet Uttar Pradesh District Level Household and Facility Survey 2007-08, India:* Mumbai: IIPS.

Ministry of Health and Family Welfare, Govt. of India and International Institute for Population Sciences (2009). *Fact Sheets Districts of Uttar Pradesh District Level Household and Facility Survey 2007-08, India,* Mumbai: IIPS.

Wikipedia (2009). List of countries by population, on webpage of Wikipedia, retrieved on 20/09/2009 http://en.wikipedia.org/wiki/List_of_countries_by_population.

13

Issues of Population and Development in Bhojpuri Speaking Region of India with Special Reference to Millennium Development Goals (MDGs)

Archana K. Roy, Neelam Tripathi, Upendra Nath Roy and *Rashmi Kant*

Introduction

Bhojpuri is one of the 12 mother tongues[1] of Hindi language which is spoken by more than 33 million people in India (Census of India, 2001). It covers a large geographic area crossing state boundaries, however, 30 million of the Bhojpuri speaking people are confined in the region between Patna (Bihar) and Varanasi (Uttar Pradesh). The region, situated in middle Ganga plain, is endowed with fertile land and rich in water resources (surface and groundwater) and has high potential for human resource development. The region has a distinct culture but is known for its poverty, migration, casteism, and feudalism/suppression (locally known as *Dabangai*). Now question arises, why language is considered for demarcation of a region? And what is the implication of cultural factor on the development of the region? Language is a system of conventional spoken or written symbols used by people in a shared culture to communicate with each other. A language both reflects and affects a culture's way of thinking, and changes in a culture influence the development of its language (Britannica Concise Encyclopaedia). Language is a medium of social interact specific to each culture passed down as legacy from one generation to the other to give identity and pride with a sense of belonging to a community or even a nation that maintains distinct culture (answer.com, 2010). According to Kaul (2011) it is a strong mark of social and ethnic identity in its natural environment. However, regarding its relation with economic development the relationship is not simple. Thompson (2001) after analyzing the theories on culture and development, found that social theorists from Karl Marx to Daniel Bell have argued that economic development brings pervasive cultural change. Others, from Max Weber to Samuel Huntington, have claimed that cultural values are an enduring and have autonomous influence on society. He further adds that the relationship between culture and economic development is extremely complex.

However, in the rapidly changing era of liberalization, privatization and globalization the culture and language is also changing. UNESCO says, "Increased migration and rapid urbanization often bring along the loss of traditional ways of life and a strong pressure to speak a dominant language necessary for full civic participation and economic advancement". It also argues that in India with its burgeoning ambition, changing demographics, widening market, nuclearization of families and massive shift in population from the rural and semi-rural to the urban—the linguistic map is rapidly changing (Soondas, 2010).

When the whole world is rapidly transforming under the influence of liberalization, privatization and globalization (LPG) it is important to know that where does Bhojpuri region stand in on different development parameters? What are the different issues and challenges of population and development in the region? What is the spatial pattern of disparity in the region? This paper tries to answer these questions with reference to Millennium Development Goals (MDGs) as India is one of the signatories to Millennium Development Goals (MDGs) of United Nations. United Nations has set certain goals for the sustainable development known as Millennium Development Goals (MDGs). The progress towards the MDGs is monitored through a set of 21 measurable and time-bound targets and 60 indicators which are to be achieved by 2015. Out of the 18 targets that the old UN framework had against 8 MDGs, 12 are relevant to India.

The goals related to issues of the universalization of primary education; reduction of poverty; gender parity on socio-economic parameters; safe motherhood; access to safe drinking water and sanitation are explored in the present paper. Since data on two time scales were not available for the above mentioned goals, therefore, the mid-term MDG India Country Report 2009 is used as milestone to measure the performance of Bhojpuri region. Different sources like DLHS-3, by IIPS, 2010, Census of India 2001, Central Statistical Organization, 2009 and other government reports have been used. Data were analysed in SPSS and maps were prepared using ArcGIS 9.3 software.

Issues of Bhojpuri Region

Geographical Location and Physical Resources

The region spreads in an area of 77, 871 sq. km across the state of UP (51,486 sq. km) and Bihar (26,385 sq. km). It includes 26 districts [17 from UP and 9 from Bihar (where Bhojpuri is spoken majority under the category of Hindi speaking people)]. Physio-graphically three-fourth of the area comes under Gangatic plain and remaining one-fourth is a plateau in south (*Vindhyan and Kaimur range*). The river Ganga is the main perennial river in which other perennial rivers from the Himalaya, namely Ghagra, Gandak, Budhi Gandak, Gomti and Rapti join from its left side and from the right side or southern plateau three seasonal rivers join to the Ganga known as Son, Belan and Rihand. With such networks of rivers, land is quite fertile and rich in underground water resources. It is not very prosperous in mineral resources but the major minerals are coal, limestone and dolomite found generally in the southern plateau region (Map 13.1).

Socio-economic and Demographic Issues

Demographically, this region is home of 26 per cent of the population and 27.7 per cent of children of states (UP and Bihar) which accounts for 6.3 per cent and 8 per cent of India's people and children (Census, 2001). This figure itself is bigger than the population of Tamil Nadu or Madhya Pradesh. The average population density of the region is

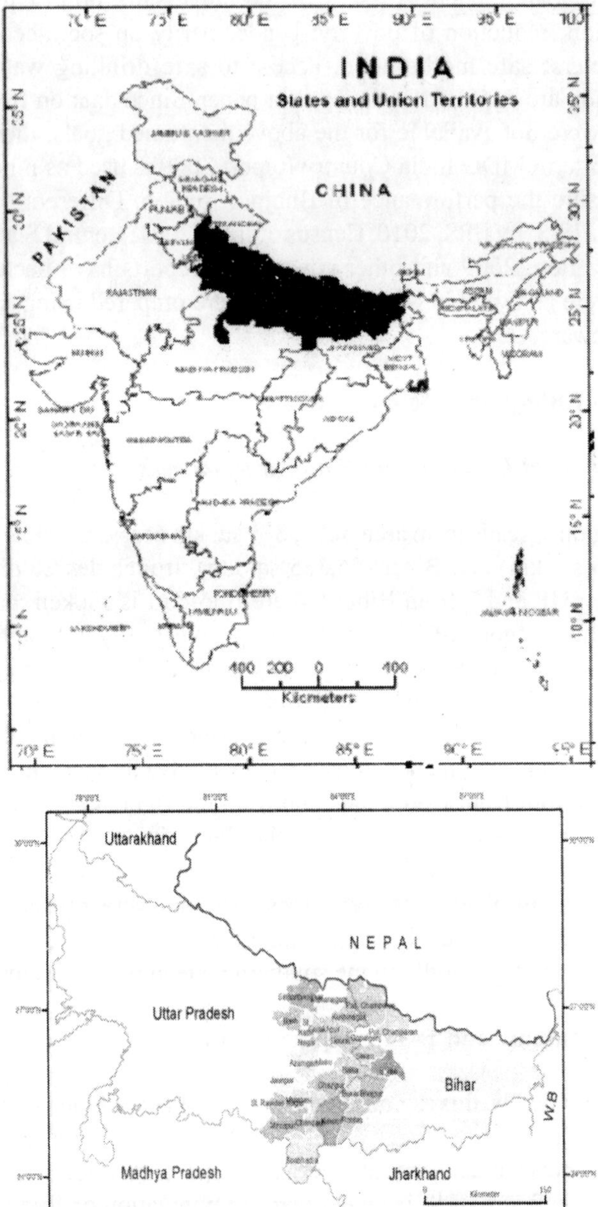

Map 13.1: Location Map of Bhojpuri Speaking Area

832 persons per sq. km with a wide regional variation. The highest population density is in Varanasi (1,989 pp sq.km) and is mainly due to high urbanization (40 per cent), while the lowest density is in Sonbhadra (215 pp sq. km) might be due to uneven terrain. The districts located in central Ganga flood plain like Gorakhpur, Deoria, Mau, Jaunpur, Ghazipur, Varanasi, St. Ravidas Nagar, Gopalganj, Siwan, Saran record population density more than 1000 people per sq. km which shows a high pressure of population on the land resource.

Because of slow urbanization, 90 per cent of the population lives in the rural area. Agriculture is a prime occupation in the region but ironically only 65 per cent of the families possess any piece of land. The land and wealth are unevenly distributed among the population. More than three-fifth (60 per cent) of households have less than one acre of land. The one-fifth of the wealth is shared by only 8 per cent of households and 60 per cent of the households share 40 of the wealth (Fig. 13.1). If we look into the social composition of the area, about 89 per cent of the population is Hindus and 11 per cent are Muslims. Among the total Hindu population, 58 per cent are OBCs, 21 per cent

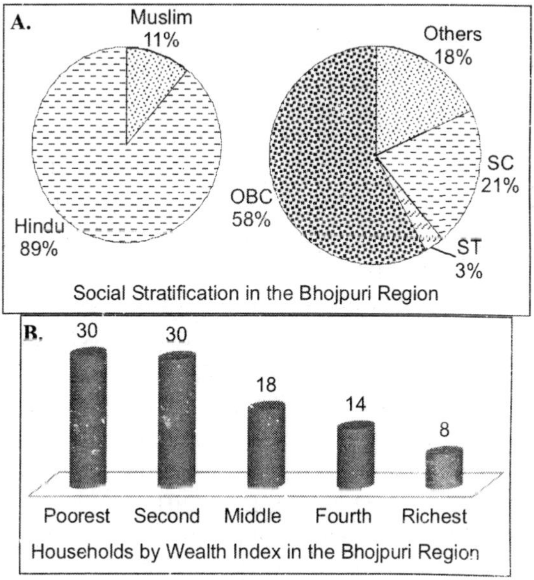

Fig. 13.1: Social and Economic Stratification in Bhojpuri Region

SCs and 18 per cent are from other categories (DLHS-3). Highest proportion (24 per cent) of Muslims in the population are found in Paschim Champaran, Bihar followed by Siddharthnagar, UP (22 per cent). The proportion Hindus was recorded highest in Ballia (97 per cent). Caste plays a very significant role in determination of economic and social status of a family. Therefore, caste groups exhibit the positive and high significant correlation with wealth index and possession of land referring that the concentration of wealth and land also increases as we move from SCs to upper castes. Similarly, regarding religion, the wealth and land are positively and significantly associated with Hindus while the case in inverse for Muslims probably because most Muslims in the region are landless artisans.

As per the statistical report of UP, the state in general and this region in particular is passing through a phase of closing down of various industries except few. The number of industrial units in UP in 1991 were 17,539 which reduced to 12,354 in 2001 (State Planning Institute, 2009).

Migration Issues

The history of migration from the region is very old. During colonial period from 1834-1916, 4,821,902 people migrated to Caribbean islands, Mauritius, Fiji, Jamaica and other island countries (Huge, 1977). Even in post-colonial period, migration from this region continued but it is within the national boundary. As per the census of India 2001, out of total 33 million Bhojpuri speaking people in India, 30 million, 1.36 million and 1.73 million were enumerated within the region, outside the region but within the state and outside the state respectively. Thus each 10th Bhojouri speaker is found outside the region which indicates high outmigration from the region mainly to the other states. However, this figure may go much higher as many Bhojpuri-speaking people living in urban area speak Hindi at home hence like many other languages it might be underreported. Study two districts Rohtas and Siwan, Roy (2011) finds that migration is the way of living for the people of this area as more than 60 per cent of the household have adopted migration as a strategy of livelihood. As per DLHS-3, out of 31,109 sample women (15-49 age group) in the region, 7,348 reported that their husband live somewhere else which accounts for 24 per cent of the sample women. Thus, migration not only with family but without family is also important characteristic of this region.

Millennium Development Goals

Goal 1: Eradicate Extreme Poverty and Hunger

According to Mid-term India Country Report, 2009, the proportion of people below the national poverty line (poverty headcount ratio or PHR) estimated for 1990 was 37.2 per cent. India is required to reduce it by half to 18.6 per cent by 2015. By the year 2004-05, the PHR has come down to 27.5 per cent. Going by the rate of change in the last 15 years, the projected PHR in the year 2015 is expected to be just short of the year's MDG-target mark (18.6 per cent) by about 3.5 percentage points. The absolute number of poor in the country has declined from about 320 million (36 per cent of total population) in 1993-94 to about 301 million (27.6 per cent of total population) in 2004-05. At this rate of decline, the country is expected to have a burden of about 279 million of people (22.1 per cent) living below the poverty line in the year 2015. The major States, namely Bihar, Jharkhand, Chhattisgarh, Madhya Pradesh, Maharashtra, Odisha, Uttar Pradesh and Uttarakhand, which are incidentally the more populated States, are among the slow-moving States in reducing the poverty and are not likely to achieve their target of halving the poverty ratio of 1990 by 2015 going at the pace they moved during 1990-2005. These States had about 193.5 million of people below poverty line in 2004-05 (64 per cent of total BPL population) and are expected to have nearly 198 million people below poverty line in 2015 (71 per cent of total projected BPL population). The contours of socio-economic development in these heartland States of India will continue to remain in focus.

As per DLHS 2007-08 data, about 34 per cent of the population in the Bhojpuri region live below poverty line which is quite higher than the national level. If this figure is projected for total population, number of poor reaches to 21.7 million, which accounts for approx. 11 per cent of national poor. At the district level Siddharth nagar recorded highest BPL population followed by St. Kabir Nagar, Basti, Kushinagar, Maharajganj and Deoria while the lowest share of BPL population was registered in Siwan (15.7 per cent) followed by Gopalganj, Varanasi and Saran. The districts in tarai region are the most poverty stricken. Surprisingly, Gopalganj and Siwan districts are better in comparison to other districts. The reason behind the better performance may be due to remittances sent by the huge number of migrants from these areas.

Table 13.1: Level of MDGs Indicators for India and Bhojpuri Region

	India			*Bhojpuri Region*			
	1990	*2008*	*2015*	*Total region 2008*	*Bihar 2008*	*UP 2008*	*2015*
Population below poverty line	36 (1993–94)	28 (2004–05)	22	34 (2008)	–	–	–
Net Enrolment Ratio (primary)	75	96 (2004)	100	76 (2004)	–	–	100
Drop out rate (Class I–Class V)	–	–	0	15 (2004)	–	–	0
Youth literacy	62	82	100 (2012)	79	76	81	100
Safe delivery	33	52	62 (Target 100)	39	41	38	100
Safe drinking water	–	84	100	95	97	93	100
Sanitation coverage	–	48	66	16	15	17	66

Source: For enrolment data status Statistical Handbook of Uttar Pradesh, other data for India, Mid-term India Country Report, 2009 and for Bhojpuri Region, derived from DLHS 2007-08.

Goal 2: Achieve Universal Primary Education

The Education for All Campaign and the Millennium Development Goals have focused the world's attention on ensuring that no child is excluded from receiving a primary education. India has already attained cent per cent gross enrolment ratio 5 (GER) in primary grades of schooling for both boys and girls. GER stands at 114.42 for the boys and 107.84 for the girls in the year 2006-07. However, the measure of net enrolment ratio (NER) is more appropriate indicator to examine the goal of achieving universal primary education by 2015 for all children aged 6-11 years. A trend based on DISE 7 data shows the country now well set to achieve cent per cent primary education for children in the primary schooling age of 6-11 years ahead of 2015.

Primary enrolment of 6-11 year old children by their NER measure has improved from 75 per cent for 1990 to 83 per cent in the year 2000 to over 95 per cent in 2007-08.

The data from household surveys indicates that NER in Bhojpuri speaking region is very low as in 2004, only 76 per cent of children in age group 6-11 years were in school. Rest of 25 per cent of children of this age-group were excluded from the education as they were not enrolled and remained out of school. Regarding spatial variation with respect NER[2], the districts of Kaimur from Bihar and Azamgarh, Maharajganj, and Kushinagar from UP had 90 per cent NER and in contrast it was below 65 per cent in Varanasi, St. Kabir Nagar, Paschim Champaran, Purbi Champaran and Rohtas. Even dropout rate at the primary school was also registered at 15 per cent. The Bihar side of the region has 28 per cent dropout compared to 9 per cent in UP. There was wide variation at district level. The dropout rate is also registered very high varying from highest in Bhojpur (50.6 per cent) followed by Rohtas (44.6 per cent) and Basti (42.3 per cent) to lowest value of 0.2 Kishanganj and 0.7 Ballia. The question arises that why children are out of school in spite of intensification of government efforts in recent years following the launch of various programmes including the District Primary Education Programme (DPEP) in 1992, the Minimum Levels of Learning (MLL) initiative and more recently the Sarva Shiksha Abhiyan (or the National Programme for Universal Elementary Education). The reason for dropout and never enrolled in school for primary school age-group children are enquired in DLHS-3 (Fig. 13.2). One-fourth of children have never been in school because their parents cannot afford education, for next 20 per cent either parents/guardian do not find education as a necessity or children are not interested in studies and for 8 per cent they are required at home for caring of siblings, households work and family business. These obstacles are approximately common for both boys and girls. For those who were enrolled but could not continue, gender difference is quite visible. For three-fourth of boys, lack of interest in studies (35 per cent), followed by cost (25 per cent), and required for households management (15 per cent) were most important reasons. While for girls, more than 25 per cent are dropped from the school because they are required for households affairs (probably when both the parents are out for working for wages), followed by no interest in study (18 per cent), cost of studies,

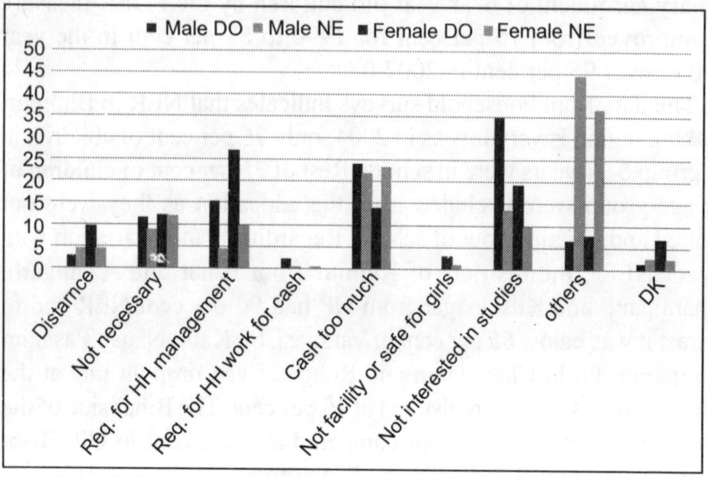

Fig. 13.2: Children in age 6-11 by reasons for dropout (DO) and never been in school (NE)

Source: DLHS-3, 2007.

not necessary to get education and far off location of schools. However, more detailed analysis is required which is beyond the scope of this paper but it may be inducted (as also mentioned by UNICEF-India) that most of the children who are out of school, mainly belong to lower socio-economic strata, where children support their families directly or indirectly, realization of value of education is low, and hence education is considered as a costly affair. Apart from this, due to structurally poverty in the region, the education system is inadequately developed—wracked by a shortage of resources, schools, classrooms and teachers, often result in, poor quality of teaching, many children drop out before completing five years of primary school and many of those who stay on, learn little.

Mean age at marriage for girls	15 yrs
Mean age of women at the time of first birth	19.6 yrs
Post natal care within two weeks of birth	27.2
Ifa tablets for 100 days	72.7
Safe delivery	38
ANC	5

Area	1992-93	2007-08	2015 expected	2015 target
India	33%	52%	62%	100
Bihar Bhojpuri		41%		100
UP Bhojpuri		38%		100
Bhojpuri region		39%		100

Goal 5: Improve Maternal Health

The rate of increase in coverage of institutional deliveries in India is rather slow. The coverage of deliveries by skilled personnel has increased from 33 per cent in 1992-93 to 52 per cent in 2007-08. Unless improved drastically the existing rate of increase in deliveries is expected to take coverage only to 62 per cent by 2015 (India report). Issues of maternal health in Bhojpuri region are presented in Fig. 13.3. The proportion of births attended by skilled health personnel was very low in the Bhojpuri region. Only 39 per cent of the births were attended by skilled health personnel, even within the region there was a great variation with maximum number of safe deliveries was reported in Rohtas (59 per cent) to the lowest value recorded in Siddharthnagar (10.7 per cent). In general the UP Bhojpuri region has lower number of safe deliveries than the Bihar counterpart (Fig. 13.3). In the region, women are at the greater risk of maternal death as the average age at marriage for women is 15 years and they deliver their first baby before age 20, full ANC coverage is just 5 per cent and more than 60 per cent of deliveries are not attended by skilled health personnel.

Goal 7: Ensure Environmental Sustainability

Under this section the indicators used are: proportion of the population using solid fuels, proportion of population with sustainable access to an improved water source and proportion of population with access to improved sanitation.

The mid-term assessment report of the MDG for water supply and sanitation shows that India is on-track in achieving the MDG target for sustainable access to safe drinking water. The overall proportion of households having access to improved water sources

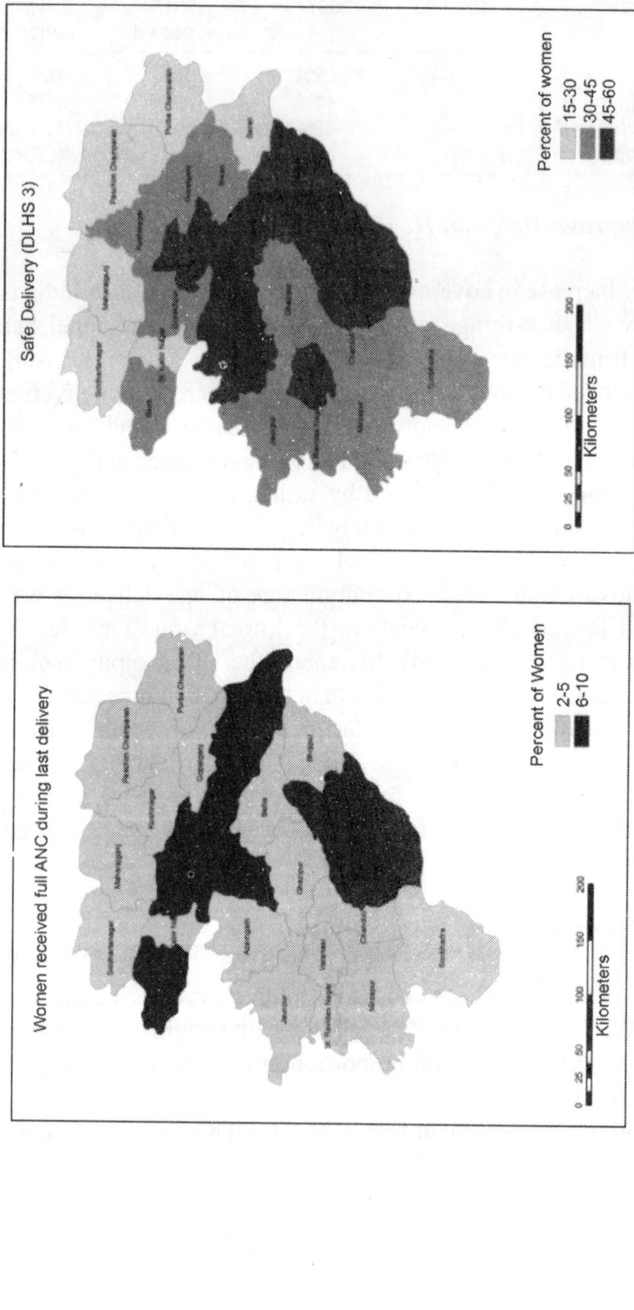

Source: For India, Country, Report and for the Bhojpuri region, DLHS-3.

Fig. 13.3: Issues of Safe Delivery and ANC in Bhojpuri Region

has increased from about 68.2 per cent in 1992-93 to 84.4 per cent in 2007-08. The access to improved sanitation facilities has not been quite impressive during the last decade. India, one of the most densely populated countries in the world, has the lowest sanitation coverage. Given the target for reducing the proportion of the households having no access to improved sanitation to 38 per cent by 2015, the proportion of households without any toilet facility declined from about 70 per cent in 1992-93 to about 51 per cent in 2007-08.

Regarding the Bhojpuri speaking region, as per DLHS-3, the condition of basic amenities are poor except for water (Fig. 13.4). It has very dense network of perennial and non-perennial channels. However, regarding fuel used for cooking, hardly 10 per cent of the households use pollution free fuel which is mainly concentrated in

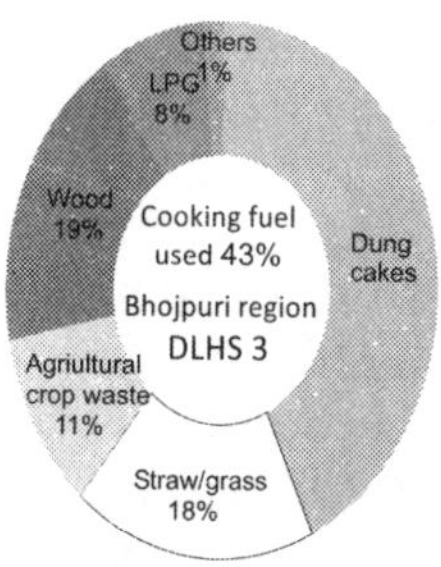

State	*Safe drinking water 2007-08*	*Sanitation coverage 2007-08*
India	84%	48%
Bihar Bhojpuri	97%	15%
UP Bhojpuri	93%	17%
Bhojpuri region	95%	16%
Target 2015	100%	66%

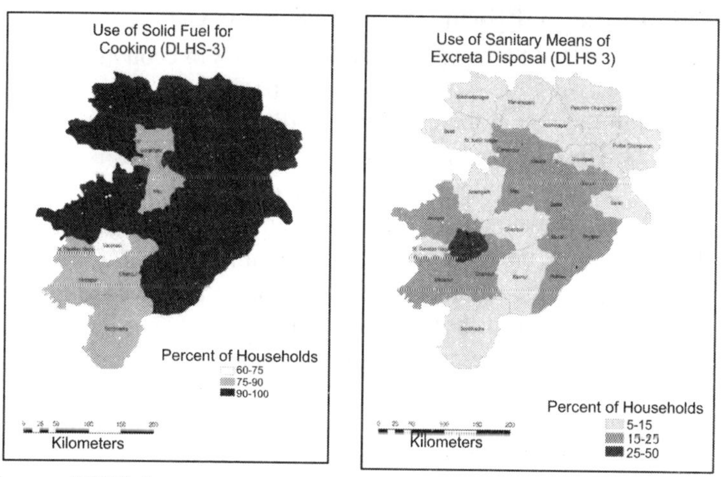

Source: *DLHS-3*.

Fig. 13.4: Status of Basic Amenities in Bhojpuri region

areas with urbanization and industrial units. Even in Varanasi where urbanization is 40 per cent, 65 per cent of the households use solid fuel. The condition of sanitation in the region is very dismal. Only 17 per cent of households have access to sanitation. However, the range varies from 46 per cent in Varanasi to lowest six per cent in Siddharthnagar. Rural sanitation coverage remains daunting. The figure is highest in case of Varanasi because it is comprised of urban municipality areas otherwise the value ranges 10-15 per cent in all other districts. The greatest threat to human health is open defecation—the root to faecal-oral transmission of diseases—lethal consequences of mortality of valuable family members of society. It leads to diarrhael diseases, stunting and under nutrition. Therefore, it should be the prime importance by the government. However, it seems to be very difficult to catch the target by 2015.

Conclusion

Bhojpuri region is characterized by underdevelopment, primitive and subsistence agriculture, lack of infrastructure, industrial growth and high population concentration. Due to its backwardness, a large numbers of youth from this region migrate seasonally and temporarily to other parts of the country and the world and it has become a way of life for the Bhojpuri youth. The literacy rate as well as the level of education is very low. The people among low socio-economic groups are not proactive about getting education particularly in rural areas. Even within the region there is a spatial disparity in socio-economic parameters. The districts in *tarai* region are the most poverty stricken and backward areas while Varanasi stands more developed district.

As far as the women are concerned, most of them are illiterate, do not have power, not representation in parliament, get early marriage and start early reproductive process. Daunting condition of sanitation and use of solid fuel further put them at the margin. The gender gap in terms of women representation in the parliament and education is very high.

This region has been one of the poorest regions of India despite its immense agricultural potential in terms food grains, vegetables and seasonal fruits; and industrial base in terms of agro based industries like sugar cane and rice, juices and cement and ceramics. The water resources can be effectively utilized for generating electricity and proving long term irrigation. Its human resource potential can only be harnessed only if the people are skilled and educated and then effectively employed or provided sustainable livelihood. Apart from this, it also has high tourism potential in the form of spiritual centre, educational centre, trade centre, history and archaelogical centre. The socio-cultural differentiation and discrimination is also high and some parts are still fall under influence of traditional feudalism and *dabangai* pose major challenge in achieving the MDGs.

If India has to become the global leader, a sincere attempt has to be made towards achievement of various socio-economic targets in general and addressing the problems of Bhojpuri speaking region in particular. The lack of political will and somewhat neglect to this region has been one of the major causes of its backwardness. The creation of a separate state as *Purvanchal* which is being demanded by some political groups is not the solution to its problem. We should learn from the cases of Jharkhand and Chhattisgarh which has been in the news for various wrong things rather than development. In order to develop this region a strong political will and honest and sincere approach is needed.

ENDNOTES

1. Mother Tongues of Hindi by number of speaking people in 2001 Census: Awadhi, Bagheli/Baghel Khan, Bagri Rajasthani, Banjari, Bhadrawahi, Bharmauri/Gaddi, Bhojpuri, Brajbhasha, Bundeli/Bundelkhandi, Chambali and Chhattisgarhi.
2. The ratio of enrolled children in 6-10/11 years to the population in 6-10/11 years.

REFERENCES

Answer.com (2010). What is the role of language in culture? retrieved on 13th December, 2010 from http://wiki.answers.com/Q/What_is_the_role_of_language _in_culture#ixzz1E Hs01F Ov.

Britannica Concise Encyclopaedia. Copyright © 1994-2008. Language and dialect, *Encyclopædia Britannica Inc., Retrieved from* http://www.britannica.com/EBchecked/topic/329791/language.

Census of India (2001). Population by Mother Tongue, C-16 Series for Uttar Pradesh and Bihar, M00147 CD-ROM.

Central Statistical Organization (2009). *Millennium Development Goals - India Country Report 2009, Mid-Term Statistical Appraisal.* Central Statistical Organization, Ministry of Statistics and Programme Implementation, Government of India.

Huge, T. (1977). *The Banyan Tree: Oversees Migration from India, Pakistan and Bangladesh*, New Delhi: Oxford University Press, 3-4.

International Institute for Population Sciences (IIPS) (2010), *District Level Household and Facility Survey (DLHS-3), 2007-08,* Mumbai: IIPS, India.

Koul, O.N. (2011). The Kashmiri language and society, Retrieved 12 February, 2011 from http://iils.org/pdf/KashmiriLanguageSociety.pdf.

Roy, A.K. (2011). *District Level Migration and Left Behind Women.* Jaipur: Rawat Publication.

Soondas, A. (2010). Are we losing mother tongue? *The Times of India,* 27 March, 2010, Retrieved on 10 February, 2011 from http://timesofindia.indiatimes.com/india/Are-we-losing-mother-tongue-/articleshow/5729796.cms#ixzz1Dyo2m56q

State Planning Institute (2009). *Statistical Abstract Uttar Pradesh 2009,* Lucknow: Economics and Statistics Division, Uttar Pradesh.

Thompson, H. (2001). *Culture and economic development: modernisation to globalisation. Theory and Science,* ISSN: 1527-5558, Retrieved on 12 Feb., 2011 from http://theoryandscience.icaap.org/content/vol002.002/thompson.html.

UNICEF (2011). Primary school years. *The Children,* UNICEF-India, Retrieved on 1 Feb., 2011 from http://www.unicef.org/india/children_166.htm

14

Industrial Growth and Deprivation: A Paradox of Developmental Strategy

Ratnawali

Introduction

Industrialization has become a synonym to development and it is generally attributed to the improvement in the economic condition, presuming to bring in all the desired changes to help in achieving a comfortable living. While economic improvement as a major constituent of development cannot be denied, the significance of social milieu as an important component is often overlooked. Such negligence thus bring to the fore the debate on the mode and direction of development and their subsequent impact on the population. Economic development fuelled by the force of industrial growth and market driven policies tend to favour the capitalist mode of production which neglect and deprive the weak.

The modern developmental measures like big dams, heavy machinery industries, road connectivity and urbanization take its toll on the lives of people who are forced to give up their habitats and settled livelihood. Displacement and uncertain future leads to a constant struggle for survival which accentuates with the loss of social support system. It thus disrupts social and cultural life of people, breeds discontent, inflicts injustice and dehumanizes the development process. The social cost of development is much higher than the gain it accrues to the population.

Since independence, India has followed the path of development that disproportionately goes against the marginalised sections of society. Apart from the industries, big hydel projects also lead to the displacement of population that is often four times the population that suffered the trauma of displacement during the partition of the country (Lobo & Kumar, 2009). It was estimated that around 60 million people have been displaced by different development projects. Among them 40 per cent are tribals (Fernandes, 2007), 22 per cent are Dalits (Mahapatra, 1994) and a vast proportion belonged to other rural poor. Irony of this problem is that they are displaced without their consent.

The miseries of the displaced or affected people have been well documented by many studies (Fernandes, 2007; Ganguly & Singh, 1995; Reddy, 1994; Fernandes & Raj, 1992). The adverse impact results in loss of prestige, impoverishment, indebtedness, alcoholism, domestic violence besides affecting the health of the ousted in a big way. As the success of developmental measures cannot be judged only by their economic output, this paper attempts to present the industrial growth in Gujarat in the context of deprivation faced by the marginalized section of society. It aims to discuss the important issues related with the acquisition of land and suggest ways to mitigate the deprivation faced by the poor.

Locating Gujarat in its Socio-Economic Milieu

Gujarat, situated on the west coast of the country, is one of the most industrialized and developed states of India. With a total area of 196.024 sq. km, it has a population of 50.15 million (Census of India, 2001). Around 15 per cent of the population is tribal spread over 11 districts of the eastern and southern part of the state. More than half of the population in the 32 talukas of these districts belongs to tribal communities. Nearly 70 per cent of the total forest area of the state comes within this region.

Gujarat has a high urban population (37.6 per cent). In terms of economic development, the state is front runner. With 6 per cent of geographical area and 5 per cent of total population, it contributes to 21 per cent of India's export and 6.42 per cent of GDP at constant price. The industrial growth is 11.5 per cent (2006-07) and is way ahead of many Indian states. There are 76 industrial clusters at different locations in Gujarat. A cluster is recognized as a group of industries manufacturing identical or complementary products, having a critical mass of minimum of 50 units within 10 km radius of a particular location (Govt. of Gujarat, 2003). Industrial development has also contributed to the rising trend of urban population which is poised to achieve half of its population as urban by 2025. However, the urbanization is likely to be associated with low productive employment, poverty, and a low quality of life for the poor in cities (Kundu, 2002).

Gujarat has witnessed a rapid pace in industrial development. With long coastal line the state has exploited the potential to the fullest by elaborate connectivity of cities with the coast which is instrumental in boosting trade in goods and services. Apart from the big hydel project to secure water availability, the rapid industrialization also led to prosperity. However, this development came at the cost of large displacements. In the initial phase of development about 6.9 lakh hectacres of land was acquired for the construction of big dams like Ukai, Madhuban and Dharoi; however, the downturn was the displacement of people from 200 villages. Acquisition for the industries took a quantum jump from the seventies, peaking at the decade of eighties and accounting a total of 1.80 lakh hectares of land affecting 15,056 families (Lobo & Kumar, 2009). Apart from the above land was also acquired for road and transportation for the development and connectivity of urban centres with remote areas.

Industrial Development and Displacement in Gujarat

Industrial development in Gujarat took place in a big way during 1981-1990. During this phase more than 87,000 hectares of land was acquired for industrial purpose and for exploration by ONGC. The Gujarat Industrial Development Corporation (GIDC) vigorously started developing the acquired land to hand them over to industrial houses at subsidized rates. Medium and small scale industries largely came up during this phase. It led to the tremendous growth of 'Golden Corridor' extending from North to South (Ahmedabad-Vadodara-Surat and Vapi) attracting heavy industrial growth. The post industrialization period (1991-95) was marked by the establishment of large scale industries that were capital intensive but their potential for employment was limited. Even the small scale industries are gradually becoming more capital intensive restricting their scope for employment generation (Lalitha, 2000). Table 14.1 presents the industrial growth and the status of employment over the years in the state.

Table 14.1: Industrial Growth at Various Periods in Gujarat

Year	*No. of registered SSIs*	*No. of working factories*	*Total employment generated*	*Average no. of workers/factory employed daily*
1960	2,169	3,649	346,462	94.95
1985	7,2479	13,067	663,614	50.79
1995	178,627	18,451	826,167	44.78
2000	251,088	20,424	866,720	42.44
2004	296,306	21,536	927,516	43.07
2006	312,782	22,480	1,038,134	46.18
2008*	NA	23,942	1,175,091	49.08

* Provisional

Source: Government of Gujarat, Socio-Economic Review of Gujarat 2009-10.

The quantum jump in the growth of small scale industries is evident from the figure. The proportion of working factories has also registered phenomenal rise which is more than 500 per cent to the base year 1960. However, a significant decline in the employment generation is observed which seems to have a faster rate after the liberalization as it dipped down to 7 percentage points (51.5 to 44.7) for the year 1990-95. During this phase more capital intensive industries were established which required less man power hence were not helping the people by way of providing employment. It is ironic that the industrial growth in Gujarat that has added net value to the tune of 15 per cent to the country also caused the displacement of the population at a massive scale. The GIDC, established for the improvement of small and medium scale industries registered a phenomenal rise with 242 (till December 2007, Socio-Economic Review 2007-08) estates and has preferred to acquire land in backward taluka near urban centre. It attracted investment by way of giving subsidies to the industries for their stated goal of generating employment. Total land with GIDC was 136 hectare in 1962-63, which rose to 26,063 hectare in 2003-04 accounting for a growth of nearly 191 per cent. It can be seen from Table 14.2 that over 1.80 lakh hectares of land were acquired during the period, affecting over 15,000 families or more than 70,000 people.

Table 14.2: Land Acquisition for Industries by Various Decades

Periods	*Land acquired (hectare)*	*Percentage share*	*No. of families*	*Percentage share*
1947-1960	2,891	1.60	369	2.45
1961-1970	15,082	8.36	2,202	14.62
1971-1980	25,659	14.23	2,642	17.54
1981-1990	87,181	48.35	6,291	41.78
1991-2000	47,710	26.46	3,323	22.07
2001-2004	1,705	0.94	218	1.44
Total (including unknown 68 hectares displacing 11 families)	1,80,296	100.00	15,056	100.00

Source: Compiled and computed from Gazette Notification cited in Lobo and Kumar, 2009.

As is apparent from the figure, the highest acquisition of land took place in 1981-1990 displacing more than 41 per cent of the total displaced families. In regional context the acquisition of land for industrial purpose was the highest in South Gujarat with nearly 51 per cent share in all acquisition, affecting a little over 45 per cent of total affected families. With the canal network to facilitate sugarcane cropping and development of string of chemical industries along the golden corridor, this region is home to chemical and hazardous industries. Being rich in forest and mineral resources, the area has been exploited for water resource and other industrial projects which together account for 9.2 lakh hectares, comprising forest and revenue land. It thus affected 1.31 lakh families and 7.2 lakh people in all (Lobo & Kumar, 2009). Incidentally these are the areas now identified as ecologically most fragile and degraded (NSSO). As we have mentioned earlier, more than half of the total population in this region belong to tribal communities. Hence loss to tribal habitat is a foregone conclusion. About 20 per cent of the households in the industrial land acquisition have lost their homestead forcing them to move out to new location. It is thus estimated that 15,000 persons or 3,000 households were forced to move out of their place of residence. The maximum acquisition of land (87 per cent) has been affected by GIDC and ONGC that has led to the displacement/deprivation of 75 per cent of the total affected families. In order to gauge the extent of socio-economic changes in the lives of people, the case of Hazira Industrial Area offers substantial insight.

The Hazira Industrial Area: A Case Study

The Hazira complex was established in early 1980 after land was acquired from 16 villages, namely Mora, Sunvali, Hazira, Bhatpor, Damka, Icchapor, Kawas etc. which comes within the Choryasi Taluka of Surat district. Located at a distance of 21 km. from the city of Surat, the complex houses 165 large and medium industries. Some of the important industries in this complex are ONGC, GAIL, KRIBHCO, Reliance, ESSAR, LARSEN & TOUBRO etc. A total of 2964.55 hectare land from 16 villages were acquired which later included two more villages for the planned development of the region under Hazira Area Development Authority (HADA). Despite protest from the villagers, the industries continued to be established, lending it a tag of hazardous zone. Land in Choryasi taluka is second most fertile land after Kheda-Anand taluka of the state. Having proximity to long coastal line, agriculture and fishing were main activities. With the acquisition of land both these activities got severely affected resulting in loss of livelihood.

The land was acquired for the above purpose by GIDC in 1980 and according to initial reports nearly 80 farmers gave away their land at a token price in the interest of national development (Lobo & Kumar, 2009). However, these lands were left unused for nearly a decade and later they were sold to industries at premium prices. The land holders who had sold their lands for a token compensation felt cheated as they had parted with the land in the interest of nation at meagre compensation whereas government sold them at reasonable profit to the industrialists. Till 2001, only 10 per cent of the total acquired land was utilized. Apart from the private land the industrial complex also consists of common property land. Table 14.3 shows

Table 14.3: Changes in land use and occupational pattern over the years

Villages	*Total area in Hect.*		*Area not available for cultivation (%)*		*Total population*		*Cultivators*		*Agricultural labour*		*Marginal labour*	
	1981	*2001*	*1981*	*2001*	*1981*	*2001*	*1981*	*2001*	*1981*	*2001*	*1981*	*2001*
Mora	2223.37	745.0	72.2	86.2	1083	6115	51	54	263	59	0	319
Sunvali	3299.05	2288.7	13.0	44.7	1939	2907	374	588	111	182	2	27
Bhatpore	824.28	425.3	39.43	69.03	2072	3379	194	16	396	313	0	66
Hazira*	–	–	–	–	3487	4793	354	80	156	195	17	214
Icchapore	912.61	317.0	2.7	100.0**	3358	8292	388	68	416	141	2	188

Note: * Included in the area of Sunvali.
** Now included in urban area.

Source: Census 1981 and 2001.

the changes in land use and occupation pattern in some of these villages since Hazira complex came into being.

The figures show that the village areas have declined over the years. Considerable land has been taken for non-agricultural purposes as is seen in the rise of area not available for cultivation. While decline in the number of cultivators is observed, an increase in the marginal labourer reflects the growing insecurity of livelihood. It is again worthwhile to mention that while two decade ago almost all the marginal workers were women; in the later years a substantial or nearly half among them were men implying the pauperisation of labour at the household level.

The socio-economic condition of these villages reveals that there has not been much gain in the socio-economic indicators (Patel *et al.*, 2003). A look at the villages (Hazira, Sunvali & Bhatpore) near the Hazira complex reveals that these villages were primarily agriculture based. Cultivation and livestock rearing was their main livelihood. It was found that the number of cultivators and workers have decreased during the 1981-1991 period. The proportion of cultivators decreased from 34 per cent to 18 per cent during 1971-1991 while in Sunvali the decrease was from 86 to 43 per cent. In Bhatpore the decline is from 34 per cent to 5 per cent. After the arrival of industries, the occupation of the people changed remarkably. While considerable loss in employment for fishermen was registered, a substantial proportion of people from other occupational groups did not find much change (Patel *et al.*, 2003). Changes have also been observed in workers profile. While there has been considerable increase in employment, it is reported that most of them were migrant. The condition of Dubla community (ST) has not improved and nearly three-forth of them live below poverty line. Studies have reported deteriorating water quality in the area with poor access to safe drinking water. Women have reportedly expressed greater difficulty in getting drinking water, procuring fire wood and grazing of animals. Domestic work became an occupation for many women. There has been no activity to support agriculture, livestock and water conservation. The nature of employment revealed that nearly 30 per cent of the total employed persons were engaged in household domestic work, which was more pronounced in case of women headed households. More than half of the people felt that industries have adversely affected their lives. As problem of air and water pollution reportedly increased, health related problems have also gone up. People were concerned about the poor health facilities in these areas which also echoed in the lack of maintenance of the infrastructure, causing problem to the people.

It is thus clear from the above that while industries have benefited people in providing some employment, it has done little to improve the quality of life. Moreover, influx of substantial migrant population is a clear indication that many jobs have gone to outsiders and the local population has not been fully absorbed as expected by the villagers. The condition of those who subsisted on agriculture based economy and was not skilled in other work, got further marginalized as employment for them was scarce in the industry. The establishment of industries, therefore, does not seem to help locals much either in terms of employment or in the improvement of the quality of life.

The Issues in Land Acquisition

The situation lead to the larger question as to how far is it judicious to take land from the people for the macro level benefit to the state by impoverishing those who have to part with their only source of living. Should the government be a facilitator in acquiring land and appropriating it to the industries without safeguarding the lives of those who depend upon them? This includes not only the agriculture land but even the common property resources that sustain the landless get transferred to the industries. Does the state have the right to deprive some for the benefit of others? Some of these issues need attention.

Implications and Acquisition of Land for Public Purpose

The inherent concept under the acquisition of land is *eminent domain*, according to which the state owns all land and is well within its jurisdiction to acquire land from the people if it is in the interest of the larger population even though people do not give consent for the same. The Land Acquisition Act, 1894 empowered the state to take over the land arbitrarily for the public purpose. Since the public purpose is still not clearly defined (Ramanathan, 1999) the very term is liable to be abused as the state alone has to decide the relevance of work that can be placed under this category. Further, legal interpretation of the above term has also varied across the cases filed under the act. Despite the caveat of '*direct use*' by the people as the rationale for acquiring land for the industries, the same was seldom taken in true spirit. According to the above Act, land can be acquired for a private company only if it is used for the purpose of erecting dwellings for its workers and providing amenities to them. It, however, does not allow states to acquire land for companies to set up factories (section 38A). In order to remove the bottle necks and facilitate the industries, Gujarat has taken advantage of the amendments of this Act (1963 & 1984) which says 'land can however be acquired for a public limited company under certain circumstances and subject to its requirement of land being justified after careful scrutiny and after all efforts by the company at obtaining the same by private negotiations at reasonable prices have failed'. Similarly, acquisition of land for cooperative industrial estate by state bodies like GIDC, GJRTC, GEB are exempted from the provisions under sections 38A to 44B that apply to the acquisition for company (Lobo & Kumar, 2009). Further to make the transfer of land from agriculture to non-agricultural use smooth and hassle free, the state is proposing to come up with a legislation that will abolish the present system of taking permission for the conversion of land use at seven different stages to give a boost to the industrial and commercial growth across the region (*Indian Express*, 2009).

By taking such measures the willingness of the state to facilitate the industrial growth even at the expense of cultivation is clear. The state owned GIDC has been instrumental in acquiring highest proportion of land over the years as we have mentioned earlier. The intention of the government favouring acquisition of land for industrial purpose seems apparent from the recent document as it clearly says that

> "Looking to the present scenario for establishment of large and medium scale industries, the corporation found that it is essential to stock land on hand to fulfil the future requirement of industries. In view of the above, corporation has further decided to acquire the land in large scale basis at various places in the state." (Govt. of Gujarat, 2010, pp. 29)

Till 2007 the corporation had acquired about 26,360 hectare of land (*Ibid.*). Many a time land is acquired much earlier and is left without use for decades. Later the land is given to the companies at premium rates leaving the landholder upset as they were given much less for their land than is gained by the GIDC in such deals. The state thus seems to be profiteering from the deals it has struck by depriving people of their resource.

The acquisition of land has differential effect on the people depending upon their economic status. The better off or big land holders seem advantageously placed as they can acquire better land or increase their asset base but the small or marginal land holders could not translate this one time gain into productive use. Ignorance, social needs and meagre amount hardly help them to improve their asset base. The heat is not only felt by the land holders but the others also get affected. Since compensation is given to the title holders, others who do not hold title but depend on agriculture are left out. The agriculture labourers, share croppers, artisans and a number of occupational groups whose livings are tied with the agrarian economy also find their economic base withering.

Loss of Common Property Resources (CPR)

Apart from the revenue land, large tracts of land owned by the government is also diverted for industrial use. These lands comes under various categories like barren and fallow land, cultivable waste land, pasture land etc. which are generally waste land not suitable for agriculture purpose but a number of families get sustenance from these lands. Thus *gauchar* land, community forest, panchayat land, ponds, rivulets etc. are utilized by the communities for various purposes. Gradually these lands have been diminishing as government usually transfers their use for various other purposes including industries. A look at land use pattern of the state from 1980-81 to 2003-04 (Socio-economic Review of Gujarat) reveals that there has been gradual decline in the forest (10.4 to 9.8 per cent), cultivable waste (10.54 to 10.51 per cent) and fallow land (4.6 to 3.1 per cent) but a steady increase in the land for non-agriculture use (5.66 to 6.09 per cent). Allocation of *gauchar* and waste land to industries and for other development projects has been documented. Studies have found that nearly 18 per cent of *gaucher* land has been transferred making living precarious for farmers and other rural population in general and pastoralists in particular (Bharwada & Mahajan, 2006). In the absence of fodder, rearing of livestock which is an important support to rural economy also gets depleted, accentuating their vulnerability to cope with the loss of livelihood. Most of CPR transfer has taken place in Saurastra and Kachcha but the transfer of *gauchar* land is the highest in south Gujarat. Loss of bio-diversity is another concern as many of the mangroves are alleged to be cut down for the Mundra SEZ. In promoting the salt industries and laying pipe lines for petrol product and optic fibres the state has also transferred substantial land from the Marine National Park for the above purpose, disregarding the ecological impact on marine life.

Increase in Rural Deprivation

One of the important notions for promoting industrial growth is reduction of poverty by generating employment and improving income. On face value it appears very promising but reality speaks otherwise. Over the years there has been a clear shift in the approach to industrial growth. From the labour intensive industries, it has now shifted to more capital intensive one, likely to generate more revenue than employment. Though, there has been some change in the economic condition, the social security among the rural masses has considerably gone down. The industries are encouraged to set up their plant in the backward region by giving them incentives like subsidies and tax exemption to develop the region by employing rural people. But such benefits are very limited as most of the rural population are unskilled and are ill suited to factory jobs. Job promises are seldom honoured and even after absorbing a few the factories do not retain them for long. It would be more appropriate to say that apart from benefiting from the low market price of these lands, industries have also been exploiting the cheap labour without giving them any social security (Hirway & Mahadeviya, 2004). Besides distorting the land and labour market to its own advantage, the effluents discharged from the industries have adverse impact on the land and environment. Pandya (1988, 1992) has pointed out the gloomy picture of pollution of air, underground and surface water and land because of poor monitoring and poor disposal of industrial waste. It may be recalled that even the High Court order for closing down 756 polluting units and directions for compensation to the population living in the vicinity of these units have not been fully implemented (rainwaterharvesting.org). Contaminated water and exposure to hazardous chemicals are resulting in increased incidents of cancer, toxicity, tuberculosis and respiratory problems, compounded by inadequate medical help from the government. The improper disposal of industrial waste thus leads to declining land productivity and poor health of the population, pushing them to further impoverishment. It was found that the land acquired by ONGC has left many lands unproductive whereas quarrying and mining has compelled the owner of adjacent plots to sell their land as access to their plot became difficult (Lobo & Kumar, 2009). The emission of smoke and particles from Cement factory led to reduction in crop produce besides affecting the health of the population (Das, Ratnawali, Sathyapalan & Pandya, 2003). The development strategy adopted by the state is thus wealth centric and non-inclusive.

The Simmering Discontent

The systematic deprivation of rural and marginalized group do not auger well for the state. With the proactive policy of promoting industries at the cost of agriculture, the food security scenario is bound to be affected. Considerable decline has already been recorded in the primary sector (Hirway and Mahadeviya, 2004). Deterioration in ecological environment and its consequences are on the rise with little effort to arrest them (Pandya, 1988, 1992). On the contrary the state is all set to claim the status of "Capital of SEZs". With around 60 SEZs[1] under its kitty the state is well poised to acquire more than 20,857 hectare of land which is likely to go up to 60,000 hectare disregarding the interest of rural and marginalized groups. It should be noted that contrary to the belief that Gujarat SEZ have been developed mainly on the barren/waste land and with mutual negotiation between the parties; reports of coercion, commissions and conversion of pastoral land to waste land have been mentioned (Manshi, 2008). The resentment among the people is brewing and they are protesting at the negation of their rights.

According to reports the people in and around a dozen villages in Mahuva Taluka of Bhavnagar district in Saurastra are agitating against the transfer of more than 4000 hectare of government land to Nirma group of company. The land include waste land, a water body spread over 100 hectare and grazing land (Dave, 2009). It is expected to uproot 15 villages and will cause salinity ingress. Similarly, farmers from Vanthali and Manvdar taluka of Junagarh district are agitating for denial of underground water for cultivation. The area is put under "Dark Zone" for the last fifteen years because of low underground water table which is below 200 feet. Though, water level has improved over the years but farmers are denied withdrawal for the agricultural purposes. There is no financial assistance by way of bank loans for the construction of new wells or tube wells neither are there any new power connection allotted for agriculture. However, the same restriction does not apply to industries and are free to use underground water (Bukhari, 2009). Many more such cases have been reported by newspaper like the resentment over transfer of 650 acres of *gauchar* land acquired by Gujarat Maritime Board (GMB) and transferred to L&T Port Sutrapada for constructing a private port in Veraval, Junagarh, 400 hectare of *gauchar* land from 14 villages in and around Mundra port to Adani group. Protest by Charan Caste who are cattle rearer, against the allotment of 1000 hectare of grazing land to Adanis in Kutch, resentment in Mocha village at Porbandar against ploy to transfer grazing land to mining bigwigs for the extraction of limestone (Sharma, 2009) reveal the growing restlessness among the people.

The Way Forward

Industrial development in Gujarat therefore presents a paradox. While it has led to the fast economic growth, the state has shown high income differentials across the region and social groups due to regional imbalance in the location of industries and their concentration in the urban or urban agglomerates. In spite of this, poverty in urban area is much higher than the rural areas. Though the reason is attributed to high rural-urban migration, casualisation of labour is yet another reason which is hitting the rural poor harder in the urban areas. Poverty even exists in the close vicinity of industrial agglomerates. A region-wise poverty analysis found out that poverty in urban areas of golden corridor is least while the rural areas of the same, is the region of second most poor people of the state (Hirway, 2002). Migration of rural folk to this area has been highest (56 per cent) in the state which though diversify the surplus rural labour, ends up giving them dismal situation of life and livelihood (Shah, 2002). The degraded environment and low agricultural resources seem to be the defining factors of the scenario that pushes rural people from their habitat. Commenting on the pattern of development adopted by the country, Kothari (1988) observes that the present pattern of development are so wasteful and destructive of natural resources like land, water and entire ecosystem, that they are producing an increasing and exponential growth in desertification, deforestation and soil degradation, which together are leading to an overall erosion of productive process itself. He thus pointed out

> "....the issue is not that of convincing policy-makers...... It is something far more fundamental; it involves a clear admission of the faulty and inappropriate, wasteful and irrational; inequitous and exploitative and in the end, highly dangerous and possibly suicidal nature of the development. It calls for nothing less than evolving a new concept of development to replace the one now commonly accepted."
>
> (*Ibid.*: 46-47)

The least we require is balanced growth so that linkage between the agriculture and industries could be strengthened and prosperity could be trickled down. Investment in agriculture and related industries will arrest the migration rate and deprivation of the community. Promoting one by depriving the other would not be sustainable as it will lead to wide income disparity and unrest. Secondly, it should be noted that acquisition of land and rehabilitation should go side-by-side. A humane approach is required towards providing necessary living condition and livelihood options to the displaced. Since loss of land tantamount to loss of livelihood, it would be more considerate to provide equity share in the industry ensuring stake of the displaced in the industry. One can expect that the new Resettlement and rehabilitation policy will take care of these issues. On the other hand the industrial houses and corporates need to be monitored for their strict adherence to the regulating laws with strong will and determination by the policy makers and administrators.

Development is multidimensional; on the one hand it espouses the poverty eradication and economic upliftment of masses, on the other it also calls for social empowerment by liberal ideas, equity and social and distributive justice. Concentration of wealth among few will always lead to discontent because of desperation and sufferings faced by the deprived. This also goes against the universal Human Rights. To make development broad based, Sen (1997a and b) has favoured GALA (getting a little assistance), a friendly process of development which celeberates people's agency and the expansion of human freedom and capabilities. He argues that development is essentially a process of expansion of people's capabilities and that development strategy should help people in acquiring these capabilities. It will open up the horizon of development to everyone and will be more inclusive in nature.

ENDNOTE

1. As at the end of September 2009, with the grant of approval from BOA in MOCI, New Delhi (GOG—Socio-Economic Review 2009-10).

REFERENCES

Manshi, A. (2008). Picture imperfect: unravelling the Gujarat SEZ model. *Down to Earth*, October 16-31, p. 52.

Bukhari, S.H. (2009). No 'dark zone' tag, cry out 35000 farmers in Junagarh. *Indian Express*, 23 May, p. 4.

Das, B., Ratnawali, Sathyapalan, S. and Pandya, K. (2003). *Socio-Economic Study of Lodhava-Singsar Region.* (Mimeo.) Surat: Centre for Social Studies.

Dave, H. (2009). Government invites Mahuva agitators to dialogue table. *Indian Express*, 28 May, p. 4.

Fernandes, W. (2007). *The human cost of development induced displacement* (Mimeo.). Guwahati: North Eastern Social Research Centre.

Fernandes, W. and Raj, A. (1992). *Development, displacement and rehabilitation in the tribal areas of Odisha.*(Mimeo.), New Delhi: Indian Social Institute.

Ganguly, T.T. and Singh, M. (1995). Dams and the displaced in India. In H.M. Mathur (ed.), *Development, displacement and resettlement: Focus on Asian Experience*, New Delhi: Vikas Publishing House, 93-221.

Government of Gujarat (2003). *Location of industries in Gujarat state: 1990 and 1998*, Vol. 1. Gandhinagar: Government of Gujarat.

Government of Gujarat (2010). *Socio-Economic Review. Gujarat State 2009-10*. Gandhinagar: Directorate of Economics and Statistics.

Indian Express (2009). Govt. plans to abolish NA Permission system for land. *Indian Express*, 23 May 2009, p. 3.

Hirway, I. and Mahadevia, D. (2004). *Gujarat Human Development Report 2004*. Ahmedabad: Mahatma Gandhi Labour Institute.

Hirway, I. (2002). Dynamics of Development in Gujarat: some issues. In I. Hirway, S.P. Kashyap, Amita Shah (eds.), *Dynamics of Development in Gujarat*, New Delhi: Concept Publishing Company.

Industry at any cost. Retrieved on 20 Oct. 2010, www.rainwaterharvesting.org/Crisis/Industrial-pollution.html.

Kothari, R. (1988). *Rethinking Development: In Search of Humane Alternatives*. New Delhi: Ajanta Publications.

Kundu, A. (2002). Globalizing Gujarat: urbanisation, employment and poverty. In I. Hirway, S.P. Kashyap, A. Shah (eds.), *Dynamics of development in Gujarat*, New Delhi: Concept Publishing Company.

Lalitha, N. (2000). *Issues in the Industrial Development of Gujarat and Policy Initiatives by the Government of Gujarat.* Occasional Paper Series, Ahmedabad: Gujarat Institute of Development Research.

Lobo, L. and Kumar, S. (2009). *Land Acquisition, Displacement and Resettlement in Gujarat 1947-2004*, New Delhi: Sage Publication.

Patel, A., Sathyapalan, J., Joshi, V. and Pandya, K. (2003). *Socio-Economic Study: Gujarat. Port Development Gujarat (PODEG)* (Momeo.), Surat: Centre for Social Studies.

Sen, A. (1997a). What is the point of a development strategy? *The development economic research programme—New series*. London: London School of Economics.

—— (1997b). Development thinking at the beginning of the 21st century. *The development Economic Research Programme—New Series*. London: London School of Economics.

Shah, A. (2002). Uneven development and migration: Insights from micro initiatives. In G. Shah, M. Rutten, H. Streefkerk (ed.), *Development and Deprivation in Gujarat, In Honour of Jan Breman,* New Delhi: Sage Publication, 273-293.

Sharma, G. (2009). Rural Gujarat girds up loins to stop industrial behemoths from eating up grazing land. *Indian Express*, 23 July, p. 3.

15

Relationship between Women Work Participation and Education in Rajasthan

Neelam Saraswat and *Mithlesh Gupta*

Introduction

Education and employment are important determinants of the development of women. There are a number of social, economic and demographic benefits of female employment outside home. Besides adding to the family income, it increases women's status and decreases human fertility. Moreover, working women, whether they are holding high government posts or are self-employed or working in the private sector, influence the socio-economic condition of a country to a great extent. Therefore, for all-round development of a country or state, it is necessary that there is appropriate arrangement of participation of women in its development and planning. And education develops a feeling self-confidence, imparts skills and makes women enlightened citizens, socially and economically secure and viable for the emerging labour market. Of course, this, to a large extent, depends on cultural environment, geographical conditions, etc.

Women being almost half of the human capital resource of the world, it may be argued that education may favourably affect both willingness and ability of females to participate in market activities. On the contrary, if discrimination approach is considered, existence of sex-discrimination would generally tend to discourage female participation and effect of education on it is not evident.

Keeping in cognizance the importance of socio-economic, cultural, geographical factors, a state-level analysis of relationship between education and work participation of women seems imperative for policy formulations for providing women with equal opportunities of education and employment.

Thus, the present work attempts to study the relationship between women's education and employment in Rajasthan, which is a predominantly desert and socially, educationally and economically backward state of India. The analysis is based on Census of India, 2001 data for rural, urban and total population of males and females by literacy rate as well as levels of education, numbers of main and marginal workers and work participation rates. Correlation between literacy rates and work participation rates has been computed and analysed using Spearman's Rank Correlation coefficient technique. If a multiple regression is used involving other variables affecting education and labour force participation, then there may arise econometric problems (e.g., multicollinearity) which may flaw the attempt. Another proplem particularly in making any inter-temporal comparison, e.g. different censuses, is due to lack of uniformity in the concept of work participation used.

Plan of the paper is as follows: After present section on the background of the study, relationship between levels of education and employment has been presented in Section 2. Combined statistical analysis of female literacy rate and work participation rate has been undertaken in Section 3. Finally, Section 4 summarizes major conclusions and policy implications of the study.

Employment and Levels of Education

In the present section, on the basis of Census 2001, total population has been classified into educated and uneducated and then numbers of main workers, marginal workers and non-workers among them have been analysed in total, rural and urban areas of Rajasthan (Tables 15.1, 15.2 and 15.3). Educated women have been further classified into five categories:

Table 15.1: Percentage of Main Workers, Marginal Workers and Non-Workers in Total Population in Rajasthan by Levels of Education, 2001

Levels of education	*Total persons*			*Main workers*		*Marginal workers*		*Non-workers*	
	Male	*Female*	*Total*	*Male*	*Female*	*Male*	*Female*	*Male*	*Female*
Total	52.06	47.94	100.00	43.65	16.97	6.30	16.52	50.05	66.51
Educated	65.15	34.85	100.00	50.18	12.57	7.30	13.24	42.52	74.19
Uneducated	39.48	60.52	100.00	33.29	19.40	4.72	18.34	61.90	62.26
(i) Educated but below Matric / Secondary	63.83	36.17	100.00	43.47	10.09	7.62	12.97	49.10	76.94
(ii) Matric / Secondary but below Graduation	75.19	24.81	100.00	62.56	8.55	6.49	6.16	31.50	85.29
(iii) Technical Diploma or Other Certificate which not equivalent to Degree	81.54	18.46	100.00	70.38	50.39	3.19	3.90	27.43	45.71
(iv) Graduation and Higher Education other than Technical Degree	73.05	26.95	100.00	82.07	22.36	4.25	2.40	14.68	75.24
(v) Technical Degree or Diploma Equivalent to Technical Degree or Post-Graduation	79.54	20.46	100.00	76.60	49.19	3.12	2.13	21.28	48.68

Source: Computed from Census of India, 2001.

(i) Educated but below Matric/Secondary;
(ii) Matric/Secondary, but below graduation;
(iii) Technical Diploma or other Certificate which is not equivalent to Degree;
(iv) Graduation and higher education other than Technical Degree; and
(v) Technical Degree or Diploma equivalent to Technical Degree or Post-Graduation.

Position of women according to these five levels of education and three levels of workers has also been discussed.

Total Population

According to Census, 2001, percentage of females among total educated persons was only 34.85 per cent, whereas that among uneducated 60.52 per cent (Table 15.1). Even among educated females, only 12.57 per cent were main workers, 13.24 per cent being marginal workers, while remaining 74.19 per cent were non-workers. That is, only one-fourth of all educated females (25.81 per cent) were participating in labour force. Among the uneducated females, the proportion of workers was higher at 37.74 per cent. Moreover, somewhat higher percentage of uneducated females were engaged as main workers (19.40 per cent) as compared to marginal workers (18.34 per cent). Among educated men, on the other hand, 50.18 per cent were main workers, 7.30 per cent marginal workers and 42.52 per cent non-workers, as against the corresponding figures for uneducated being 33.28 per cent, 4.72 per cent and 61.9 per cent, respectively. Thus, the proportion of main workers and marginal workers was higher and that of non-workers lower among educated men, as compared to uneducated men. Thus, it can be said that education is positively correlated with employment for men and negatively with employment for women. This lends support to the discrimination approach of education-employment relationship.

To find further evidence on whether educated females are subject to greater degree of discrimination than uneducated ones, we now examine data on employment by five categories of levels of education (Table 15.1).

Table 15.2: Percentage of Main Workers, Marginal Workers and Non-Workers in Rural Population in Rajasthan by Levels of Education, 2001

Levels of education	*Total persons*			*Main workers*		*Marginal workers*		*Non-workers*	
	Male	*Female*	*Total*	*Male*	*Female*	*Male*	*Female*	*Male*	*Female*
Total	51.80	48.20	100.00	43.57	20.08	7.17	20.54	49.26	59.38
Educated	67.41	32.59	100.00	49.66	15.99	8.79	19.41	41.55	64.60
Uneducated	39.36	60.64	100.00	35.25	21.84	4.95	21.30	59.80	57.13
(i) Educated but below Matric/Secondary	66.23	33.77	100.00	44.42	12.99	8.66	17.70	45.92	69.31
(ii) Matric/Secondary but below Graduation	84.39	15.61	100.00	63.58	14.33	9.18	16.73	22.09	68.94
(iii) Technical Diploma or Other Certificate which not equivalent to Degree	84.01	15.99	100.00	73.07	59.60	5.51	9.84	21.42	31.56
(iv) Graduation and Higher Education other than Technical Degree	88.45	11.55	100.00	73.03	27.87	8.19	10.55	18.78	61.68
(v) Technical Degree or Diploma equivalent to Technical Degree or Post-Graduation	92.00	8.00	100.00	81.25	54.55	6.16	8.00	12.59	37.45

Source: Computed from Census of India, 2001.

An examination of figures for levels of education and employment reveals that a larger proportion of technically educated women was engaged as main workers [50.39 per cent for category (ii) and 49.19 per cent for category (v)] as compared to other categories, [i.e., 10.09 in category (i), 8.55 in category (ii), and 22.36 in category (iv)]. Similarly, the percentage of non-workers was lower among technically educated women as compared to other categories. This shows that technical education is helpful in providing employment to women. However, education below graduation has in fact lower proportions of main workers as compared to uneducated women (19.40 per cent). In the same way the proportion of non-workers was extremely high for all non-technical categories of educational levels [76.94 per cent for category (i), 85.29 per cent for category (ii), and 75.24 per cent for category (iv)] than those for uneducated females (62.26 per cent), which is very close to uneducated males (61.90 per cent). Thus, non-technical education does not seem to help in increasing work participation among women. To some extent, family responsibilities, unfavourable working conditions, lack of awareness, lack of mobility, lower level of achievement motivation, sex-discrimination at work place etc. are all responsible for this. This also partly manifests for technical education in the form of segmentation of labour market, as evident from the comparable data for males, where technical education had a much more positive effect on employment. Moreover, proportions of main workers among males were higher for non-technical educational categories also as compared to uneducated males. Data show wide discrepancy between male and female education. Thus, it is necessary to give utmost importance to technical education among women.

Next we examine education and employment relationship for rural and urban areas of Rajasthan separately.

Rural Population

In rural areas, picture similar to the overall Rajasthan emerges. The proportions of main workers lower and non-workers higher among educated women as compared to uneducated women are found (Table 15.2). More than half (62.61 per cent) of the educated women are non-workers in rural areas. Hence, education does not seem to have

Table 15.3: Percentage of Main Workers, Marginal Workers and Non-Workers in Urban Population in Rajasthan by Levels of Education, 2001

Levels of Education	*Total Persons*			*Main Workers*		*Marginal Workers*		*Non-Workers*	
	Male	*Female*	*Total*	*Male*	*Female*	*Male*	*Female*	*Male*	*Female*
Total	52.92	47.08	100.00	43.90	6.51	3.52	3.04	52.58	90.45
Educated	60.03	39.97	100.00	51.49	6.27	3.51	1.85	45.00	91.88
Uneducated	40.11	59.89	100.00	23.43	6.80	3.54	4.46	73.01	88.74
(i) Educated but below Matric/Secondary	56.73	43.27	100.00	40.21	3.42	4.02	2.00	55.77	94.58
(ii) Matric/Secondary but below Graduation	65.61	34.39	100.00	61.20	5.82	2.89	1.16	35.91	93.02
(iii) Technical Diploma or Other Certificate which not equivalent to Degree	79.95	20.05	100.00	68.57	46.20	1.64	0.86	29.79	52.94
(iv) Graduation and Higher Education other than Technical Degree	66.13	33.87	100.00	77.40	21.52	1.88	1.15	20.72	77.33
(v) Technical Degree or Diploma Equivalent to Technical Degree or Post-Graduation	73.34	26.66	100.00	73.70	48.39	1.22	1.26	25.08	50.35

Source: Computed from Census of India, 2001.

positive effect on employment of rural women also. Data for rural men, on the other hand, show that among educated men, proportion of main workers in higher and that of non-workers lower. Thus, education shows a positive effect on employment of men. There is, however, need to mention here that employment is not an outcome of education alone.

Analysing the data on levels of education for rural area, it is found that proportion of main workers is much higher in technically educated women than the non-technical categories. Thus, in rural area also, technical education seems to be helpful in providing employment to women. Among rural men, proportions of main workers are equally high for non-technical higher education also. On the whole, it is also commensurate with general qualifications.

Among main workers, the proportions for women workers are found to be much lower than those for men, whether educated up to any level or are uneducated, whereas among marginal workers and non-workers they are higher. This shows that relatively larger proportions of women are in the marginal and non-worker categories which suggests segregation of work against women and lower status of women in spite of attaining education.

Urban Population

Data for urban females displayed in Table 15.3 show that, again, the proportions of main as well as marginal workers lower and that of non-workers higher in the case educated ones than those for their uneducated counterparts. Thus, even in urban areas, which are expected to be having better facilities for women's educated employment, such women are lagging behind. Data for men, on the other hand, show a positive correlation between education and employment.

Analysis of information on levels of education and employment shows that in urban areas, though general higher education is also directly correlated to employment, technical education seems to be much more effective in providing employment to women. For men, on the other hand, proportion of main workers is highest for the category of 'graduation or higher education other than technical degree' (77.40 per cent).

We may conclude from the above analysis for total rural and urban areas that in main as well as marginal workers, proportions for uneducated females were higher than those for educated females. Thus, if appears that there is no positive correlation between education on employment. Educated male main workers, on the contrary, were in larger proportions than uneducated male main workers, showing positive effect of education on unemployment. On the basis of different educational levels among females, it is found that women having technical qualifications, followed by graduation and higher education other than technical degree, had higher proportions of main workers than other normal categories of educated women. This shows that technical education is helpful in providing women employment. However, in all categories of levels of education percentage of educated females in main workers is much less than that of educated males in main workers, and much more in marginal and non-workers. This suggests segmentation of labour market even against educationally qualified women. This results in their lower wages and socio-economic status. To some extent, family responsibilities and unfavourable working conditions are responsible for overwhelmingly marginal and lower levels of employment of educated women than educated men particularly so in general category.

Literacy Rate and Work Participation Rate: Rank Correlation Coefficient

Correlation between literacy rate and work participation rates of women has been statistically analysed in the present section through Spearman's rank correlation coefficient technique, using districtwise Census of India data for Rajasthan, as given in Tables A1, A2 and A3. Following formula has been used:

$$r_R = 1 - \frac{6\Sigma d^2}{N\left(N^2 - 1\right)}$$

Where, r_R = rank correlation coefficient;
Σd^2 = sum of squares of rank differences;
N = number of pairs of items

For tied ranks, following correction is made in the formula:

$$r_R = 1 - \frac{6\left\{\left(\Sigma d^2\right) + \frac{1}{12}\left(m_1^3 - m_1\right) + \frac{1}{12}\left(m_2^3 - m_2\right) + \ldots\right\}}{N\left(N^2 - 1\right)}$$

Where m_1, m_2 — are number of items with tied ranks.
If $r_R > 0$, the two variables change in the same direction.
If $r_R < 0$, the two variables change in opposite directions.

Table 15.4: Spearman's Rank Correlation Coefficient between Literacy Rate and Work Participation Rate in Rajasthan

Area	*Males*	*Females*
Total	-0.62	-0.48
Rural	-0.50	-0.28
Urban	+0.27	-0.014

Source: Self-computed from Tables A1, A2 and A3.

The results obtained by using the above formula on data given in Appendix Tables A1, A2 and A3 are shown in Table 15.4. For overall population of all the 32 districts of Rajasthan, Spearman's rank correlation coefficient is found to be –0.48 for females, indicating, moderate negative correlation. It is also negative for rural females, though lesser in magnitude (–0.28). It is extremely low negative in case of urban females (–0.014). Thus, female literacy rate does not seem to positively affect work participation among women. Hence this result supports the outcome of the analysis in the previous section.

Relationship between the two variables for overall male population is highly negative (–0.62) and to a somewhat lesser degree for rural areas (–0.50). However, rank correlation between male literacy rates and work participation rates is moderately positive (+0.27) for urban areas of the 32 districts of Rajasthan. Thus, only for urban males, positive effect of literacy on work participation rate is observed.

Conclusion

From the above discussion on levels of education and employment of women in Rajasthan, it is observed that educated female main workers are much less than uneducated female main workers, indicating greater participation of uneducated women in long-term economic activities as compared to educated women. Women's activity rates in a state like Rajasthan are influenced by innumerable set of factors such as changing nature of employment opportunities, propensity to work, social status, family decision, sex discrimination at work place, mobility, awareness etc., apart from capabilities. Better infrastructure facilities and motivation, if provided, can encourage employment and self-employment among educated unemployed women. Also, evidence exists that proportion of main workers was higher in women attaining technical education. Hence, apart from providing general education to women, it is necessary to give importance to their technical education and skill development.

Proportion of educated female main workers is found to be much less than that of educated male main workers. Among marginal workers, both educated and uneducated, female workers are more than male workers. This suggests segmentation of labour market against even the educationally qualified women in the state. As a result, wages of women are low, resulting in their lower socio-economic status. Whereas the percentages of non-workers were almost identical for uneducated males and females, educated women have predominance of non-workers as compared to educated men. To some extent, family responsibilities and unfavourable working condition are responsible for this.

Spearman's rank correlation coefficient indicated a negative relationship between female literacy rates and work participation rates for 32 districts of Rajasthan. Hence it supports the conclusion drawn on the basis of levels of education and employment. In fact, for only urban males, a positive correlation between literacy rates and work participation rates is manifested.

Thus, a strategy needs to be evolved where, on the one hand, women are motivated to seek higher qualifications, particularly technical skills, along with facilities provided for them to do so, and on the other, not only have additional employment opportunities be generated, but distributive justice must be done.

APPENDIX

Table A1: Districtwise Literacy Rate and Work Participation Rate by Sex in Rajasthan—Total Population, 2001

Sl. No.	*District*	*Literacy rate*				*Work participation rate*			
		Male		*Female*		*Male*		*Female*	
		%	Rank	%	Rank	%	Rank	%	Rank
1.	Ganganagar	75.49	15	52.69	7	53.5	6	25.0	30
2.	Hanumangarh	77.41	11	52.71	6	51.7	9	29.9	25
3.	Bikaner	70.78	26	42.55	15	50.2	16	27.5	28
4.	Churu	79.52	9	53.87	5	50.0	18	38.4	10
5.	Jhunjhunu	86.61	1	60.1	2	46.5	29	32.6	19
6.	Alwar	78.91	10	43.95	11	53.1	7	43.9	4.5
7.	Bharatpur	81.39	5	44.12	10	47.1	28	33.0	17
8.	Dholpur	75.85	14	42.36	16	51.7	9	34.0	16
9.	Karauli	80.93	6	45.44	9	44.8	32	34.2	15
10.	Sawai Madhopur	76.75	13	35.44	26	47.7	26	35.6	13
11.	Dausa	80.37	7	43.15	14	45.8	30.5	36.1	12
12.	Jaipur	83.58	4	56.18	4	47.4	27	22.3	31
13.	Sikar	85.20	3	56.40	3	45.8	30.5	31.5	21
14.	Nagaur	75.33	16	40.45	18	48.1	24	32.9	18
15.	Jodhpur	73.86	20	39.18	20	48.4	23	27.1	29
16.	Jaisalmer	66.89	29	32.25	29	51.7	9	29.4	26
17.	Barmer	73.64	21	43.91	12	51.3	13.5	41.8	6
18.	Jalore	65.10	31	27.53	32	54.1	3	46.1	2
19.	Sirohi	70.58	27	37.37	23	50.1	17	30.1	23
20.	Pali	73.06	22	36.70	24	48.5	22	30.9	22
21.	Ajmer	79.96	8	49.10	8	49.8	19	28.0	27
22.	Tonk	71.25	25	32.30	28	49.3	20	30.3	11
23.	Bundi	72.17	23	37.76	22	53.8	5	40.5	7
24.	Bhilwara	68.12	28	33.47	27	54.4	2	38.6	9
25.	Rajsamand	74.05	19	37.89	21	51.4	12	30.0	24
26.	Udaipur	74.47	17	43.71	13	51.5	11	31.8	20
27.	Dungarpur	66.19	30	31.22	30	51.3	13.5	44.7	3
28.	Banswara	60.24	32	27.86	31	50.5	15	43.9	4.5
29.	Chittorgarh	71.82	24	36.45	25	56.7	1	46.2	1
30.	Kota	86.25	2	61.25	1	48.0	25	19.4	32
31.	Baran	76.86	12	42.18	17	49.2	21	35.6	13.5
32.	Jhalawar	74.29	18	40.39	19	53.4	4	39.6	8
	Rajasthan	76.46		44.34		50.0		33.5	

Source: Census of India, 2001, Rajasthan, Population Totals.

Table A2: Districtwise Literacy Rate and Work Participation Rate by Sex in Rajasthan—Rural Population, 2001

Sl. No.	*District*	*Literacy rate*				*Work participation rate*			
		Male		*Female*		*Male*		*Female*	
		%	Rank	%	Rank	%	Rank	%	Rank
1.	Ganganagar	72.00	17	47.27	6	54.5	6	30.8	32
2.	Hanumangarh	75.97	10	50.01	5	52.5	9	35.8	26
3.	Bikaner	61.92	31	28.53	25	51.2	15.6	38.6	18.5
4.	Churu	78.63	8	51.45	3	52.0	10.5	49.7	2
5.	Jhunjhunu	86.36	1	59.80	1	48.1	27	39.1	16
6.	Alwar	76.54	9	39.16	13	53.8	7	49.2	3
7.	Bharatpur	79.95	6	39.62	11	47.4	28	38.8	17
8.	Dholpur	75.29	11	39.37	12	52.9	8	38.6	18.5
9.	Karauli	80.45	4	43.84	8	45.1	32	38.2	20
10.	Sawai Madhopur	74.13	13	29.69	24	48.8	23.5	42.0	12
11.	Dausa	79.19	7	40.83	10	46.0	31	39.2	15
12.	Jaipur	79.96	5	44.42	7	46.1	30	34.3	28
13.	Sikar	84.74	2	55.70	2	46.3	29	37.8	22
14.	Nagaur	73.66	14	37.58	15	48.8	23.5	38.0	21
15.	Jodhpur	66.94	25	25.10	31	48.6	25	36.8	23
16.	Jaisalmer	63.09	29	27.45	27	52.0	10.5	33.2	29
17.	Barmer	72.15	16	42.43	9	51.5	15.5	44.5	10
18.	Jalore	63.52	28	25.88	29	54.8	5	48.6	4
19.	Sirohi	65.94	26	31.47	22	50.3	19	34.4	27
20.	Pali	69.79	21	31.76	21	48.5	26	35.9	25
21.	Ajmer	72.60	15	32.72	19	51.2	17	39.9	14
22.	Tonk	68.48	23	25.62	30	50.2	20	44.3	11
23.	Bundi	68.99	22	32.41	20	55.4	2	47.0	5
24.	Bhilwara	62.85	30	29.09	28	55.2	3	44.8	8
25.	Rajsamand	71.23	19	33.22	18	51.7	13.5	33.1	30
26.	Udaipur	69.52	20	35.46	17	51.8	12	36.4	24
27.	Dungarpur	64.12	27	28.19	26	51.7	13.5	46.4	6
28.	Banswara	57.49	32	23.78	32	50.8	18	52.3	7
29.	Chittorgarh	67.91	24	29.98	23	58.0	1	31.0	1
30.	Kota	82.56	3	50.60	4	49.9	21	40.1	31
31.	Baran	74.81	12	38.21	14	49.8	22	44.6	13
32.	Jhalawar	71.46	18	35.51	16	55.0	4	44.6	9
	Rajasthan	72.96		37.74		50.7		40.6	

Source: Same as in Table A1.

Table A3: Districtwise Literacy Rate and Work Participation Rate by Sex in Rajasthan—Urban Population, 2001

Sl. No.	*District*	*Literacy rate*				*Work participation rate*			
		Male		*Female*		*Male*		*Female*	
		%	Rank	%	Rank	%	Rank	%	Rank
1.	Ganganagar	85.34	25	68.54	10	50.9	2	7.3	26
2.	Hanumangarh	83.00	28	63.40	15	48.5	11	5.9	32
3.	Bikaner	85.38	24	65.62	12	47.7	15	7.2	27
4.	Churu	81.71	30	59.95	24	44.9	25	9.1	18.5
5.	Jhunjhunu	87.51	14	61.28	20	40.7	32	7.0	28
6.	Alwar	91.50	3	71.24	4	48.8	8	11.6	10
7.	Bharatpur	87.08	15	61.47	19	44.3	27	9.2	17
8.	Dholpur	78.35	32	54.87	29	46.1	23	13.2	7
9.	Karauli	83.81	26	54.65	30	42.9	31	10.7	12.5
10.	Sawai Madhopur	87.54	13	59.17	25	43.3	30	8.1	23.5
11.	Dausa	90.12	8	62.54	16	43.7	29	8.8	21
12.	Jaipur	87.03	16	67.89	11	48.7	9	9.7	14
13.	Sikar	86.91	17	60.60	22	43.8	28	6.7	30.5
14.	Nagaur	83.06	27	54.48	31	44.6	26	8.1	23.5
15.	Jodhpur	86.12	20	65.28	13	48.1	13	7.6	25
16.	Jaisalmer	85.70	21	58.33	26	50.3	4	6.9	29
17.	Barmer	90.52	7	61.54	18	48.4	12	6.7	30.5
18.	Jalore	82.61	29	47.97	32	46.9	18	14.1	3
19.	Sirohi	89.76	10	64.44	14	49.4	7	9.1	18.5
20.	Pali	85.40	23	55.27	28	49.6	10	12.0	9
21.	Ajmer	89.89	9	72.58	3	47.7	15	9.6	15
22.	Tonk	81.65	31	56.89	27	45.6	24	15.5	2
23.	Bundi	85.53	22	60.15	23	46.7	20	12.2	8
24.	Bhilwara	86.81	18	62.29	17	51.5	1	13.6	4.5
25.	Rajsamand	91.11	4	69.24	7	49.8	6	8.7	22
26.	Udaipur	93.35	1	78.29	1	50.3	4	10.8	11
27	Dungarpur	89.25	12	69.03	9	47.2	17	15.9	1
28.	Banswara	92.13	2	77.03	2	46.8	19	10.7	12.5
29.	Chittorgarh	91.06	5	70.19	6	50.3	4	13.6	4.5
30.	Kota	89.29	11	70.30	5	46.5	21	9.3	16
31.	Baran	86.77	19	61.11	21	46.3	22	13.5	6
32.	Jhalawar	90.58	6	69.09	8	47.7	15	9.0	20
	Rajasthan	87.10		65.42		47.4		9.5	

Source: Same as in Table A1.

REFERENCE

Census of India (2001). *Rajasthan, Population Totals*. New Delhi: Census Commissions of India.

16
Demographic and Health Composition of Tribal Population in Rajasthan

B.L. Nagda

Introduction

Scheduled Tribes are officially recognized by the Government of India, as socially and economically backward and need special protection from injustice and exploitation. They constituted separate socio-cultural groups having distinct customs, traditions, marriage, kinship, property inheritance system and living largely in agricultural and pre-agricultural level of technology. They are food gatherers, hunters, forestland cultivators, and minor forest product collectors. The tribes are inhabited widely varying ecological and geo-climatic conditions in different concentrations. They have developed strong magico-religious health care system and wish to survive in their own style.

The state of Rajasthan situated in northwest part of India. It is largest state of the country in term of geographical area. It has great geographical diversity, northern west part of the state is desert, southern part is hilly, north-east part is plateau and central part is plain. There are several types of tribal inhabited in the state and constitutes around twelve per cent of the total population of the state but major portion of tribal population belongs to Bhil and Mina. In fact, they are original inhabitants of this area. Apart from these main tribes, number of smaller tribes also inhabited. Around 39 per cent of tribal are Bhil inhabited in the Udaipur, Banswara, Dugarpur and

Chittorgarh districts of Rajasthan, Bhil is believed to be fine archers and they called as bowmen. The Bhil was originally forest food gatherers and hunters. However, during the course of development, they engaged in other developmental work. The second largest tribal group of state is Minas. They found in majority in region of Shekhawati and eastern Rajasthan. Minas solemnize marriage in the younger years of the children. Garasias is in small number, inhabiting Abu Road area of southern Rajasthan. Sahariyas considered as the jungle dwellers is the most backward tribe, inhabited in Kota and Baran districts of south-east of Rajasthan. The main occupation of this tribe is shifting cultivation, hunting and fishing. Damor tribe is mainly cultivators and manual labourers lived in Dungarpur and Udaipur districts.

Sources of Data

The main sources of data about health and population of tribes are decennial census and National Family Health Survey (NFHS-2) Rajasthan.

Population Size and Growth

According to 2001 census, tribal population of the state was around 71 lakh, which forms 12.6 per cent of total population of the state. The population has grown at rapid rate for the last ninety years and at an almost flooding rate during the last four decades. During 1901 total population in the state was 103 lakh, which rose to 565 lakh in 2001. During the decades 1961-71, 1971-81 and 1981-91 the growth rate of tribal population in the state was 35.33 per cent, 34.46 per cent and 31 per cent respectively. During these periods the growth rate of tribal population has decreased. The growth rate of tribal population was reported 29.8 per cent during 1991-2001. It is higher than the growth rate of total population of the state (28.33).

Distribution of Tribal Population

Tribal population is unevenly distributed in the state. The highest concentration of tribal population is noticed in Banswara and

Dungarpur districts, and its proportion to total population of the districts comes to 72.3 per cent and 65.1 per cent respectively. Other districts with higher proportion are Udaipur (47.9 per cent), Sirohi (24.8 per cent), Sawai Madhopur (21.6 per cent), Bundi (20.2 per

Table 16.1: Distribution of Scheduled Tribes in Rajasthan (in %)

Sl. No.	Name of District	District Population	Tribal population	Proportion of tribal population to total population of the state	Proportion of tribal population to district population
1.	Ganganagar	1789423	14,744	0.21	0.80
2.	Hanumangarh	1518005	10,029	0.14	0.70
3.	Bikaner	1674271	5,945	0.08	0.40
4.	Churu	1923878	10,063	0.14	0.50
5.	Jhunjhunu	1913689	36,794	0.52	1.90
6.	Alwar	2992592	2,39,905	3.38	8.00
7.	Bharatpur	2101142	47,077	0.66	2.20
8.	Dhaulpur	983258	47,612	0.67	4.80
9.	Karauli	1209665	2,70,630	3.81	22.40
10.	Sawai Madhopur	1117057	2,41,078	3.40	21.60
11	Jaipur	5251071	4,12,864	5.82	7.90
12.	Dausa	1317063	3,53,187	4.98	26.80
13.	Sikar	2287788	62,512	0.88	2.70
14.	Ajmer	2181670	53,634	0.76	2.40
15.	Tonk	1211671	1,45,891	2.06	12.00
16.	Jaisalmer	508247	27,834	0.39	5.50
17.	Jodhpur	2886505	79,540	1.12	2.80
18.	Nagaur	2775058	6,497	0.09	0.20
19.	Pali	1820251	1,05,814	1.49	5.80
20.	Barmer	1964835	1,18,688	1.69	6.00
21.	Jalore	1448940	1,26,799	1.79	8.80
22.	Sirohi	851107	2,10,763	2.97	24.80
23.	Bhilwara	2013789	1,80,556	2.54	9.00
24.	Udaipur	2633312	12,60,432	17.76	47.90
25.	Rajsamand	987024	1,29,198	1.82	13.10
26.	Chittorgarh	1803524	3,88,311	5.47	21.50
27.	Dungarpur	1107643	7,21,487	10.04	65.10
28.	Banswara	1501589	10,85,272	15.29	72.30
29.	Bundi	962620	1,94,851	2.75	20.20
30.	Kota	1568525	1,51,969	2.14	9.70
31	Baran	1021653	2,16,869	3.06	21.20
32	Jhalawar	1180323	1,41,861	2.00	12.00
	Rajasthan	56473122	70,97,709	100.00	12.60

Sources: Census India, Final population totals, Rajasthan, Directorate of Census Operation, Rajasthan, 2001.

cent), Chittorgarh (21.5 per cent) and Baran (21.2 per cent). The dispersal pattern of tribal population in various districts in terms of absolute number observed that the highest proportion of the tribal population has resided in Udaipur district (17.76) followed by Banswara (15.29 per cent), Dungarpur (10.1 per cent), Jaipur (5.81 per cent) and Sawai Madhopur (3.4 per cent) districts. These five districts together contributed more than half of the state's total tribal population. The districts, which have less than one per cent of the state's total tribal population are Ganganagar, Churu, Jhunjhunu, Sikar, Ajmer, Jaisalmer, Jodhpur and Nagaur (Table 16.1).

Sex Ratio

Sex ratio is number of females per thousand males. Sex composition of a population is an indicator of the socio-economic and health conditions of male and female in society. The sex ratio of tribal population (944) was higher than the general population (922) of the state or say twenty-two females per thousand males was more in tribal population. It shows no discrimination of sex at birth in the tribes. Table 16.2 shows that sex ratio of tribal population was reported 900 and less females per 1000 males in 16 districts. Sex ratio in between 900-950 found in 11 districts and sex ratio 950 and above females per 1000 males reported in 5 districts. It is observed that those districts having higher concentration of tribal population had higher sex ratio.

Table 16.2: Districtwise Sex Ratio of Tribal Population, 2001

Sex Ratio	*Districts*
900 and less	Ganganagar (880), Bikaner (817), Churu (884), Alwar (878), Bharatpur (868), Dholpur (834), Swai Madhopur (876), Jaipur (897), Dausa (894), Karoli (854), Jaisalmer (869), Hanumangarh (868), Nagaur (854), Barmer (885), Bundi (880), Kota (885)
900 – 950	Jalore (903), Jhunjhunu (929), Pali (926), Sikar (923), Ajmer (925), Sirohi (953), Bhilwara (939), Baran (918), Jhalawar (916), Jodhpur (918), Tonk (915)
950 and above	Udaipur (985), Rajsamand (968), Chittorgarh (967), Dungarpur (1028), Banswara (983)

Sources: Census India, Final population totals, Rajasthan, Directorate of Census Operation, Rajasthan, 2001.

Major portion of Bhil live in southern districts of the state where sex ratio was more than 950 females per thousand males. Whereas, majority of Mina concentrated in Sawai Madhopur and Jaipur districts and they are socio-economically well-being, having sex ratio similar to general population of the respective districts.

Literacy

Literacy is one of the powerful instruments of social change. Rajasthan has the lowest literacy rate except the state of Arunachal Pradesh in the country. According to 2001 census over all literacy rate of the country was 65.38 per cent whereas, this figure for Rajasthan was 60.4 per cent. Table 16.3 shows that the position of literacy among tribal of Rajasthan is extremely poor, and more so in the case of female literacy. It is very painful and even disgusting to note that the literacy among tribal in Rajasthan still very low. More than two-third of tribal female population is continued to be illiterate. However, there appears to be slight improvement in 2001.The big portion of tribal female children is away from School because they tend to use children for ancillary services (Nagda, 1992).

Tribal Culture and Fertility

Ethno-demographic characteristics of tribes bring out wide variation in the level of fertility within the community. It is mainly because of variation in social, cultural and economic practices. The socio-cultural practices vary from one tribe to another. The factors, like traditional religious systems, value of children, polygamy, remarriage, bride price and poor literacy pave the way for high fertility among the tribes.

Table 16.3: Tribal literacy rate in Rajasthan 2001

Year	*Total*	*Male*	*Female*
1961	3.98	7.42	0.28
1971	6.46	12.02	0.20
1981	10.27	18.85	1.20
2001	44.7	63.4	29.3

Source: Census of India, Rajasthan, Final Population Totals Directorate of Census Operation, Jaipur, 2001.

The average living children of Bhil was 6.7 followed by Christian Bhil 6.2 and Bhagat Bhil 5.4. The age at effective marriage of boys and girls was 18.2 and 17.4 years respectively. More than 50 per cent of the Bhil lived in nuclear family consisting of 5 to 6 members. Almost all mothers are illiterate. Marriage is usually solemnized after the couple has had the opportunity to know each other intimately over number of months. Divorce is granted easily on grounds of adultery, carelessness in the upkeep of the house etc. (Nagda, 1992).

The results of National Family Health Survey, (NFHS-2) India, Rajasthan indicates the silent issues of fertility pattern of tribes as evident in Table 16.4. About 12 per cent women were currently pregnant and mean number of children ever born to ever married women in age 45-49 was 5.72. More than half of mothers had 3 and more live births. The median age at first birth was reported at 19.4 years. Tribal women produced 4.31 children during her reproductive life but they want only 2.7 children. Tribes considered on an average 2.9 children as ideal in their family. The tribal maintain sufficient space between births of two successive children. Median interval since previous birth was 29.5 months. It shows that tribes are aware about adverse implications of rapid growth of population. Son preference strongly reported among tribes, 97 per cent of tribes want at least one son.

Level of Mortality

Like fertility, mortality was also higher among tribes in comparison to general population of the state. Table 16.5 reveals that among the

Table 16.4: Fertility Indicators of Tribes of Rajasthan

Indicators	*Value*
Percentage of currently pregnant women	12.4
Mean no. of children ever born to ever married women age 45-49 years	5.72
Percentage of third and above birth order of children	55
Median age of first birth women in age group 20-49	19.4
Total fertility rate	4.3
Wanted fertility	2.7
No. of children considered as ideal	2.9
Percentage of women wanted at least one son	97
Median months interval since previous birth	29.5

Source: National Family Health Survey, (NFHS-2) India, Rajasthan, 1998-99.

tribes 58 births per 1000 live births have not survived the first month of life, about 37 per cent of infants die in between age of one month to 12 months and 95 per cent infant die before reaching their first birthday. Child mortality (1-5 years) was calculated to be 155. The major causes of infant and child mortality in tribes were acute respiratory infections, fever, diarrhoea and anaemia. It is suggested that for reduction in mortality especially infant mortality, existing infrastructure of health and medical facilities should be improved and National Rural Health Mission should be popularized more among them through modern and traditional methods of motivation and communication.

Table 16.5: Mortality Indicators of Tribes of Rajasthan

Indicators	*Rate*
Neonatal mortality rate	58
Post neonatal mortality rate	36.7
Infant mortality rate	94.7
Child mortality (1-5 years) rate	155

Source: National Family Health Survey, (NFHS-2) India, Rajasthan, 1998-99.

Health Scenario of Tribal Women

Health can be defined as complete physical, mental and social well-being and not merely the absence of diseases or infirmity. Tribal population in India inhabited widely varying socio-cultural and geographical condition and exposed differently to the various geo-cultural stresses and strain. Socio-cultural milieu of the tribal includes diverse factors such as reproductive health practices, health seeking behaviour, practice of traditional and modern systems of medicine etc. The tribal has own life style and cultural practices, it includes economic subsistence, diets, occupation, habits, ceremonial, ritualistic and customs. On the one hand reproductive health of tribal women generally governs by poverty, powerlessness and lack of control over resources. Majority of tribal women suffered from malnutrition and, infections. On the other hand, lack of access of health care with poor quality of delivery system and its responsiveness of women's need exacerbate the reproductive health status.

Reproductive health condition is one of the indicators of quality of life; quality of life of women can be assessing through demographic indicators, like maternal and infant mortality rate. The report of Sample Registration System (SRS 1997-2003) indicates that nearly 24 per cent of maternal mortality declined in the country during 1997-2003. Report further highlight the fact that about two-third of the maternal deaths occur in Empowered Action Group (EAG) states—Bihar, Chhattisgarh, Rajasthan, Jharkhand, Odisha, Madhya Pradesh, Uttar Pradesh, Uttarakhand and Assam. Maternal Mortality Ratio declined from 398 in 1997-98 to 301 in 2001-03. Infant mortality rate is also considered as one of the most sensitive indicators of reproductive status of women, Although Scheduled Tribes have a lower infant mortality (62) than Scheduled Castes (66) (NFHS-3). Malnutrition is widespread among the tribal girls and women. Women have equal rights to enjoy the standard health services, but in reality, women are different and unequal access to basic health services. The awful poverty, lack of nutritious food and safe drinking water, lack of sanitation and unhygienic accommodation, hostile life etc. create problem of the reproductive health of women in tribal areas. Early child bearing, miscarriages and multiple pregnancies create serious health hazards for tribal women.

Socio-culture Issues of Health

Every culture has its own belief and practice concerning disease and evolves its own system of medicine. Tribal practices different types of diagnosis and treatments during illness of person, the interference of supernatural agency is strong in context of health and disease. Tribal believed that different deities and spirits created various diseases. The tribal communities have specific gods for treatment of diseases, calamities, treatment of diseases of cattle, bites of snakes and dogs and so on. All these deities have their own respective sphere and field. The tribal priest is known as Bhopa whose services are sought after finding out the cause of illness. The Bhopa worships the deities at the sacred place (*Devru*). He is mainly entrusted with the benevolent deities and control the malevolent deities. The dependency and believe on Bhopa are often responsible for the non-acceptance of modern medicine. The Bhopa share the common cultural beliefs and practices of the patients, naturally they have more faith in them.

Traditional healers also occupied prominent place in tribal community for the treatment of various diseases. If the reason of illness is identified as evil-eye, sorcery or witchcraft, the tribal always would call their Bhopa instead of consulting a doctor, as they strongly feel that the doctor are quite helpless against such evil forces which can only be counteracted by Bhopa (Nagda, 1992). The traditional health care system and treatment are based on their deep observation and understanding of nature. The Tribal healers used different part of plants not only for treatment of various diseases, but also for birth control. This knowledge can be fruitfully utilized in a wider context.

Nutritional Status of Tribal Women

Body Mass Index (BMI) can be used to assess both thinness and obesity. The BMI is defined as the weight in kilograms divided by the height in metres squared (kg/m^2). The mean BMI for women in Rajasthan is 19.9. About 40 per cent of Scheduled Tribe women had BMI below 18.5, indicating higher nutritional deficiency among tribal women. More than half (58 per cent) tribal women and 80 per cent of children were suffering from anaemia (NFHS-2). The problem of malnutrition in tribes is related to their poverty, illiteracy, lack of nutritional food, and eating food habits. Several traditions are prevailing among tribes about food intake by pregnant and lactating mother. The pregnant mother is prohibited to eat ghee, oil seeds, groundnuts, curd and hot foods. Newly delivered mother eats several types of herbal products and gum with ghee. It is believed that during pregnancy rich foods containing ghee and fat is injurious to health of womb and after the delivery it is good for the health of mother as well as child. The tribal people of southern Rajasthan especially Kathodi and Garasia believe that during pregnancy use of forest foods and animal foods are good for health of mother and child. Kathodi women eat monkey meat after delivery (Nagda, 1996).

Health Care

Antenatal and post-natal care is one of the indicators of health status of children as well as mother. Results of NFHS-2 show that 58 per cent tribal mothers were not accepted ante-natal care during their

Table 16.6: Health Care Indicators in Tribal of Rajasthan

Indicators	*Percentage*
Women received ante natal care	42.0
Women received two injections of T.T.	42.0
Women received 100 IFA Tablets	35.8
Home delivery	86.0
Delivery conducted by *Dais* and others	76.0
Children Immunization	
Three doses of DPT	15.7
Three doses of Polio	31.5
Measles	19.0
One dose of BCG	39.7
One dose of vitamin A	12.9
Complete vaccination	10.3
Started breast feeding within one hour of birth	03.7
Mother Squeezed first Milk from breast	73.8

Source: National Family Health Survey, (NFHS-2) India, Rajasthan 1998-99.

pregnancy. They stated that it is not necessary and not customary. About 42 per cent of mothers were given two doses of tetanus toxide injection and 36 per cent were given IFA tablets. About 86 per cent deliveries performed at their home and three-fourth of deliveries conducted by *Dais* and other untrained persons (Table 16.6). Initiation of breast-feeding immediately after childbirth is important because it benefits both mother and infant. As soon as the infant starts suckling the breast, the Harmon Oxytocin is released, resulting in uterine contraction that facilitate expulsion of the placenta and reduce the risk of postpartum hemorrhage. Small number of mothers (3.7 per cent) in tribes started breast-feeding within one hour of birth and 74 per cent of the mothers squeezed first milk from breast, this practice create the problem of natural immunity in the children.

Immunization of children is concerned, about 16 per cent children were given three doses of DPT, 31.5 per cent three doses of polio, 19 per cent of measles, and 13 per cent one dose of vitamin A. The complete immunization of children was only 10.3 per cent among the tribes. The low practices of various immunizations create the problem of higher incidence of infant mortality among tribes.

Traditional Methods of Conducting Delivery

The tribal believe that when the menstrual period is delayed by a month, assumed to be woman pregnant. She does not take *mahua* liquor during pregnancy. There is no restriction on her daily routine work. The delivery (*Japa*) is conducted in hut called *Jopada*. The hut is cleaned and pasted with cow dung in advance. In case of any problem, during the pregnancy, call traditional *Dai* of the community and elderly ladies of the community help in conducting the delivery. The naval cord is cut by mother herself with the help of a Bamboo strip, knife and stone.They prefer to cut the naval cord with a bamboo strip because it is more safe from infections (Nagda, 1992). After cutting the naval cord, the stump is tied and *mahua* oil is applied. The mother is given a small quantity of *mahua* liquor with water mixed *haldi* (turmeric) and *gour* (Jaggary) for taking after two days of delivery. These drinks are bringing out all harmful substances from the body of the mother. Mother do not take sour foods and pluses because it could adversely affect health of mother and child. The baby is breast fed after one day of delivery. The mother breast fed the child upto 2-3 years and supplementary food given to child after 5-6 months. They do not stop breast-feeding during the sickness of the child.

Birth Control Method

Table 16.7 shows that about 29 per cent of currently married women using any method of birth control. About one-fourth of the women using any modern method of birth control. Female sterilization was more popular among tribes. The unmet need for family welfare was 18 per cent. It is almost same for limiting and spacing methods. About

Table 16.7: Practicing Birth Control Methods in Scheduled Tribes in Rajasthan

Indicators	*Percentage*
Any methods of birth control	29.3
Any modem method of birth control	26.5
Female Sterilization	21.8
Male Sterilization	1.0
Not using any method	70.7
Unmet needs of family welfare	
For spacing	8.9
For limiting	9.1

Source: National Family Health Survey, (NFHS-2) India, Rajasthan 1998-99.

71 per cent of the women were not using any method of birth control. About 2 per cent women were using traditional methods of birth control. It indicates that modern method of birth control is not popular among the tribes.

Indigenous Practices of Birth Control

The tribal is an ancient social community. Since long they have knowledge about indigenous methods of birth control. They used several types of forest products and *Jadibuties* (Herbals) for controlling the birth as well as removing the sterility. They thought that traditional methods of birth control are easily available with no side effects. The family welfare practice among the tribes is very ancient as a part of their cultural practices. For removing the unwanted pregnancy, they use boiled water of bamboo pills, old *gour*, leaf of adhatoda, vasica (Adusa), clove basil, carsia, black-pepper, *sunth* and thymol. This mixture can be used in any period during the pregnancy but it believes that if it should be used within three months of the pregnancy would be not harmful for health of the mother. The tribes also know the treatment of permanent and safe methods of birth control. For permanent method they used juice of gurbell, seed of castor and citron, mixing it with *gour* and give it to the women for 15 days with water. During this period a woman has to take some precautions such as to avoid use of milk, ghee, sour things, and allowed to take only *roti* of barley with *mug dal* without chilies and salt. For maintaining space between two children, they used boiled water of cottonseeds, cannabis sativa seeds (*Ganja*) along with picrorhiza (*Kutki*). This mixture has to be taken by the female every month for four days after the menstruation period. They even know the herbal treatment for the cure of infertility. In this treatment they use roots of Appamarga plant, seeds of Nagkeshar, stem of *Shatawari* plant and dry leaf of *Sisham* and *Banyan* tree. All these are to be grind together in form of powder and prepare tablets, dry them at cool and dry place. The sterile woman has to take tablets twice in a day with milk of cow for the period of three months. Tribal people also know the treatment of choice of birth, for birth of son they used *Shivlingi* seeds with milk after two months of pregnancy. They believed that this herbal treatment is very effective for those women who want birth of male child. The local heath healers have knowledge about the herbal medicines and they don't want themselves to be exposed it. They generally keep it a trade secret. The tribal *Dai* know these medicines but they also have it secret (Nagda, 1992).

REFERENCES

International Institute for Population Sciences (2001). *National Family Health Survey (NFHS-2), India, Rajasthan 1998-99*. Mumbai: IIPS.

International Institute for Population Sciences (2007). *National Family Health Survey, India (NFHS-3) Vol-2, India 2005-2006,* Mumbai: IIPS.

Nagda, B.L. (1992). *Social Correlates of Fertility*, Himanshu Publication, Udaipur.

Nagda, B.L. (1996). *Good Nutrition of Healthy Mothers*, Social Welfare, No. 7, New Delhi: Central Social Welfare Board.

Ministry of Home Affairs, Govt. of India (2004). *Sample Registration System (SRS) 1997-2003, Maternal Mortality in India (MMR),* New Delhi: Registrar General of India.

Census of India (2001). *Final Population Totals, Rajasthan, Directorate of Census Operation, Rajasthan 2001*, New Delhi: Census Commissioner of India.

17
Development Induced Displacement: A Case Study of Odisha

Ranjana Bajpai

Introduction

The Indian government has been 'planning' development to improve the living standards of its citizens through successive Five-Year Plans under a centralized planning mechanism. Such plans comprise a wide variety of projects like dams, bridges, power, mining, industries, transport networks, urban development, info-tech cities, etc. Unfortunately, some of these projects have brought adverse affects in the form of displacement of people from their original place of habitation due to large-scale land acquisitions. Development-induced displacement in the country has brought socio-economic and environmental problems to the displaced people. In India, around 20 million people displaced since 1947 and 75 per cent of these became impoverished, causing disturbing psychological and socio-cultural consequences. These include shattered long-established production systems, violation of inherited sacred zones, graves and places of worship, disruptions of family and kinship systems and informal social networks (Kothari, 1995). Cernia (2000) lists landlessness, joblessness, homelessness, marginalization, food insecurity, increased morbidity and mortality, loss of access to common property and social disarticulation as major impacts of displacement.

This leads one to question whether these development projects harm the very people they are meant for? If yes, then whom does this development cater to? Who are its real beneficiaries: the state, the market or the big business? And finally, is the concept of development being compartmentalized (social or human development, economic development and ecological development) for the comfort of the policy makers who use the term in the context of a given situation? These are key questions that the government of the day cannot shy away from, especially in the world's biggest democracy. This paper discusses development-induced displacement in Odisha, one of the poorest states of India, currently experiencing a phenomenal growth of FDI in industrialization and service sector; at the same time the state has witnessed massive protests against land acquisitions for industrialization and resultant degradation of environment.

A Brief Socio-Economic Profile of Odisha

Located on the east coast of India, Odisha is blessed with a rich mineral wealth, forests, lakes and rivers, besides being home to a large number of indigenous people (classified as Scheduled Tribes) who have harmoniously lived in their traditional lands for centuries. But, with around 85 per cent of Odisha's 37 million people living in the rural areas, the state is the second poorest in the country having one of the lowest levels of urbanization. Ever since independence, Odisha has been one of the most poor states and currently over 45 per cent of the state's population live below the poverty line, despite its rich natural and cultural heritage (India: Transition in Odisha, 2006). With an annual Per Capita Income of Rs. 13,329 in 2004-05, Odisha is ranked 17th among all states. The brunt of poverty is borne by the Scheduled Tribes most of whom live in geographical isolation in remote village communities with little access to basic services and employment opportunities. These communities make up a large segment of the population, constituting 22 per cent of Odisha's people compared to 8 per cent in the rest of the country. As far as Human Development Index is concerned, though Odisha has achieved a significant progress in terms of its Human Development Index (HDI) –51.3 per cent during 1981-2001and the absolute value of HDI has gone up from 0.345 in 1991 to 0.404 in 2001, it is well below middle-income states. Health indicators of the state like MMR and IMR are much higher than the national average (See Table 17.1).

Table 17.1: Human Development Indicators

Indicator	*Odisha*	*All India average*
Maternal Mortality per 100,000 live births	738	408
Infant Mortality per 1000 live births	81	67.6

Source: Human Development Report of Odisha, 2004.

In terms of overall literacy, Odisha has significant gains but inter-district and gender disparities in literacy are worrisome. The state is also prone to many natural calamities like cyclones, floods and droughts which aggravate mass poverty. Regional disparities in social indicators are pronounced and so are the gaps between rural-urban, interior-coastal and also between STs and rest of the population (Human Development Report of Odisha, 2004; Ravishankar, 2008). According to the District-Level Household and Facility Survey-3 (DLHS) 2007-08, 15 of the 30 districts in Odisha have over 80 per cent of households with low standards of living, while 18 of the 30 districts have more than 50 per cent BPL card holders (Patro, 2009). The eight districts of Malkangiri, Kandhamal, Kalahandi, Rayagada, Koraput, Nuapada, Nabarangpur and Boudh, most of them with a high per cent of tribal population, have a very low HDI in almost all the yardsticks of the UNDP. Khurda and Ganjam districts on the other hand have performed well in terms of HDI (Human Development Report of Odisha, 2004).

Development Policies of the Odisha Government

The government of Odisha endorsed Central Government's liberalization policy of economic reforms and followed a similar track in order to realize the true potential of the state's mineral wealth, economic growth and overcome mass poverty. One way to do this was to encourage broad-based private sector development which the government perceived would lead to employment generation and substantial spill-over effects in the economy. As a result, a good number of development projects were proposed in the economically poor but resource rich remote, hilly regions of the state. Besides the government, a good number of big businesses were also attracted to mineral resources of Odisha. The state attracted huge investments by the private ventures in a single year 1995-96, amounting somewhere between Rs. 60,000 crore in industrial projects alone, excluding power and mining sectors (Sengupta, 2001).

In 2001, the State government adopted a New Industrial Policy to improve the investment climate through targeted regulatory reforms and passed legislation for the same in 2004. A single window system to facilitate and streamlining the setting up of new businesses in the state were the key features of this policy. Since then, the economic environment of Odisha has transformed swiftly. The economic growth rate measured in terms of gross state domestic product (GSDP) went up from 4 per cent in the 1990s to about 8.5 per cent over the Tenth Plan period (2002-2007), compared to an all-India average of 7.8 per cent. The manufacturing sector is growing at 11 per cent a year, and a large number of investment projects are underway. In short, Odisha has been following an aggressive path of economic development over recent years. From 2005-2008, the state has been consistently ranked number one among Indian states in terms of private investment projects under implementation, according to Centre for Monitoring Indian Economy (CMIE) and there are around 470 projects (costing US $125 billion or about seven times of the GSDP) under different stages of implementation (Ravishankar, 2008), out of which 45 are the steel plants only (Shrivastava, 2007). In 2005, the State government signed as many as 43 MoUs with various corporate bodies for setting up of their industrial units, bringing an investment of Rs. 1,60,000 crore to the state.

Going by the investment figures, one can conveniently say Odisha's economy is marching ahead, but there are other key aspects of these investments which the government seems to have comfortably put into the backburner: the large-scale forced displacement of the already marginalized sections of the society and the issue of land acquisition by the government for these private ventures. Land acquisition refers to the process by which the government forcibly acquires private property for public purpose without the consent of the land owner, which is different from a market purchase of land. The Land Acquisition Act, 1894 addresses the process of land acquisition in India and was last amended by the Land Acquisition

Amendment Act, 1984. The Act takes a broad definition of 'public purpose' permitting a diverse range of projects. A number of Supreme Court cases have highlighted concerns related to fair compensation, valuation of land, definition of 'public purpose' and other issues related to land acquisition (Parker and Vanka, 2008). Citing problems with the principal Act, the government recently introduced the Land Acquisition (Amendment) Bill, 2007, which was passed by the Lok Sabha in February 2009. It expands the rights of those displaced by land acquisition, and limits the ability to acquire land for public purpose. The Bill also establishes the Land Acquisition Compensation Disputes Settlement Authority at the state and national levels to arbitrate all disputes resulting from land acquisition proceedings. This Bill was introduced in conjunction with the Rehabilitation and Resettlement Bill, 2007 to address land acquisition, compensation, and resettlement of displaced persons. As far as Odisha is concerned, its R&R Policy of 2006 is one of the most advanced in the country, but its effective implementation is yet to take final shape, because most of the controversial projects in the state were conceived much earlier when the amended version of the core Land Acquisition Act of 1894 was in place. Also, as the case studies in the later section point out, most of the projects are already wired in some or the other form of controversy with respect to environmental impacts, compensation packages, loss of jobs, etc.

Impact of Government Policies on People

Large-Scale Displacement

The government policies to develop Odisha have led to large-scale displacement of the people. Unfortunately, the government equated the growth with development, assuming the spillover effects of the growing economy to fill the necessary gaps elsewhere in the state. What the government could not foresee was the large-scale displacements happening particularly in the remote hilly and poor regions of the state which have been traditionally occupied by the Scheduled Tribes over centuries. Not only were their lands encroached upon, but also their sources of livelihood degraded, social fabric disturbed and cultures invaded by the modern systems.

Volume of Displacement in Odisha

According to UNDP around 100,000 families have been displaced since independence and about two million have been affected in varied degrees on account of development projects in Odisha. As most of Odisha's natural resources lie in the hilly and forest areas, a majority of the development projects have come up in such regions. These forests are home to indigenous tribes, one of the most poor and disadvantaged of Odisha's population (Jena, 2006) and these are the very people affected by adverse impacts of land acquisition for mining and industries.

Till 2000, about 20 lakh people were directly affected by development projects in varying proportions out of which about 5 lakh were physically displaced losing their traditional habitat. Statistical figures further indicate that while dam/irrigation projects alone have displaced nearly 3.5 lakh people which is roughly 70 per cent of the total displaced, whereas the Mining Projects, Urban Development Projects, Thermal Projects and Wild Life Sanctuaries have displaced 3.37 per cent, 12.86 per cent, 2.60 per cent and 0.5 per cent of the total displaced people in the State of Odisha. These figures account only for the already completed projects and there are many other projects which are either ongoing or are in the pipeline about to displace 2 lakh more people (Ota, 2001).

Impact of Displacement

Displacement of human populations is forced, involuntary and therefore, very disturbing. Not only are the people physically displaced but also suffer from immeasurable and irreparable environmental, socio-economic and cultural losses, something which is very hard to reimburse and restore after the displacement has actually taken place. Usually displacement unravels the social fabric of a society, especially of indigenous groups. Social networks get fragmented as families and sometimes the entire community is torn apart. It can destabilize law and order and interfere with the self-sustaining socio-economic and ecological systems. Some explicit problems linked to the displacement are (Downing, 1998):

- Loss of livelihoods, homes and land,
- Increased occurrence of disease and reduced life expectancy,
- Lack of food security and loss of access to common property resources, and
- Loss of a sense of cultural identity and denial of human rights.

In most of the cases, these problems are made worse by poor implementation of resettlement programmes and inadequate training of personnel. On the other hand, corrective measures to successfully execute the rehabilitation and resettlement of displaced families can help better management of natural resources and strengthen economies through capacity building and employment generation. Otherwise, international as well as national experiences have shown that faulty rehabilitation and resettlement policy frequently give rise to socio-political unrest and international disrepute to the government for alleged human rights abuses. Moreover, local people resisting the development efforts may involve colossal losses and premature closure of the industrial units which could otherwise be providing new jobs, relieving poverty and promoting social integration. The next section takes up few glaring instances of development induced displacement and people's resistance to the same in prominent regions of Odisha.

Case Studies

Utkal Alumina at Kashipur, Rayagada District

The Utkal Alumina International Ltd. (Henceforth, UAIL) bauxite project proposes to mine the Baplimali hills for bauxite and transport the mined ore to a distance of 22 km to a refinery at D. Karol in Kucheapadar village limits. The project is a joint venture between Canada's Alcan and Hindalco. The proposed alumina production is 1 million tonne of Alumina per annum to begin with, to be expanded to 3 million tonne during the course of the project. No smelter is planned since the alumina is to be exported straight from Vishakhapatnam port (Bhargava, 2005).

The villages likely to be affected by the UAIL project have been resisting the project for the last 12 years on the grounds that they would be deprived of their livelihood and will in no way benefit from the development brought about by the project. The displaced people are protesting against unacceptably low compensation rates; exclusion of many impacted communities; substantially under-estimated assets, resulting in conversion of independent and self-reliant small-farmers into paupers now living in slums. The use of force by the police in order to pressure the locals into accepting compensation through means of lathi-charges, tear-gassing, imprisonment, torture and the three shooting deaths near Maikanch village and twelve deaths in Kalinga Nagar have led to social unrest in the region (Goodland, 2007).

The company acknowledges that a total of 3 villages will be displaced by the refinery site but also estimates that 26 villages will be affected by the project. Although the company claims that 26 villages would be affected, in fact at least 42 villages will actually be affected by the refinery and the waste disposal ponds (red mud and ash ponds), causing a severe environmental crisis (Kohli, 2005). Moreover, this project is highly subsidized and the royalty of Rs. 62 per tonne to be obtained by the government of Odisha was nominal compared to fiscal incentives and the expenditures on maintaining law and order situation by the government in the region. Goodland (2007) notes that the Government of Odisha has been trying to clear all the people out of the areas needed for Utkal ever since it was conceived in 1994.

Most impacted people in this case are the Scheduled Tribes, extremely poor indigenous people. The local environment will be seriously affected. Hills will be flattened and rains can sweep silt into a huge reservoir that provides water for the region. Several streams that feed the Indravati river would be destroyed. The most visible impact of bauxite mining on the environment is effluent discharge. Dumping discharge into the streams and rivers would raise the *p*H level causing severe contamination. Several perennial sources of water would dry up and thousands of people will lose their land and livelihood to displacement. Agricultural land will no longer be cultivable (Bhargava, 2005).

The Kalinga Nagar Industrial Complex

Located in the Jajpur district, the Kalingnagar Industrial Complex comprises iron ore and chromites rich regions of Dangadi, Duburi and Sukinda. It includes about 13 mineral-based industries like iron and steel, stainless steel, sponge iron and ferrochrome etc. The three regions together have an annual capacity of producing 15 million tonnes of steel. The State government has already signed an MoU with major corporate houses such as the Tatas, Jindal, Visa, Mesco who have already started their projects and a few of them are nearing completion. The state government would grant mining lease to the concerned industrial units and provide them with land for setting up their industries.

These efforts for industrialization have been met with violent protests from the displaced people over the payment of compensation. The compensation package announced in 1994 was to be paid through the Industrial Development Corporation (IDCO), a corporation of the Government of Odisha. Mediating between the State Government and the displaced people, the IDCO obtained the land at the rate of Rs. 35,000 per acre from the ousted and sold the land to the industrial houses at a rate of Rs. 1,00,000 per acre. The compensation package was later revised to Rs. 50,000 per acre. However, the ambiguous compensation package continues to be main irritant for the local people who had to lose their means of livelihood. Another private venture of the Tata Steel is set to displace around 700 families from their land.

On 9th May, 2005, the ADM of Jajpur and some policemen were attacked during the Bhumipuja of Maharashtra Seamless Limited (MSL). Finally, over two dozen incidents of industrial violence forced the MSL to withdraw. On 2nd January, 2006, the local tribal protests against the Tata Steel resulted in loss of life of 13 persons and one policeman and serious injuries to other four policemen in the encounter that ensued between the police and the displaced people (*Prajatantra Saptahiki*, 2006). Since January 2, 2006, the tribals of Kalinga Nagar launched an indefinite economic blockade on the National Highway No. 200 at Madhuban Chhak. Not prepared to work out any compromise with the State government, the people have formed a Bisthapan Birodhi Janamanch and demanded a complete halt to displacement in the area. The place where the 13 victims of police firing were cremated has been named as Bir Bhumi and on 23rd May, 2006, about 5,000 women from Odisha, Chhattisgarh,

Jharkhand and Andhra Pradesh took out a rally and resolved not to lift the road blockade (*The Times of India*, May 24, 2006).

Sterlite Industries Limited at Lanjigarh in Kalahandi

Sterlite Industries Limited in collaboration with the Odisha Mining Corporation has set up an alumina refinery at Lanjigarh in the Kalahandi district of Odisha, based on bauxite mining from Niyamgiri Hill complex which has a bauxite deposit of 73 million tonnes. The plant is estimated to produce 3 million tonnes per annum of bauxite. Drawing in around Rs. 4,500 crores of investment, the stakes in the project seem to be high.

The district of Kalahandi is one of the least industrially developed parts of the country, having attracted a lot of attention of the media due to outbreak of famine conditions, drought and incidence of extreme poverty of the masses leading to sale of babies. It has a sizable population of the SCs, STs and OBCs. While efforts are being made to change a predominantly rural economy and hilly area into a major industrial complex, the people have been strongly resisting the work of setting up of the plant and mining of bauxite. A mass movement is going on since 2004. The tribal people of the area have formed a local group Niyamgiri Suraksha Samiti supported by another organization called Green Kalahandi, to protect the forest, land and water of the area, giving the slogan, 'Vedant Hatao' (Remove Vedant).The people are resisting leasing out of the hills to the company in order to protect their local environment (*The Times of India*, 28 May, 2006).

The Niyamgiri region has a number of perennial hill streams and some of the last remaining forests in the whole Kalahandi District, which are useful for drinking water and local economy of the tribal people living in the foothills. It is a source of the river Bansadhara and is even proposed as an Elephant Reserve. Rivers play an important ecological function in forest areas, and are critical for the survival of the forests. They are integrally connected to the lives of the local people as well, both for drinking water and irrigation. For drought prone Kalahandi, mining could spell disaster. The hill is also a place of worship of the people belonging to the Dongaria and Khond tribes who depend upon fruits grown by them in this area. The diversion of the Niyamagiri forests may push these tribes to extinction (Kohli, 2005).

The displaced people of two villages, namely Kinari and Bolbhatta resettled in the foothills of Niyamgiri are yet to adjust to their new environment. The use of ground water as well as the use of water from the river Tel by the company is a major area of contention, because it directly affects the cultivating class in an area that is already drought prone. Besides, people are also protesting the fact that the company has a plan of giving employment benefit to only 286 persons (*Bikalpa Bichar*, May 2004; *The Sambad*, 17 May, 2006).

A report by the Central Empowered Committee (CEC) of the Supreme Court has not only warned about the socio-environmental implications of the project, but also highlighted the ambiguous status of clearances applied by the company (See Kanchi Kohli, 2005). The CEC report confirmed several violations pertaining to mining and processing of bauxite. It also pointed out the detrimental impacts of the mining activity on the flora/fauna and the water resources of the region. The report went on to recommend against the diversion of the forestland for entire project (both refinery and mining). Furthermore, the CEC highlighted that area allocated to company forms part of Schedule V area as specified by the Constitution. Schedule V provides protection to the *adivasi* people living in these areas. No land in these areas is allowed to be transferred to non-tribals.

In spite of this, the economic interests continue to flout environmental laws to convert forest land into mining landscapes. The Sterlite has already cut considerable portions of the Niyamgiri reserve forests and begun work in non-forest areas. This includes construction of dwelling units, preliminary constructions and excavation. Worse still, the authority meant for the protection of the environment, the Ministry of Environment and Forests, is the one which granted clearance to the refinery, even though the clearance for the mining part remains pending.

Posco-India Steel Plant at Paradip

The project: Posco is a South Korean company planning to set up its Rs. 51,000 crore steel plant since June 2005. The company also plans to open a new private port of its own and avail the status of SEZ (Special Economic Zone). The MoU with the South Korean major signed on 22nd June, 2005 by the State government proposes a mining lease of 600 million tonnes that the company would be given along with the extent permission to the company to export iron ore. Further, the State government would acquire 435 acres of private land for the plant which involves displacement of 20,000 to 25,000 people (*The Sambad*, 22 July, 2006). While the government insists only 435 acres of private land out of 4,000 would be acquired, most of the land has been the source of livelihood of the indigenous communities who cultivate betel, cashew, etc., besides being the source of ancillary employment like making baskets for *Paan* leaves (Asher, 2007). The project has failed to take off as the company could not acquire land for it due to opposition from the local people.

The controversy: This is a big project expected to displace the people living in 22 villages of three gram panchayats, namely Kujang, Dhinkia and Nuagaon of Ersama block. Affected villagers demand Rs. 20,000 per decimal of land on which the betel vines have been planted, a demand not heeded to by the administration (Mohanty, 2010). The people of Jagatsinghpur district in coastal Odisha have formed the POSCO Pratirodh Sangram Samiti that also includes children and women of the locality, to a protest against the company (*The Times of India*, 25 June, 2006). The Samiti has raised a brigade of 1200 people who are ready to keep the movement alive. At times there have been violent clashes between the supporters of the proposed plant and the people protesting the project, calling for intervention of the police (*The Times of India*, 16 June, 2006). In one such incident which took place on 11 April, 2006, at Dhinkia village 11 persons were injured and nine activists were rounded off by the police (*The Times of India*, 21 June, 2006).

The people also complain against police excesses committed at the behest of the State government. More than 30,000 people have pledged not to obey any rule which the State government may enforce, pressing for their displacement. The Sangram Samiti has put up barricades at the entry points to these villages which are guarded by women and children to prevent entry of government and POSCO officials (*The Sambad*, June 22 and 24, 2006). As a result of people's agitations, the setting up of the plant is running behind the schedule.

Although the POSCO officials insist that the proposed plant would recruit 97 per cent Indians in its labour intensive project and would extend employment benefits to 48,000 people directly (*The Sambad*, June 30, 2006), the protests have continued and resistance to land acquisition is strong.

The project has got the environmental clearance but the forest clearance is still pending and people have passed resolution for regularization of their lands under the Forests Rights Act, 2006. The objection to the setting up of the plant is on several grounds. The POSCO steel plant and captive port will have massive impact on the coastal ecosystem and the Gahirmatha sanctuary along the east coast—breeding ground of the endangered Olive Ridley turtle, as the environmentalists have highlighted. Apart from the plant and the port, the mines of the Sundargarh district over 6,000 hectares would affect the Paudi Bhuian tribes dependent upon Khandadhar forests. The question of giving mining lease to a foreign company at a cheap rate has stirred up patriotic feelings among the people. The huge quantity of water which would be utilized by the proposed plant from the river Mahanadi would not only affect the cultivators but also rob the fishing folk of their only means of livelihood. It is estimated that the number of project affected people is likely to cross one lakh (Asher, 2007).

Resettlement and Rehabilitation

First and foremost, it is the prime responsibility of the government to acknowledge the constitutional rights like the rights to land resources, culture, education, development, etc. of the people affected (Sahoo, 2005), especially the women and Scheduled Tribes and avoid as far as possible any development plan which denies them their traditional ownership of their lands over which they have depended for centuries for their livelihoods. Secondly, even if such development plans are carried out there has to be an implementing agency whether the Government or Private/Public Sector in place to undertake preventive measures so that the displaced families are rehabilitated, resettled and adequately compensated for their acquired lands.

Many activists and organizations working for the local people believe the government's role remains confined to that of a facilitator to acquire lands for big projects (Asher, 2007; Bhargav, 2005; Shrivastava, 2007). In any case, Rehabilitation and Resettlement (Henceforth R&R) remains the responsibility of the government/ company which has to acquire the land. Till recently, the government of Odisha followed at least 11 different R&R for various development sectors. However, in June 2005 the government brought a comprehensive R&R policy for project-affected people. It has various components to separately deal with displacement caused due to industrial and mining projects; irrigation projects, national park and sanctuaries; linear projects like roads and railways. Moreover, it proposes to make displaced people the stakeholders in the industry that displaces them, by giving them 50 per cent of the compensation money in the company scrip. While it bars "physical displacement" before the "actual rehabilitation" is complete, it has also increased the compensation amount to Rs. 1.5 lakh from Rs. 37,500 per acre (0.4 hectare) (Odisha's Draft Rehabilitation and Resettlement Policy, 2006). The draft policy has strongly recommended that all R&R need to take into cognizance the needs of weaker sections, including women affected by displacement. It says that "displacement and resettlement will be treated as a development opportunity to improve living conditions of the affected people, especially those of Scheduled Castes and Scheduled Tribes, Below Poverty Line, women-headed households and other vulnerable groups in ways culturally familiar to them" (Jena, 2006).

The draft, however, fails to ensure any employment guarantee to the displaced; it carries just a stipulation that the industries give job "preference" at least to one nominated member of each affected family. It remains silent about the government's role in cases where people don't want to be displaced by the industrial projects. Besides, the document indicates that the State government will gradually withdraw from the land acquisition process and act merely as a facilitator (Odisha's Draft Rehabilitation and Resettlement Policy, 2006).

Here it is significant to note that Odisha has a history of protest movements not only against the development policies but also its R&R policy which by and large recognized the physical displacement only and even after the introduction of the new draft R&R policy, the industrial complex of Kalingnagar saw a violent tribal protests which brought back the state in national land acquisition news. Such protests have the potential to drive away investors in the face of what appears to be a resurgence of the movement against big industrial and infrastructure projects. The Utkal Alumina project, a joint venture of the Aditya Birla Group and Alcan, a Canadian company, is a case in point. Work on this project had been stalled for almost a decade because the tribals who were getting displaced opposed it. It is high time the government ponders as to why despite best of its efforts to placate the displaced people, its development plans are stiffly rejected.

The four cases cited here are tips of the iceberg. People's protests against displacement are widespread all over the state and gaining momentum against the private development projects. The displaced people of the Mahanadi Coalfields Limited (MCL), a subsidiary of the Coal India have been strongly pressing their demands for rehabilitation, taking resort to agitational methods (*The Times of India*, 22 May, 2006). While the State government considers all these resistance movements as law and order problems, and has geared up the security arrangements in and around the industrial hub, the people's voice has put a big question mark on the fate of many such projects. The questions of rehabilitation, resettlement and improving of the quality of life of the project-affected people, are closely associated with all these protests. As observed here all these protest movements are mainly directed against displacement of people and for protection of the local environment. The people complain against poor rate of compensation and gross inadequacies in matters of resettlement and rehabilitation. There are several instances in the past in which people have suffered heavily due to apathy of the state government wherever major projects were undertaken. As regards employment benefits to the displaced people, it is found that the PSUs have been lagging far behind in the state. It is estimated that the MCL is yet to provide jobs to nearly 2,017 displaced families and the Neelachal Ispat Nigam at Kalinga Nagar to at least 450 in recent years ('PSUs top rehab default roster', *The Times of India*, May 22, 2006).

Future Directions

Industrialization and development projects in Odisha leave behind many unanswered questions. One is that of depletion of mineral deposits. The state has 5,428 million tonnes of iron ore deposits of which the government can recommend mining lease for 1,529 million tonnes. The rest has already been leased out to various companies. If all the 43 projects start working, they will need 2,785 million tonnes of ore creating a shortfall of 1,256 million tonnes. The mega POSCO plant alone needs 600 million tonnes. The situation would be still more complicated if Arcelor-Mittal goes ahead with its 12 million-tonne-project, its requirement being 600 million tonnes of raw material. The picture seems to equally gloomy with regard to alumina projects (Shrivastava, 2007).

Another question relates to scarcity of water for the purpose of irrigation. Widespread mining activities are heavily dependent upon the natural sources of water. The mining activities in Koraput, Keonjhar, and Sundargarh districts are severely affecting the water resources of the region. The mega industrial units would divert water from the paddy fields of the state. The cultivators of Sambalpur have been raising their voice against supply of water from the reservoir of the Hirakud Dam to the industrial units located in and around Sambalpur and Jharsuguda. Unless proper care is taken, industrialization may have a great adverse impact on agriculture and ruin the rural economy.

A World Bank report on Odisha notes that 43 per cent of the rural population in Odisha is very poor due to fall in agricultural production which has an adverse impact on rural economy. The report makes it clear that there is a declining trend in agricultural production. While the growth rate during 1980 to 1992 was 3.2 per cent, it came down to 2.4 per cent during 1992 to 2003 (India Development Policy Review, 2006). Thirdly, it may be pointed out that the average life span of a plant is 35 to 50 years. What would happen to the state after all the mineral deposits are used up barely after 50 years? Widespread mining operations are sure to result in destruction of forests and green belts. The devoured land mass may not be suitable for agriculture nor for any productive purpose.

Development through foreign investors inducing displacement of local people may be an issue with which people may not reconcile easily. A Conference on Development Induced Displacement and Impoverishment, held in Oxford, UK, in January 1995 suggested priority strategies the national (state) governments could adopt to improve matters relating to development induced displacement (Downing, 1998):

- finding alternatives to the sorts of projects that give rise to displacement. If they are unavoidable, action should be taken to ensure that the displaced people at least regain their earlier living standards and feel some benefit from the project. They should be fully compensated for their loss and be helped to re-establish their lives and communities,
- training those responsible for managing resettlement to avoid socially counter-productive wrong turns,
- building institutional capacity among resettled groups as well as in prospective host population,
- keeping resettled, as well as host communities, informed about decisions likely to affect their future and providing fair opportunities for both to participate in decision making processes, and
- incorporating emerging international standards on resettlement into their own policy and law so that the social, economic and cultural rights of displaced people are guaranteed.

Therefore, the R&R policy needs to be formulated in such a way that the people feel that they are not the losers. The offer needs to be made attractive and make the people willing to accept it. Attachment to one's home, community assets and local resources form a natural bond among the people in a locality. It is a social reality closely knit with the psycho-social behaviour of the individual and community developed over a period of time. In all matters of development induced mass displacement government intervention is desirable with utmost care.

Conclusion

The development policies of the Odisha government have certainly not addressed the larger interests of the people. The current investment climate of Odisha ensures economic benefits to the powerful industrial lobbies and poverty and alienation for the masses, so much so that the new R&R policy of the government too has not been able to garner people's support for big mining and industrial projects. Rehabilitation and compensation packages are only the mechanisms of curing the disease. The government is yet to put its head into the causes and preventive measures related to large-scale displacements in Odisha which have adversely affected the marginalized sections, causing physical displacement, economic deprivation and destruction of socio-cultural fabric of the people, widening the already prevalent regional and social disparities. Not only are the people displaced and their livelihood affected, the mining and industrial activities pollute their air, water and soil, spelling disaster for the project-affected people. The process of development in Odisha has been reduced to manufacturing and service sectors, agriculture rarely making it to the realms of policy making, although 85 per cent population of the state is rural, dependent upon cultivated land for its immediate livelihood. Economically, the current development policy of Odisha stands unviable in long run, with a majority of projects getting stalled in legal matters relating to R&R or facing stiff resistance from the local people; socially they have not done justice—forcefully evacuating people from their traditional lands and disturbing their family and kinship ties and so on; environmentally, they are polluting the common property resources of the tribal people with their erroneous waste disposal methods. Such a mess leaves a big question to ponder upon: development for whom?

REFERENCES

Bhargava, S.N. (2005). *Panel assessing Odisha mining impacts*. Retrieved on 04 January, 2010, http://www.indiatogether.org/direct/2005/cdr-000080.html.

Bikalpa Bichar, Bargarh, No. 25. May 2004.

Cernia, M.M. (2000). Risks, safeguards and reconstruction: a model for population displacement and resettlement. *Economic and Political Weekly*, October 7, 2000, 25.

Downing, T.E. (1998). Rights for the world's evicted. Are development projects harming people they're meant to help? *Findings of the Conference on Development Induced Displacement and Impoverishment*, 3-7 January 1995, Oxford, UK. http://www.eldis.org/id21ext/10cmw2.html, Retrieved on 4th January, 2010.

Goodland, R.J.A. (2007). India: Odisha, Kashipur: Utkal bauxite and alumina projects: human rights and environmental impacts. *Library of Congress Cataloguing*, Washington D.C.: Publication Data, p. 138.

Human Development Report of Odisha (2004). *Planning and Coordination Department*, Bhubaneswar: Government of Odisha.

India Development Policy Review (2006), Retrieved on 27th December, 2009 http://siteresources.worldbank.org/SOUTHASIAEXT/Resources/DPRGraphs.pdf.

India: Transition in Odisha (2006), Retrieved on 27th December 2009 www.worldbank.org.in/./0,contentMDK:21877366~pagePK:1497618~piPK:217854~theSitePK:295584,00.html,.

Jena, M. (2006). *Odisha's Draft Resettlement Policy is Promising*. http://www. indiatogether.org/2006/feb/rlf-resettle.htm, Feb. 2, Retrieved on 27 December, 2009.

Kohli, K. (2005). *Odisha's aluminium mining costs are steep*. http://www.india together.org/2005/mar/env-alumina.htm, Retrieved on 04 January, 2010.

Kothari, S. (1995). Development, displacement and official policies: a critical review. *Lokayan Bulletin*, March-April, No. 11.5, New Delhi.

Manshi Asher, SEZs: Stirring up a storm along the Indian coast, *Info Change News and Features*, July 2007.

Mohanty, D. (2010). Odisha renews talks to enlist locals, support for Posco. *The Indian Express*, 19 January.

Odisha's Draft Rehabilitation and Resettlement Policy (2006). *Down to Earth*, June 30.

Odisha Social Infrastrucure, Retrieved on 27 December 2009, http://www.teamOdisha.org/social-infrastructure.asp

Ota, A.B. (2001). Reconstructing livelihood of the displaced families in development projects: causes of failure and room for reconstruction. Paper presented at the Conference on Livelihoods and Poverty Reduction: Lessons from Eastern India, 25-27 September 2001.

Parker, Priya and Vanka, Sarita (2008). *New rules for seizing land.* Retrieved on 04 January 2010. http://www.indiatogether.org/2008/may/law-land.htm.

Patro, S.K. (2009). Odisha poor in human development index. http://www.worldbank.org.in/WBSITE/EXTERNAL/COUNTRIES/SOUTHASIAEXT/INDIAEXTN/0,,contentMDK:20952860%7EpagePK:141137%7EpiPK:141127%7EtheSitePK:295584,00.html - Human_Development_Indicatorshttp://www.expressbuzz.com/edition/story.aspx?Title=Odisha+poor+in+human+development+index&artid=|1dW04jWBXA October 12th, Retrieved on 27th December 2009.

Prajatantra Saptahiki, Cuttack, January 15-21, 2006.

Ravishankar V.J. (2008). Transition in Odisha: A glass half full or half empty? Opinion Piece Special to *The Telegraph and Samaj*, August 14.

Sahoo, S. (2005). Tribal Displacement and Human Rights Violations in Odisha. *Social Action: A Quarterly Review of Social Trends*, April-June, 55(2).

Sengupta, J. (2001). State, market and democracy in the 1990s: liberalization and the politics of Oriya identity. In Jayal, Niraja Gopal and Sudha Pai, *Democratic Governance in India: Challenges of Poverty, Development and Identity*, New Delhi: Sage Publication.

Shrivastava, Aseem (2007). Industrialisation or environmental colonialism? *Info Change News and Features*, March.

The Sambad, Bhubaneswar, 17 May, 2006.

The Sambad, Bhubaneswar, 30 June, 2006.

The Sambad, Bhubaneswar, 22 July, 2006.

The Times of India, PSUs top rehab default roster, Bhubaneswar, 22 May, 2006.

The Times of India, Bhubaneswar, 22 May, 2006.

The Times of India, Bhubaneswar, 24 May, 2006.

The Times of India, Bhubaneswar, 28 May, 2006.

The Times of India, Bhubaneswar, 16 June 2006.

The Times of India, Bhubaneswar, 21 June, 2006.

The Sambad, Bhubaneswar, 22 and 24 June, 2006.

The Times of India, Bhubaneswar, 25 June, 2006.

Part IV
Environmental Issues

18
Earth Democracy, Living Democracy: Rebuilding Security in an Age of Insecurity

Vandana Shiva

Humanity seems to be in a free fall towards disaster. The ecological fabric of our existence is being torn apart, as the violence of corporate globalisation combines with the violence of war on a planetary scale. Alternatives beyond war, non-sustainability and social and economic injustice are becoming a survival imperative. These alternatives need to combine our making peace with the planet and our making peace among people from diverse cultures. One is not possible without the other because the roots of terrorism, violence and war lie in environmental and economic exclusion and the insecurity it generates. People's security does not lie in larger military budgets, bigger bombs and stronger police states. It lies in ecological security in economic security, in cultural and political security. Rebuilding these multiple securities can recreate peace, justice and sustainability.

Economic Globalisation and the Threat to Ecological Security

Climate change, species extinction, the destruction of water resources are threatening the very basis of our existence. Even as we attempt to grapple with the pollution and externalities of the first two industrial revolutions based on fossil fuels and chemicals, in the form of climate change and toxic pollution, new threats of genetic pollution are being unleashed by the third industrial revolution in the form of biotechnology and genetic engineering. The fundamental principles of precaution and polluter pays are being given a go by to rush untested, unnecessary, potentially hazardous GMOs to the market place for release in ecosystems.

It is the poor whose ecological security is most threatened by corporate led globalisation. While an attempt is being made to project resource conservation as anti-poor, it is the poor who pay the highest economic costs for resource destruction and resource privatisation.

Globalisation at the most fundamental level, is rewriting our relationship with the earth and her species, alienating land, water and biodiversity from local communities, transforming commons into commodities to be traded freely for profits—with total indifference to the ethical, ecological and economic impacts of this commodification of life.

Globalisation is a break from all earlier stages of human relationships with the earth and her resources.

1. It is based on enclosures of the remaining ecological commons—biodiversity, water and air, and the destruction of local economies on which people's livelihoods and economic security depends.
2. The commodification of water and biodiversity is ensured through new property rights built into trade agreements like the WTO which are transforming people's resources into corporate monopolies, viz. TRIPs and trade in environmental goods and services. The corporatisation of food and agriculture is ensured through the W.T.O. Agreement on Agriculture and Structural Adjustment Programmes of the World Bank.

   In Odisha, one of the poorest Indian states, World Bank and DFID are using northern tax payer money to privatise irrigation water which now costs 10 times more and destroying agriculture, the only livelihood of the poor.

   In Delhi, a $ 2.5 million World Bank loan for water privatisation has largely financed the consultancy fees for an international accounting firm, Price Waterhouse.

   In Bolivia, communities resisting water privatisation have been shot at by police. In South Africa, protestors have been arrested.

3. The transformation of commons and basic needs to commodities is ensured through shifts in governance with decisions moving from communities and countries to global institutions, and rights moving from people to corporations through increasingly centralised and unaccountable states acting on the principle of eminent domain—the absolute sovereignty of the ruler.

This corporate takeover of vial resources and vital economic sectors has in turn led to anti-democratic formations and constellations. Instead of acting on the public trust doctrine and principles of democratic accountability and subsidiarity, globalisation led to governments usurping power from parliament, regional and local governments, and local communities.

For example, the TRIPs agreement was shaped by corporations and allows five corporations to control seeds, plants, food and health. It forces central government to hijack the rights to biodiversity and knowledge from communities and assign them as exclusive, monopolistic rights to corporations who see a 1 trillion $ market in seeds. Centralised decision making is being used by Monsanto to push its GM crops and seeds, bypassing democratic decision making and all concerns for Biosafety.

The Agreement on Agriculture was shaped by global grain giants like Cargill and is based on taking decisions away from farming communities and regional governments. Governments are being forced to sacrifice their farmers and food security for export led agriculture and dumping of subsidised products on others' markets, creating a regime of mutually assured destruction, instead of systems of food security.

The General Agreement on Trade in Services (GATS) takes decisions and ownership over water from the local and public domain to the privatised, global domain. Water trade is assessed at a $ 1 trillion economy by the World Bank. Water privatisation is carried out through water policies that make water the property of the State. This has happened in the case of India. Land is privatised by using the government's right to acquire land often in violation of the constitutional and community rights as in the case of Bastar.

This undemocratic process of privatisation and deregulation has led to increased political bankruptcy and corruption and economic bankruptcy. It has also led to concentration of power and ownership.

Globalisation has relocated sovereignty from people to corporations, through centralising, militarising States. Rights of people are being appropriated by States to carve out monopoly rights of corporations over our land, our water, our biodiversity, our air. States acting on the principle of eminent domain or absolute sovereignty of the State are undermining people's sovereign rights and their role as trustees of people's resources on the public trust doctrine. State sovereignty, by itself, is therefore not enough to generate countervailing forces and processes to corporate globalisation.

The reinvention of sovereignty has to be based on the reinvention of the state so that the state is made accountable to the people. Sovereignty cannot reside only in centralised sate structures, nor does it disappear when the protective functions of the state with respect to its people start to wither away. The new partnership of national sovereignty needs empowered communities which assign functions to the state for their protection. Communities defending themselves always demand such duties and obligations from state structures. On the other hand, TNCs and international agencies promote the separation of the community interests from state interests and the fragmentation and divisiveness of communities.

Corporate Globalisation and the Growth of Economic Insecurity

Both ecology and economics have emerged from the same roots—"oikos"—the Greek word for household.

As long as economics was focused on the household, it recognised and respected its basis in natural resources, the limits of ecological renewal. It was focused on providing for basic human needs within these limits. Economics as based on the household was women centred.

Today, economics is separated from, and opposed to both ecological processes and basic needs. While the destruction of nature has been justified on grounds of improving human welfare, for the majority of people poverty and dispossession has increased. While being non-sustainable it is also economically unjust. While being promoted as "economic development", it is leading to underdevelopment, while projecting growth, it is causing life threatening destruction.

The dominant model of "economic development" has in fact become anti-life.

And it is this non-sustainability model, which is being globalised. Corporate globalisation is creating economic insecurity worldwide as livelihoods are destroyed and basic needs are denied to millions.

Globalisation of agriculture has allowed dumping of subsidized artificially cheap products which are robbing small farmers of livelihoods, markets and incomes. Giant corporations controlling production and distribution, are using subsidies to push up costs of production and push down prices farmers receive for their product.

It is this negative economics that is being globalised by the rules of W.T.O. and the conditionalities of structural adjustment.

The dominant economy is negative both in ecological terms and in financial terms when viewed from the perspective of the poor and women.

It is ecologically negative in a dual sense—it is destructive of the resource base of our survival, and it consumes more resources than it produces. The destruction and resource waste is not taken into account, and an essentially unproductive, non-sustainable economy is pushed on grounds of improving welfare.

A usual argument used in promoting industrial agriculture like the Green Revolution earlier and genetic engineering in agriculture now is that only industrial agriculture and industrial breeding can keep up with increased food productivity for feeding a growing population. However, increased mouths to feed implies more efficient resource use so that the same resources can feed more people. Since resources, not labour, are the limiting factor in food production, it is resource productivity, not labour productivity which is the relevant measure. A sixty fold decrease of food producing capacity in the context of resource use is not an efficient strategy for using limited land, water and biodiversity to feed the world.

Further, since food security is based on food entitlements, and entitlements in peasant societies are based on livelihoods and work, increase in labour productivity based on displacement farmers cannot reduce hunger. It decreases food entitlements by destruction of livelihoods.

A myth promoted by the one-dimensional monoculture paradigm is that bio-diversity reduces yields and productivity, and monocultures increase yields and productivity. However, since yields and productivity are theoretically constructed terms, they change according to the context. Yields usually refers to production per unit area of a single crop. Planting only one crop in the entire field as a monoculture will of course increase its yield. Planting multiple crops in a mixture will have low yields of individual crops, but will have high total output of food.

The negative economy of agriculture destroys biodiversity, soil and water. It also destroys rural livelihoods and creates hunger and starvation.

When farmers spend more on production costs than they earn from selling their produce, the economics is negative. It is artificially kept afloat by subsidies. Rich countries give $370 billion as subsidies. Corporate farmers of California receive more than $500,000 in subsidies. The new farm bill increases U.S. subsidies to agriculture by $73.5 billion to $180 billion over the next 6 years. This will artificially bring down prices worldwide, destroying millions of peasant livelihoods. It is not efficiency but subsidies which are lowering prices of farm commodities.

It is the poor farmers and small farmers who are destroyed by the globalisation of a negative economy. Thousands of Indian peasants have committed suicide because of increasing costs of seeds and chemicals. The poor consumers also suffer, starvation deaths have returned to India for the first time since independence. As 500 million Indians face starvation, 60 million tonnes are rotting in storage because poor people cannot afford to buy food because of a four fold increase in prices because of removal of food subsidies. Food denied to people is now being exported with subsidies. While the poor in India were paying Rs. 7,000 per tonne of wheat and of Rs. 11,300 per tonne of rice because of withdrawal of subsidies, exporters were getting food at a subsidized rates of Rs. 4,300 per tonne and Rs. 5,650 per tonne respectively.

Negative economics push people out of production. They create economic displacement and disposable people, people whose very right to survival is denied.

A negative economy is based on pseudo productivity and pseudo surpluses. It is based on an illusion of growth, which hides the poverty and destitution it creates. The growth is of course real for the Monsantos—the seed, chemical and biotechnology industry which have expanding markets even as farmers take their lives, sell their kidneys, or their land. The growth is real for the Cargills who buy cheap from farmers and sell processed food at high cost to consumers. The trade rules are aimed at benefiting the Monsantos and Cargills. An economics for creating scarcity and poverty is transformed into an economics for creating surpluses and growth by a jugglery of statistics. Food output is replaced by "high yield" categories, food security in the household and community is replaced by agricultural trade figures. Yields go up while output goes down, trade goes up while consumption by the poor goes down. Pseudo surpluses hide starvation, superprofits hide a crisis of distribution.

The economy that generates superprofits for the global corporations also generates ecological poverty for the planet and economic poverty for the poor. It is an economy that produces insecurity for all. And it is this economics of insecurity that is creating politics and culture of insecurity. Globalisation is fuelling extremism, terrorism, xenophobia and fundamentalism and deepens destruction of nature, people's rights and democracy.

Globalisation, the Decay of Democracy and the Rise of Terrorism, Fundamentalism and Police States

An Economics of Exclusion Creates a Political Culture of Exclusion

Non-democratic economic systems that centralize control over decision making and resources and displace people from productive employment and livelihoods create a culture of insecurity. Fundamentalist, extremist and excluding politics emerges to fill the void left by the collapse economic democracy and the growth of economic security. Instead of identifying the root causes of economic insecurity in the global economy, every policy decision is translated into the politics of "we" and "they." "We" have been unjustly treated, while "they" have gained privileges. Globalisation is creating a global culture of insecurity and insecure people can be made afraid of the "other" more easily.

Destruction of resource rights and erosion of democratic control of natural resources, the economy, and means of production undermine cultural identity. With identity no longer coming from the positive experience of being a farmer, a craftsperson, a teacher, or a nurse or from the security of national sovereignty, culture is reduced to a negative shell where one identity is in competition with the "other" over scarce resources that define economic and political power. Positive identities mutate into negative identities—"I" am not the "other" and annihilation and extinction of the other is necessary for my security and survival. Politics emptied of an economics of welfare becomes a negative political and hatred and division on basis of race, caste and religion.

The rise of Jean Marie LePen in France is an illustration of how the culture of insecurity created by globalisation fuels xenophobia. LePen has a slogan of "French First" and he sees insecurity as the most severe problem, but he does not identify the erosion of French security with globalisation as Jose Bove does. Instead his attempt is to "link insecurity and immigration." It is not the removal of borders for foreign investment, foreign corporations which he sees as creating threats to French identity, culture and jobs. His exclusive focus is on immigrants. If barriers against people are not erected, according to LePen France will be submerged, it will disappear.

It is the W.T.O. and World Bank that create the LePens in France, Fortuyns in Netherlands, Haiders in Austria. and the Narendra Modis in India. There is in fact a comfortable partnership between the supporters of globalisation and the promoters of the politics of hate. By keeping people's energies diverted from issues of basic needs and economic democracy, xenophobia and communalism help keep an unjust and inequitable economic order in place. During the worst violence since India's Independence, both GM crops and a new Patent Act were cleared. Corporations gain from globalisation and its fall out in terms of social disintegration.

I believe a new politics of hatred and intolerance is arising from the growing economic insecurity and a sense of shrinking space for survival. Simultaneously, representative democracy is loosing its base in economic democracy as decisions move out of countries into global institutions like World Bank, IMF and WTO and boardrooms of global corporations. Politicians robbed of mandates to ensure people's basic needs are provided for are moving to political agendas of exclusion to garner votes and stay in power. People's economic security and democracy are both being systematically killed by the forces unleashed by globalisation.

Democracy emptied of economic freedom and ecological freedom becomes a potent breeding ground for fundamentalism and terrorism.

Over the past two decades, I have witnessed conflicts over development and conflicts over natural resources mutate into communal conflicts, culminating in extremism and terrorism. My book *Violence of the Green Revolution* was an attempt to understand the ecology of terrorism. "The Harvest of Rage" drew similar links between the Oklahoma bombing and the U.S. farm crisis, and how it creates fertile ground for extremist fundamentalist ideologies and actions.

Our international solidarities, our national sovereignties are evaporating and being replaced by the narrowest conceivable sense of who we are.

Centralized economic systems also erode the democratic base of politics. In a democracy, the economic agenda is the political agenda. When the former is hijacked by the World Bank, the IMF, or the WTO, democracy is decimated. The only cards left in the hands of politicians eager to garner votes are those of race, religion, and ethnicity, which subsequently give rise to fundamentalism. And fundamentalism effectively fills the vacuum left by a decaying democracy. Economic globalisation is fuelling economic insecurity, eroding cultural diversity and identity, and assaulting the political freedoms of citizens. It is providing fertile ground for the cultivation of fundamentalism and terrorism. Instead of integrating people, corporate globalization is tearing apart communities.

Globalisation is creating negative economies, negative politics and negative identities. Rejuvenating, deepening, widening democracy has become a survival imperative for the human species. Reinvention of freedom in our times requires freedom from fear, freedom from violence, freedom from denial of basic needs, freedom from non-sustainable and unethical production, trade and consumption patterns. It requires a shift from negative economies, politics and culture to positive life enhancing positive economies, positive politics and positive identities.

Instead of addressing the root causes of terrorism and fundamentalism in the growth economic insecurity and collapse of economic democracy by ensuring people's needs are met and their livelihoods protected, states across the world are equipping themselves with laws to shut down democracy and freedom in the name of fighting terror.

Whether it is the Patriot Act in the U.S., Prevention of Terrorism Act in India (POTA), or the Anti-Terrorism, Crime and Society Act (ATCSA) of the UK, new laws created after September 11, 2001 are not just laws against terrorists—they are also laws against citizens democratic defence of their fundamental freedoms that are being trampled upon by the forces of globalisation.

Fear and violence are currently the dominant forms of human expression. Rule through fear and violence are becoming the dominant mechanism for governance. In another period, it would have been described as the rise of facism and totalitarianism with the totalitarianism of corporate control over markets combining with totalitarianism of militarised states, taking away from people their fundamental rights and freedoms.

Earth Democracy

From Life Destroying to Life Preserving Cultures

Why are we as a species destroying the very basis of our survival and existence. Why has insecurity been the result of every attempt to build security? How can we as members of the earth community reinvent security to ensure the survival of all species and the survival and future of diverse cultures? How do we make a shift from life annihilating tendencies to life preserving processes? How do we from the ruins of the dominant of culture of death and destruction build cultures that sustain and celebrate life?

A major source of threat to our ecological survival and growing culture of insecurity is the rule and domination by abstract constructions, and a denial of the real, the concrete, the life giving. This "escape" from our biological species being is historically rooted in the Cartesian/Newtonian divide between mind and matter, between objective and subjective, between primary and secondary qualities.

In the reductionist mechanistic worldview, the constructed has primacy and is more real than reality itself. This domination of abstract construction over lived reality is based on division and separation, decontextualisation and disembeddedness and blindness to the disintegration that result from separation of that which is interconnected. Atomisation has become the dominant thought and dominant practice. The false assumption of separability based on abstraction allows the emergence of life threatening technologies like genetic engineering, and terminator seeds. It allows the dominance of corporations, a legal fiction, over the lives of real people and citizens. It allows the destruction of real wealth is soil, water, biodiversity, good food, rich cultures for the creation of fictitious wealth of more than 3 trillion dollars a day which is nearly a hundred fold more than the real goods and services it can command.

Closely linked to the rule and reification of abstraction are dominant traits which threaten life in its diversity self-organisation and self-renewal—the monoculture of the mind and the law of the excluded middle.

The monoculture of the mind pushes to oblivion and extinction biological and cultural diversity which are the very preconditions of ecological and cultural security.

The law of the excluded middle becomes the basis of the legitimisation of exclusion of ecocide and genocide. It shuts out in between species in nature and culture. It denies the existence of biodiversity on farms, and food from forests.

Once we break free of the mental prison of separation and exclusion and see the world in its interconnectedness and non-separability, new alternatives emerge. Despair turns to hope. Violence gives way to non-violence. Scarcity transforms into abundance and insecurity to security.

Globalisation has shrunk our humanity. It has replaced the diverse cultures of conservation and compassion with a global monoculture of consumerism. It has pitted community against community, neighbour against neighbour, faith against faith. Globalisation, a paradigm based on competition and war, has created a world of many wars and multiple conflicts.

We need once more to feel at home on the earth and with each other. We need a new paradigm to respond to the fragmentation caused by various forms of fundamentalism. We need a new movement which allows us to move from the dominant and pervasive culture of violence, destruction and death to a culture of non-violence, creative peace and life. That is why in India we started the Earth democracy movement.

Earth democracy is based on creating living economies that protect life on earth and provide basic needs and economic security to all. It is based on living democracy which is inclusive.

The Earth Democracy Movement is a commitment to go beyond the triple crisis of economic injustice and inequality, ecological non-sustainablility and the decay of democracy and the rise of terrorism.

Earth democracy provides an alternative worldview in which humans are embedded in the Earth Family, we are connected to each other through love, compassion, not hatred and violence and ecological responsibility and economic justice replaces greed, consumerism and competition as objectives of human life.

In earth democracy economy, politics and society move from negative systems that benefit a few in the short run to positive systems which ensure that the fundamental right to life of all species and all peoples are defended and guaranteed. The maintenance of life in its diversity and integrity rather than limitless extraction of profits through monopolies over biodiversity and water and all vital resources is the basis of relationships in Earth Democracy.

Earth Democracy transforms our mind and our actions, and liberates us from patterns of thought and paradigms which have pushed us to our contemporary predicament. It helps address the common roots of problems that are defined separately as economic, ecological and political. Exclusivist monocultural modes of dominant thinking are leading to:

1. the destruction of resources and the creation of monopolies over land, biodiversity, water and food;
2. the deepening of poverty and the exclusion of millions from livelihoods and economic security; and
3. the destruction of democracy, peace and cultural diversity.

Earth democracy enables us to make mind shifts that are conducive to meeting our needs without destroying other species and improving human welfare while ensuring welfare of all beings. In India, we pray: "Let all beings be happy".

Earth democracy embodies principles that enable us to transcend the polarisation, divisions and exclusions that are putting the economy against ecology, development against environment, people against the planet and against one another in a new culture of hate. It is symbolised in Hindus visiting a Muslim shrine on their pilgrimage to Sabarimala, and Wasiffudin Dagar a Muslim singer praying to Durga and Shiva, keeping the ancient Dhrupad tradition alive for a composite culture. It is symbolised in farms rejuvenating biodiversity, and species acting in mutuality to benefit one another.

Earth democracy recontextualises humans as one member of the Earth family (*Vasudhaiva Kutumbakam*), and diverse cultures in the mosaic of cultural diversity which enriches our lives.

Re-embedding humans in the ecological matrix of biological and cultural diversity reopens spaces for sustainability, justice and peace by reorganising relationships restructuring constellations of power and revitalising freedom and democracy.

We are being ruled by terror and greed, fear and insecurity. As we face the double closure of spaces by corporate globalisation and militarised police states, by economic facism aided by political facism, our challenge is to reclaim our freedoms and the freedoms of our fellow beings. Reclaiming and recreating the indivisible freedom of all species is the aim of the Earth Democracy Movement. The Earth democracy movement embodies two indivisibilities and continuums. The first is the continuum of freedom for all lives on earth, and all humans without discrimination on the basis of gender, race, religion, class and species. The second is the continuum between and indivisibility of justice, peace and sustainability—without sustainability and justified share of the earth's bounties there is no justice, and without justice there can be no peace.

Corporate globalisation ruptures these continuities. It establishes corporate rule through a divide and rule policy, and creates competition and conflict between different species and peoples and between different aims. It transforms diversity and multiplicity into oppositional differences both by breeding fundamentalisms through spreading insecurity and then using these fundamentalisms to shift humanities focus and preoccupation from sustainability and justice and peace to ethnic and religious conflict and violence.

Earth democracy is the democracy of all lives, not just humans, definitely not just humans privileged through class, race, gender and religion.

Since other species do not vote, they cannot lobby, and they have no purchasing power in the market place, earth democracy creates an obligation on us as humans to take their well-being into account. As His Holiness the Dalai Lama said on his 60th birthday,

> All beings have a right to well-being and happiness. We have a duty to ensure their well-being.

This creates human responsibility as trustees and stewards, instead of the dominant notion of mastery, control and ownership.

Earth democracy privileges diversity in nature and society in form and in function. When the intrinsic worth and value of every life form and even human is recognised, biological diversity and cultural diversity flourish. Monocultures result from exclusion and dominance of species, one variety, one race, one religion. Monocultures are an indication of coercion and loss of freedom. Freedom implies diversity. Diversity signifies freedom.

Earth democracy also nourishes diversity by going beyond the logic of exclusion of apartheid, of "us" and "them" of either/or, of the law of the excluded middle. It is in the included middle that diversity and creativity flourishes in nature and in culture.

The law of the included middle also implies multifunctionality, the logic of "and", of inclusion. It transcends the false polarisation of wild *versus* cultivated, nature *versus* culture, or even the false clash of cultures. It allows for the forest farm and the farmed forest, it recognises that biodiversity can be preserved and also support human needs.

Through diversity replacing monocultures, and multi-dimensionality replacing one dimensional systems, the negative economics of scarcity creation can be replaced by the positive economics of mutually shared abundance and guaranteed provision of basic needs and access to vital resources.

Gandhi's concept of Swadeshi—of economic freedom and economic democracy—is at the core of Earth democracy.

Earth Democracy puts responsibility and duties at the core of our relationships, with rights flowing from responsibility instead of dominant paradigm of rights without responsibility and responsibility without rights. The separation of rights and responsibility is at the root of ecological devastation and gender, class inequality. Corporations that earn profits from chemical industry, or from genetic pollution resulting from GM crops do not have to bear the burden of that pollution. The social and ecological costs are externalised and born by others who are excluded from decisions and from benefits.

Earth democracy is based on those who pay the price having a say those who carry responsibility having the rights. This creates the need for direct or basic democracy. On the one hand this implies decisions moving downwards, from global institutions and centralised governments to local communities. On the other hand it implies a shift in our interpretation of sovereignty.

Rights to natural resources for maintaining life are natural rights. They are not given by States, nor can they be extinguished by States, the W.T.O., or by corporations, even though under globalisation, attempts are being made to alienate people's rights to vital resources of land, water and biodiversity.

Earth democracy shifts the constellation of power from corporations to people, and rebalances the role and functions of the state which is increasingly taking on undemocratic forms.

In the dominant constellation, non-sustainable and unjust economics are being kept in place by undemocratic governments. The erosion of economic and political democracy is giving rise to fundamentalist cultures: Old divisions between have and have-nots are mutating into new polarisations of the "lives and live nots", as millions are robbed of their very right to stay alive. The economic "apartheid" between the "lives and live nots" is being translated into the genocidal tendencies of ethnic cleansing.

Earth democracy creates sustainable and just economics, governed by democratic and accountable systems. Cultural and biological diversity is protected.

Earth democracy is about life, and natural rights to the conditions of staying alive. It is everyday life and decisions and freedoms related to everyday living—the food we eat, the clothes we wear, the water we drink. It is not just about elections and casting votes once in 3 or 4 or 5 years. It is a permanently vibrant democracy. It combines economic democracy with political democracy and ecological democracy. It creates positive economies, positive politics, positive identities. It creates security, and hence the conditions of peace.

Earth democracy is not dead, it is alive. Under globalisation, democracy even of the shallow representative kind is dying. Governments everywhere are betraying the mandates that brought them to power. They are centralising authority and power, both by subverting democratic structures of constitutions and by promulgating ordinances that stifle civil liberties. The September 11 tragedy has become a convenient excuse for anti-people legislation worldwide. Politicians everywhere are turning to xenophobic and fundamentalist agenda to get votes in a period when economic agenda have been taken away from national levels and are being set by World Bank, I.M.F., W.T.O. and global corporations, and parties loose elections because of the economic betrayal of their people.

The Earth democracy movement is about living rather that dead democracy. Democracy is dead when governments no longer reflect the will of the people but are reduced to anti-democratic unaccountable instruments of corporate rule under the constellation of corporate globalisation as the Enron and Chiquita case make so evident. Corporate globalisation is centred on corporate profits.

Earth democracy is based on maintaining life on earth and freedom for all species and people.

Corporate globalisation is based on privileging the logic of competition. Earth democracy is based on cooperation.

Corporate globalisation operates to create rules for the global, national and local markets which privilege global corporations and threaten diverse species, the livelihoods of the poor and small, local producers and businesses.

Earth democracy operates according to the ecological laws of nature, and limits commercial activity to prevent harm to other species and to people.

Corporate globalisation creates poverty and threatens peace. Earth democracy creates an inclusive prosperity and through it creates peace.

Corporate globalisation destroys cultural and biological diversity. Earth democracy is the protection of diversity in nature and society.

Corporate globalisation is exercised through centralising, destructive power.

Earth democracy is exercised through decentralised power and peaceful coexistence.

Corporate globalisation globalises greed and consumerism. Earth democracy globalises compassion, caring and sharing.

This shift is also an ecological imperative. As members of the earth family, *Vasudhaiva Kutumbakam*, we have a share in the earth's resources. Rights to natural resources for needs of sustenance are natural rights. They are not given or assigned. They are recognised or ignored. The eminent domain principle inevitably leads to the situation of "all for some"—corporate monopolies over biodiversity through patents, corporate monopolies on water through privatisation and corporate monopolies over food through free trade.

The most basic right we have as a species is survival, the right to life. Survival requires guaranteed access to resources. Commons provide that guarantee. Privatisation and enclosures destroy it. Localisation is necessary for recovery of the commons. And Earth democracy is the movement to relocate our minds, our production systems and consumption patterns from the poverty creating global markets to the sustainability and sharing of the earth community. This shift from global markets to earth citizenship is a shift of focus from globalisation to localisation of power from corporations to citizens.

Resisting the privatisation of biodiversity through IPRs, the privatisation of water through structural adjustment programmes and trade in services and corporatisation of food and agriculture needs a combination of strategies at local, national and global levels and a shift in the constellation of the power and rights of states, corporations and citizens.

Privatisation based on exclusive rights of corporations to vital resources like biodiversity and water is an enclosure of the commons. Reversal of this enclosures requires a recovery of the commons through a combination of actions at their levels.

1. At the local level, the recovery of the commons needs the strengthening and assertion of local community rights and people's sovereign and natural rights to vital resources such as water and biodiversity.
2. At the national level, the recovery of the commons requires a recognition of local community rights and a reinvention of sovereignty, and a shift from the state functioning on the doctrine of eminent domain to states functioning on the public trust doctrine.

3. At the global level, the recovery of the commons requires a movement of people to keep water and biodiversity beyond monopoly, ownership and commodification, which in turn prevents patenting of life forms and privatisation of water. These principles need to be enshrined in international law and policy. They become the source of popular democratic pressure reform W.T.O. and the World Bank—to implement the review of TRIPs in which countries of the South are calling for an exclusion of life forms from patentability and to review structural adjustment programmes which imposing export-led agricultural strategies and policies of water privatisation.

The citizen treaties to defend water and biodiversity as commons will have democratic power and substance to the extent they reflect the recognition and strengthen local community rights at the global level. Global commons not built from or based on local commons is an ecologically and democratically fraudulent category. The global commons is merely a recognition and reinforcement for local community rights. It is not the level at which rights are exercised or assigned.

	Commons	*Privatisation regimes*
Global	Treaties on commons reflect commitment to share, recognition of local community rights as highest level of rights to water and biodiversity. Reform of TRIPs and GATS and AoA of W.T.O. to ensure the people's sovereignty over biodiversity, food and water.	TRIPs for privatisation of biodiversity, GATS and SAPs for privatisation of water. Private monopoly rights of corporations as highest level of rights. No recognition of community rights.
National	States function on the doctrine of public trust. Respect and recognise local communities as the highest competent authority, recognise community rights and people's sovereignty.	States function on doctrine of eminent domain. Transfer common property resources to corporations through privatisation laws and policies.
Local	Local communities have sovereign rights conserve, share and sustainably use commons, excluding no one, letting no one exclude. Localisation of economics improves livelihood security and ecological security.	Local communities left politically, ecologically and economically disenfranchised and their access to their own resources block by IPRs and privatisation regimes.

Since sovereignty based on the doctrine of eminent domain is becoming the conduit for global usurpation of people's resources, and undermining of people's sovereign rights, reclaiming the biodiversity and water commons must go hand in hand with reclaiming sovereignty, and redefining a new partnership between people and governments on the basis of subsidiarity and the public trust doctrine.

The reclamation of sovereignty embodies Gandhi's concept of Swaraj or self rule.

	Corporate globalisation	*Earth democracy*
Aim of human life	Profits, Accumulation, Consumerism.	Maintenance of life on earth, now and in the future. Improvement of well being and happiness of all beings.
Axis of global integration	Global Markets	Ecological processes. Our common humanity and planetary consciousness.
What disintegrates as a result	Disintegration of local economies, local cultures, ecosystems, ethical basis of life.	Unjust rule, non-sustainable lifestyles.
Security	Military	Ecological, economic, social and cultural.
Dominant tendencies	Cruelty	Compassion
	Greed	Giving
	Fear of everything alive and autonomous ownership.	Love for everything alive
	Control of everything and everyone.	Belonging, not owning
	Exclusion	Inclusion
	Exploitation	Conservation
	Live and live not	Live and let live
	Monocultures	Diversity
	Monopolies	Sharing
	Separation	Inter-relationship
	Facism of market and fundamentalism	Earth democracy

The economic, ecological and social crises resulting from corporate globalisation are inviting a new way of thinking and being on this planet. A new worldview in which it is not greed but compassion which is globalised, a new consciousness in which we are not reduced to consumers of globally traded commodities if we are privileged and to narrow fragmented one-dimensional identities based on colour, religion or ethnicity, but experience our lives as planetary beings with planetary consciousness, mindful and aware of what our actions, our consumption, cost other humans, other species and future generations and connected to all in a common fabric of life.

Beginning with people's everyday actions Earth Democracy offers a potential for changing the way governments, intergovernmental organisations and corporations operate. It creates a new paradigm for global governance while empowering local communities. It creates the possibility of strengthening ecological security while improving economic security. And on the foundations of ecological and economic security it makes societies immune to the virus of communal hatred and fear.

Earth democracy offers a new way of seeing in which everything is not at war with everything else but through which we can cooperate to create peace, sustainability and justice in our violent and volatile times.

Rio+20: Will green be the colour of life or money?

> "The significant problems we face cannot be solved at the same level of thinking we were at when we created them."
>
> —*Albert Einstein*

In 1992, the citizens and movements of the world gathered in Rio for the Earth Summit.

In 2012, the world community will gather again in Rio. On 24th December, 2009, the General Assembly of the United Nations adopted a Resolution (A/RES/64/236) to hold a conference 20 years after the Earth Summit. Member States agreed that the Rio+20 Summit would focus on—

> "Green Economy within the context of sustainable development and poverty" and "Institutional framework for sustainable development."

But what is the "green economy" and what is the "institutional framework for sustainable development"? If the answers are offered in the old paradigm of market driven solutions which have failed to protect the earth, "green economy" will mean more of the same. It will mean more carbon trading which has failed to reduce emissions. It will mean more commodification of food and water, land and biodiversity, which has failed to reduce hunger and thirst, poverty and ecological degradation and has instead increased it.

If the "institutional framework" creates a World Environment Organisation like a World Trade Organisation based on commodification and trade in nature's gifts, and trade wars as global environment management, we will further impoverish the earth and local communities, and we will further destroy democracy.

If on the other hand the answers are offered in the context of the emerging paradigm of the Rights of Mother Earth, then the green economy is Gaia's economy, and the institutional framework is Earth Democracy, democracy from the bottom up, democracy rooted in the Earth. The world order built on the economic fundamentalism of greed, commodification of all life, and limitless growth and the technological fundamentalism that there is a technological fix for every social and environmental ill is clearly collapsing.

The collapse of Wall Street in September 2008 and the continuing financial crisis, signal the end of the paradigm that put fictitious finance above real wealth created by nature and humans that put profits above people and corporations above citizens. This paradigm can only be kept afloat with limitless bailouts that direct public wealth to private rescue instead of using it to rejuvenate nature and economic livelihoods for people. It can only be kept afloat with increasing violence to the earth and people. It can only be kept alive as an economic dictatorship. This is clear in India's heartland, as the limitless appetitive for steel and aluminium for the global consumer economy, and the limitless appetitive for profits for the steel and aluminium corporations is clashing head on with the rights of the tribals to their land and homes, their forests and rivers, their cultures and ways of life. The tribals are saying a loud and clear "no" to their forced uprooting. The only way to get to the minerals and coal that feed the "limitless growth" model in the face of democratic resistance is the use of militarized violence against the tribals, operation "Green

Hunt" has been launched in the tribal areas of India with precisely this purpose, even though the proclaimed objective is to clear out the "Maoists". Under operation Green Hunt, more than 40,000 armed paramilitary forces have been placed in the tribal areas which are rich in minerals and where tribal unrest is growing. Operation Green Hunt shows clearly that the current economic paradigm can only unfold through increased militarization and the undermining of democratic and human rights.

The technological fundamentalism that has externalized costs, both ecological and social and blinded us to ecological destruction has also reached a dead end. Climate chaos, the externality of technologies based on the use of fossil fuels, is a wakeup call that we cannot continue on the fossil fuel path. The high costs of industrial farming is running up against limits, both in terms of the ecological destruction of the natural capital of soil, water, biodiversity and air, as well as in terms of the creation of malnutrition, with a billion people denied food and another two billion denied health because of obesity, diabetes and other food related diseases.

The green economy agenda for Rio+20 can either deepen the privatization of the earth, and with it the crisis of ecology and poverty or it be used to re-imbed economies in the ecology of the earth.

Green economics needs to be an authentic green. It cannot be the brown of desertification and deforestation. It cannot be the red of violence against nature and people, or the unnecessary conflicts over natural resources—the land and water, seeds and food. As Gandhi had said "the Earth has enough for everyone's needs, but not for some people's greed".

To be Green, economics needs to return to its home to Oikos. Both ecology and economics are derived from "Oikos" which means "home". Ecology is the science of household, economics is supposed to be the management of the household. When economics works against the science of ecology, it results in the mismanagement of the earth, our home. The climate crisis, the water crisis, the biodiversity crisis, the food crisis are different symptoms of this crisis of mismanagement of the earth and her resources.

We mismanage the earth when we do not recognize nature's capital as the real capital and everything else as derived. If we have no land, we have no economy. When we contribute to growth of nature's capital, we build Green Economies. And the richer nature's capital is, the richer human society is.

The Great Reducing: From Earth Community to Humans, from the Human Community to Corporations as Legal Persons

The ecological and economic problems we face are rooted in a series of reductionist steps, which have shrunk our imagination and our identity, our purpose on the earth, and the instruments we use to meet our needs. We are first and foremost earth citizens. And our highest duty is to maintain the living systems of the earth that support our life. Earth citizenship needs earth centred cultures, earth based democracy, and earth centred economies.

However, society and culture have been reduced to economies, and economies has been reduced to market economics, and market economics has been reduced to finance and finance has been reduced to abstract instruments like derivatives, securitization, collateral debt obligation.

Simultaneously, humans as earth citizens, with duties and rights have been replaced by corporations, with no duties to either the earth or society, only limitless rights to exploit both the earth and people. Corporations have been assigned legal personhood, and corporate rights are now extinguishing the rights of the earth, and the rights of people to the earth's gifts and resources corporate rights are premised on maximization of profits.

There are three tools corporations use to maximize profits. The first is the use of technologies that transfer production from local communities to distant corporations, substitute biodiversity with toxic products, and reduce everyone to be consumers of toxic, non-renewable products whose cost is high but price is cheap.

The second instrument is the creation of tools for wealth accumulation. These tools include measuring wealth as capital, ignoring nature's wealth and society's wealth. They also include measuring wealth as growth of GNP and GDP. At the core of the second instrument of profit maximization is the privatization of the earth's gifts—of biodiversity of water, even the atmosphere.

More often than not, the tools of technology and economic tools of wealth appropriation go hand in hand, reinforcing each other. Thus genetic engineering goes hand in hand with patents on life and privatization of biodiversity. Dams for long distance water transfer go hand in hand with water privatization.

Earth Rights: The New Unfolding

We need a new paradigm for living on the Earth because the old one is clearly not working. An alternative is now a survival imperative for the human species. And the alternative that is needed is not only at the level of tools, it is at the level of our worldview. How do we look at ourselves in this world? What are humans for? And are we merely a money-making and resource guzzling machine? Or do we have a higher purpose, a higher end.

I Believe We do

I believe that we are members of the Earth family—of *Vasudhaiva Kutumbakam*. And as members of the Earth family, our first and highest duty is to take care of Mother Earth—*Prithvi, Gaia, Pachamana*. And the better we take care of her, the more food and water, health and wealth, we have. "Earth rights" is first and foremost the rights of Mother Earth, and our corresponding duties and responsibilities to defend those rights. Fartn rights is also the rights of humans as they flow from the rights of Mother Earth—the right to food and water, the right to health and a safe environment, the right to the commons—the rivers, the seeds, the biodiversity, the atmosphere.

I have given the name Earth Democracy to this new paradigm of living as an Earth Community, respecting rights of Mother Earth.

Earth Democracy enables us to envision and create living democracies. Living democracy enables democratic participation in all matters of life and death—the food we eat or do not have access to; the water we drink or are denied due to privatization or pollution; the air we breathe or are poisoned by. Living democracies are based on the intrinsic worth of all species, all peoples, all cultures; a just and equal sharing of this earth's vital resources; and sharing the decisions about the use of the earth's resources.

Earth Democracy protects the ecological processes that maintain life and the fundamental human rights that are the basis of the right to life, including the right to water, the right to food, the right to health, the right to education, and the right to jobs and livelihoods. Earth Democracy is based on the recognition of and respect for the life of all species and all people.

Ahimsa, or nonviolence, is the basis of many faiths that have emerged on Indian soil. Translated into economics, nonviolence implies that our systems of production, trade, and consumption do not use up the ecological space of other species and other people. Violence is the result when our dominant economic structures and economic organization usurp and enclose the ecological space of other species or other people.

According to an ancient Indian text, the *Ishopanishad*

> The universe is the creation of the Supreme Power meant for the benefits of [all] creation. Each individual life form must, therefore, learn to enjoy its benefits by forming a part of the system in close relation with other species. Let not anyone species encroach upon's other rights.
>
> Whenever we engage in consumption or production patterns which take more than we need, we are engaging in violence. Non-sustainable consumption and non-sustainable production constitute a violent economic order.

In the *Ishopanishad* it is said:

> A selfish man over utilizing the resources of nature to satisfy his own ever increasing needs is nothing but a thief, because using resources beyond one's needs would result in the utilization of resources over which others have a right.

The rights of corporations extinguish the rights of the Earth and all her children, including humans. The economy as currently structured is centred on corporations and corporate profits. Corporate profits are based on destruction of the earth and dispossession and uprooting of people.

The technological and economic systems that impoverish the earth, also impoverish local communities.

The rights of the Earth are ultimately intertwined with the rights of the people. The rights of corporations to appropriate or contaminate the earth's resources undermine both the rights of the Mother Earth and the human rights of people to livelihoods and basic needs of food and water. That is why the rights of Mother Earth are the very basis of the human rights of people to land and natural resource, food and water, to livelihoods and basic needs.

Earth Rights are the basis of equity, justice and sustainability. On Earth Day 2010, the President of Bolivia, Juan Evo Morales Ayma organized a conference on Rights of Mother Earth. The idea was to start a process for adopting a Universal Declaration of the Rights of Mother Earth on the lines of the Universal Declaration of Human Rights. Without Earth Rights, there can be no human rights. It is time to deepen human rights by deepening the recognition that humans depend on the Earth.

Earth Rights are Human Rights

Humanity stands at a precipice. We have to make a choice. Will we obey the market laws of corporate greed or Gaia's laws for maintenance of the earth's ecosystems and the diversity of her beings?

The laws for maximizing corporate profits are based on—

1. Privatising the earth
2. Enclosing the commons
3. Externalising the costs of ecological destruction of hazards

The laws for protecting the rights of Mother Earth are based on—

1. Respecting the integrity of the earth's ecosystems and ecological process
2. Recovery of the commons
3. Internalising ecological costs

The corporate ideology has presented corporate profits as growth and growth as beneficial to all. In a false way, corporate greed that appropriates resources that meet people's needs. Peoples needs for food and water can only be met if nature's capacity to provide food and water are protected. Dead soils and dead rivers cannot give food and water.

Defending the rights of Mother Earth is therefore the most important human rights and social justice struggle of our times.

19
Declining Biotic Cover and Strategies for Conservation

Sudha Srivastava

Introduction

A forest, with its varied vegetation, animals and micro-organisms, is an ecosystem which has evolved over thousands, may be millions of years. Forests regulate the local and global climate as well as the hydrological cycle, maintain the compositional balance in air, act as a vital carbon sink and control pollution by absorbing noise and dust. They check soil erosion, protect watersheds, control floods and help recharge groundwater. Forests are home to innumerable species and a vast store of genetic information, many of which we do not yet know about. All our crop plants and domesticated animals were once a part of forests. Their wild relations and all the other species that inhabit forests may hold possibilities that we cannot yet imagine. Forests are of great economic importance too as sources of timber, fuel, bamboo, cane, essential oils, resins, tannin, latex, medicines, honey and many other things. For the tribal and others living near forests, they are the main source of livelihood.

Among the key findings of FAO (2005) are several facts about forests:

- Forests cover about 30 per cent of the total area

- Primary forests account for 36 per cent of forest area—but 6 million hectares are lost or modified each year
- There exists a wide variation in number of native tree species—from 3 in Iceland to 7780 in Brazil
- 84 per cent of the world's forests are publicly owned, but private ownership is on the rise
- 11 per cent of the world's forests are designated for the conservation of biological diversity
- One-third of the world's forests are primarily used for production of wood and non-wood products
- More than 300 million hectares of forests are designated for soil and water conservation
- Use of forests for recreation and education is increasing, but difficult to quantify
- The value of wood removals is decreasing, while the value of NWFPs is increasing—and underestimated
- Around 10 million people are employed in forest management and conservation
- Forest disturbances can be devastating—but they have been severely underreported
- Up to 50 per cent of pharmaceuticals in the market today have an origin in natural products, and 42 per cent of the top 25 selling drugs worldwide are derived from natural products.

Deforestation

However, there has been large-scale destruction of forests, resulting in a number of problems. Deforestation is the conversion of forested areas to non-forest land for use such as arable land, pasture, urban use, logged area, or wasteland. Deforestation results from removal of trees without sufficient reforestation, leading to a decline in habitat, biodiversity, wood for fuel and industrial use and quality of life. Since about the mid-1800s, the Earth has experienced an unprecedented rate of destruction of forests worldwide. In addition to the direct effects brought about by forest removal, indirect effects caused by edge effects and habitat fragmentation can greatly magnify the effects of deforestation. Generally, the removal or destruction of significant areas of forest cover has resulted in a degraded environment with reduced biodiversity. In many countries, massive deforestation is ongoing and is shaping climate and geography. Shrinking forest cover lessens the landscape's capacity to intercept, retain and transport precipitation. Instead of trapping precipitation, which then percolates to groundwater systems, deforested areas become sources of surface water runoff, which moves much faster than subsurface flows. Deforestation also contributes to decreased evapotranspiration, which lessens atmospheric moisture which in some cases affects precipitation levels.

Improving Quality and Area of Our Forest Cover is a Centrepiece of Our Strategy to Combat Climate Change

Forests in India

The forest and tree cover of our country is about 78.37 m ha (i.e. 23.84 per cent of geographical area). The net increase over a two-year period is 0.18 m ha (0.23 per cent), the mangrove cover increasing by 5800 hectares. Between 1997 and 2007 the increase in forest cover was 3.13 m ha (4.75 per cent) (SFR, 2009). There has been a significant increase in the hill and tribal districts of the country. Forest conservation efforts include protection of forests, check on diversion, farm forestry/private area plantations, expansion of protected area network (PAN) to include 4 per cent of entire country and efforts to control forest fires. Institutional arrangements are in place, but their implementation leaves a lot to be desired. The National Forestry Action Programme, Joint Forest Management, National Forest Policy, 1988 are some of the attempts in this direction.

The protected area network (PAN) comprises 95 National parks, 500 wildlife sanctuaries and 14 biosphere reserves to protect representative ecosystems. There is also a management plan for 20 wetlands with Coral Reefs and Mangroves receiving top priority.

In addition, the National Wasteland Development Board regenerates private, non-forest and degraded land, the National Afforestation and Eco-Development Board is responsible for regenerating degraded forests, land adjoining forests and ecologically fragile areas. India's Forest Conservation Act, 1980, reported as one of the most progressive forest conservation legislations in the world, puts severe restrictions on the diversion of forest land for non-forestry purposes.

In Maharashtra, most of the forest cover is restricted to the slopes of Sahyadri range and the eastern belt. While the forests of the Western Ghats comprise both the dense and open varieties, those in Eastern Maharashtra are more of the dense type. Scrubs (indicating degraded forest land) have canopy density < 10 per cent and are not considered part of forest cover. However, they are ideal for indicating areas where successful reforestation programmes can be initiated.

Maharashtra is also developing a good network of protected areas. There are 6 national parks, 35 wildlife sanctuaries, covering an area of 1.5 m ha, 3 tiger reserves covering 0.26 m ha (at Melghat, Tadoba—Andhari and Pench), and a wetland in Ujni in Sholapur. Much of endemic wildlife has been adequately conserved and protected in 35 sanctuaries. The State of Forest Report 2009 gives the following data:

Table 19.1a: Area under Varied Landuse/Landcover Categories

Land use	*Area (in '000 ha)*	*%*
Total Geographical Area	30,771	
Reporting Area for land utilisation	30758	100
Forests	5214	16.95
Not Available for Cultivation	3131	10.18
Permanent pastures and other grazing lands	1252	4.07
Land under miscellaneous tree crops and groves	249	0.81
Culturable wasteland	914	2.97
Fallow lands other than current fallows	1199	3.90
Current Fallows	1325	4.31
Net-Area Sown	17475	56.81

Table 19.1b: Per cent Coverage of Forested and Non-Forested Areas

Landuse/ landcover type	*Coverage (%)*
Very Dense Forest	2.84
Moderately Dense Forest	6.77
Open Forest	6.85
Scrub	1.35
Non-Forest	82.19

Source: Land Use Statistics, Ministry of Agriculture, GOI, 2006.

It can be clearly seen from Tables 1a and 1b that the forest cover (16.4 per cent) is far less than the stipulated norm for the country. Moreover the extent of dense forests, which is an important indicator of its quality, is also a cause for concern.

Table 19.2: Forest Cover Change Matrix

2005 Assessment (data of Oct-Dec 2004)	*2007 (data of Oct-Dec 2006)*					*Total of 2005*
	VDF	MDF	OF	Scrub	NF	
Very Dense Forest	8739	0	2	0	6	8747
Moderately Dense Forest	0	20834	5	0	8	20847
Open Forest	0	0	21063	0	4	21067
Scrub	0	0	0	4157	0	4157
Non-Forest	0	0	7	0	252888	252895
Total of 2007	8739	20834	21077	4157	252906	307713
Net Change	-8	-13	10		11	

Source: SFR 2009.

There has been a loss of 11 sq. km of forest cover compared to the previous assessment. Data reveal that there is a loss of 8 sq. km in the very dense forest, 13 sq. km in moderately dense forest and an increase of sq. km in open forest cover of the state (Table 19.2). This is attributed to coupe working in Chandrapur, clearance of private areas in Konkan, and increase in water level of some water bodies. The increase in forest cover has been mainly on account of forest plantations by the Maharashtra Forest Department and Forest Development Corporation of Maharashtra as well as private plantations and orchards of cashew, mango, etc., in Konkan (Forest Survey of India, 2009).

As per data from Ministry of Environment and Forest, the Joint Forest Management which was initiated as a World Bank project in 1990, gathered momentum in 1996-97, and today there are 10,242 committees managing 2.5 m ha of forest area (40 per cent of total forest area of state). More than 18 lakh families are involved in this activity, of which 5 lakh belong to the Scheduled Tribes (2005).

Much of the endemic wildlife has been adequately conserved and protected in 35 sanctuaries—4 in Konkan, 2 in Khandesh, 4 in Marathwada, 15 in Vidarbha and 10 in Western Maharashtra region. National Parks (6) belong to the protected area network too and are areas dedicated by status for all time. National parks serve to conserve the scenery, wildlife therein, the natural and historical objects of national significance and to provide for the enjoyment of present and future generations. Case studies of a national park and a sanctuary have been taken up to explore their status and the impact of proximity to a metropolitan area.

Case Study 1

The forested central hill complex that includes the Sanjay Gandhi National Park (SGNP, popularly known as Borivli National Park) and a Reserved Forest in Mumbai is a green oasis in the centre of an urban sprawl; SGNP is the only such park within the limits of a metropolis in Asia (Map 19.1). The hill complex has an elevation of 480m. As it lies in the region of heavy rainfall, i.e. Konkan, it receives 2487 mm of rain annually. The area has been designated as a No Development Zone. Its importance to the environment is tremendous, primarily as a carbon sink, forming the lungs of Mumbai. It is a rich and diverse forest consisting of more than 1000 species of plants, 40 species of mammals, 251 species of birds (migratory, land and water birds), 38 species of reptiles, 9 species of amphibians and a large variety of fish. It forms an important catchment area and releases water into the rivers of the city like Oshiwara, Mithi and Dahisar rivers. Three lakes—Tulsi, Vihar and Powai—nestle in the verdant hills. The Park was established in 1969, covering an area of 104 sq. km.

The number of visitors to this park is around 2 million per year. Only a small section is open to public. This section comprises a mini zoo and safari, a narrow gauge train, the celebrated Kanheri Caves, nature trails and treks, spectacular Karvi flower blooms (once in 7 years), rock climbing and boating facility.

In a growing metropolis, the survival of such a massive area under green cover is difficult as land is an invaluable asset and unless strict vigil is observed and stringent measures are taken, encroachments are bound to invade. As land is taken up for development inch by inch, especially for accommodating the burgeoning population, residential colonies are mushrooming right up to the park boundary. Constructing a sturdy, well-built fencing or boundary wall is a gigantic task; hence the fringe areas are exposed

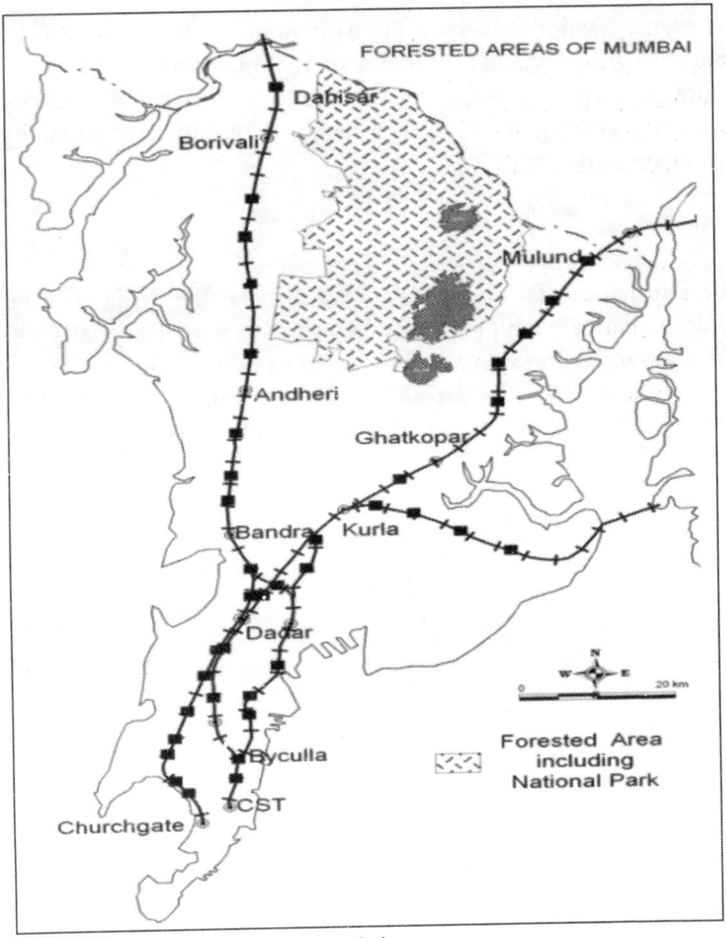

Map 19.1: The Forested Area of Mumbai

to attacks by the feline population. As the city expands its limits and makes inroads into the forests, there is a marked increase in the man-animal conflict for habitat occupation. In recent times leopards have been responsible for the deaths of 20 humans (as recorded in June 2004), while accidental deaths of leopards or their cubs have been reported on the western express highway. Several unauthorized slums have occupied the land around and at times into park. There is hence a perceived shrinking of the park (Maps 19.2a and 19.2b). Tree cover has steadily declined due to construction activities, as the area becomes a major target of real estate developers and rivers have dwindled into inconspicuous *nalas*. With increasing concretization of the natural environment, flooding is frequent during heavy rains. The changing microenvironment probably resulted in the suburbs being the worst affected during 26/7/05 (extreme rainfall) disaster. Disturbances to the natural environment include fires that are common

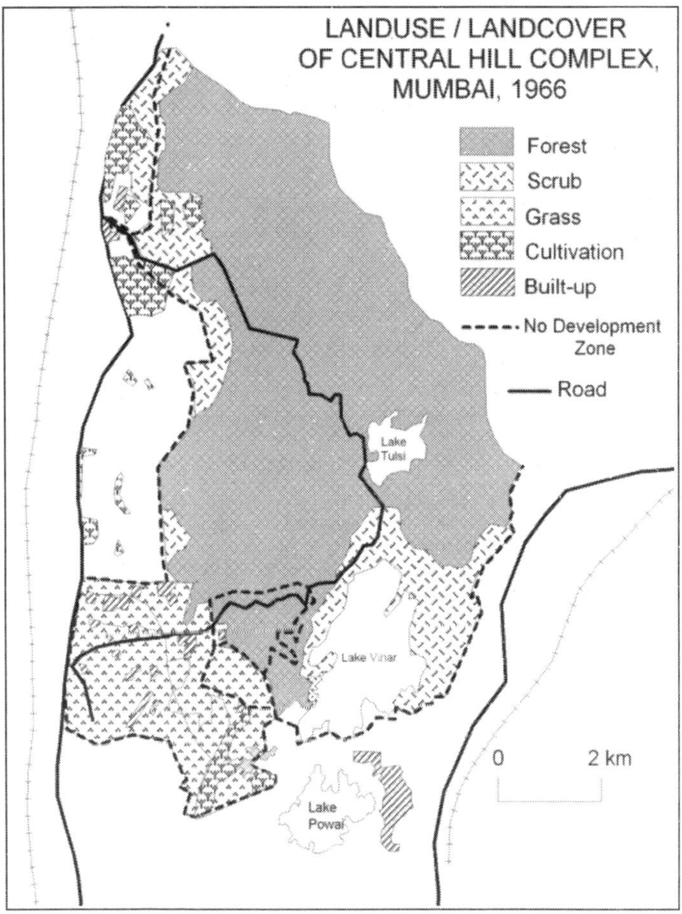

Map 19.2a: Landuse/Landcover of Central Hill Complex, Mumbai, 1966

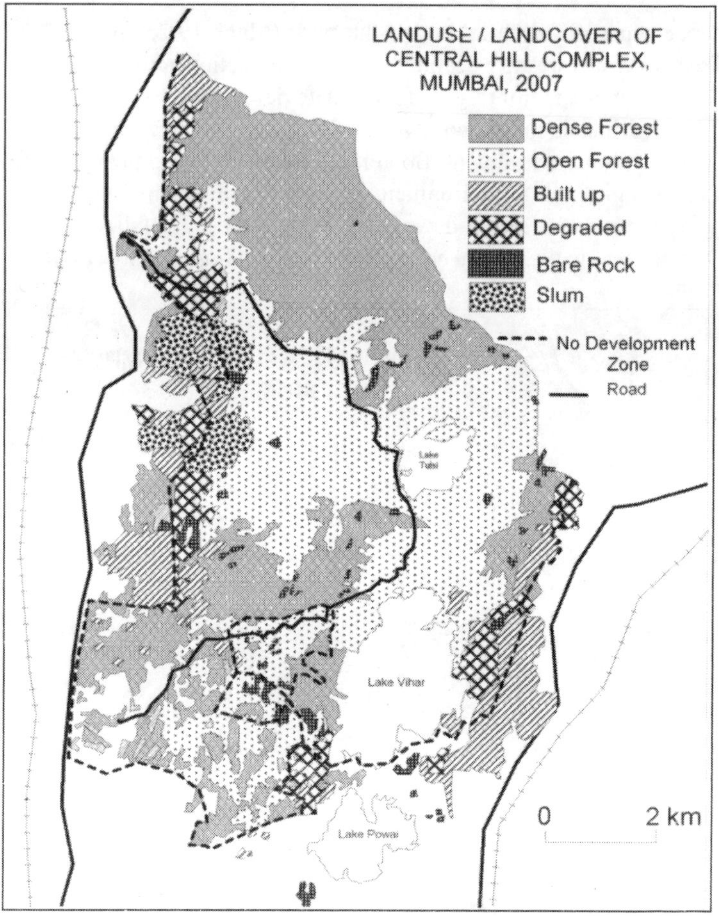

Map 19.2b: Landuse/Landcover of Central Hill Complex, Mumbai, 2007

especially during the festival of Shivratri when people throng the Shiva Temple at Kanheri in uncontrollably large numbers.

The linear pattern of the city coupled with the existence of the National Park on the central hill complex puts constraints on east-west connections. To improve the accessibility transport links are suggested across the park area. Though naturalist Humayun Abdulali in early 1970s halted a project of a road cutting through the park, the pressures may be too strong to hold back such projects in the future.

The green stretch that begins with the Aarey Milk Colony and gradually blends into the National Park has already been subjected to encroachments and development projects with the relaxation of norms in the skirting No Development Zones, with apprehensions of fragmentation of the habitat and loss of corridors for the biotic communities. The satellite imagery (Fig. 19.1) depicts the pressures along both the western and eastern limits and a creeping in of freshly built-up areas.

In the clash between people's interests and the need to protect the flora and fauna, the balance of public and official sympathy would naturally shift in favour of the former.

Fig. 19.1: Satellite Imagery showing the SGNP and its Environs

Case Study 2

The second case is that of Karnala Bird Sanctuary. It comprises an area of 4.27 sq. km, within easy reach of Mumbai. It has 150 species of resident and 37 species of migratory birds, with a very rich variety of flora. The 12th Century Karnala Fort stands 475 m high within its

Fig. 19.2: Karnala Fort

precincts (Fig. 19.2). The most interesting feature about this sanctuary is that a major road runs through it (Map 19.3). The road is now the National Highway from Mumbai to Goa. The disturbance that this very busy road creates can well be imagined. A study conducted in the sanctuary with respect to its vegetation cover reveals a profound impact of the highway. Profiles were constructed across the area and draped with the prevailing vegetation. The profiles showed that closer to the highway trees occur in patches, in fact there are more shrubs. Better quality forests are within the Reserved Forest area. There is a gradual change from deciduous to semi-evergreen as one moves away from the highway up the slope. Accessibility is a major deterrent of a good forest cover. Interference by the passing vehicles, in terms of air and noise pollution has resulted in original dense forest cover to be far from the roadway, while near it the vegetation has changed from dense to the open variety with a diminishing tree cover and increasing shrubs.

The Forest Department has tried to improve the situation but often is unable to do so due to constraints, like lack of manpower for vigilance. Hence unauthorized felling of trees goes on relentlessly,

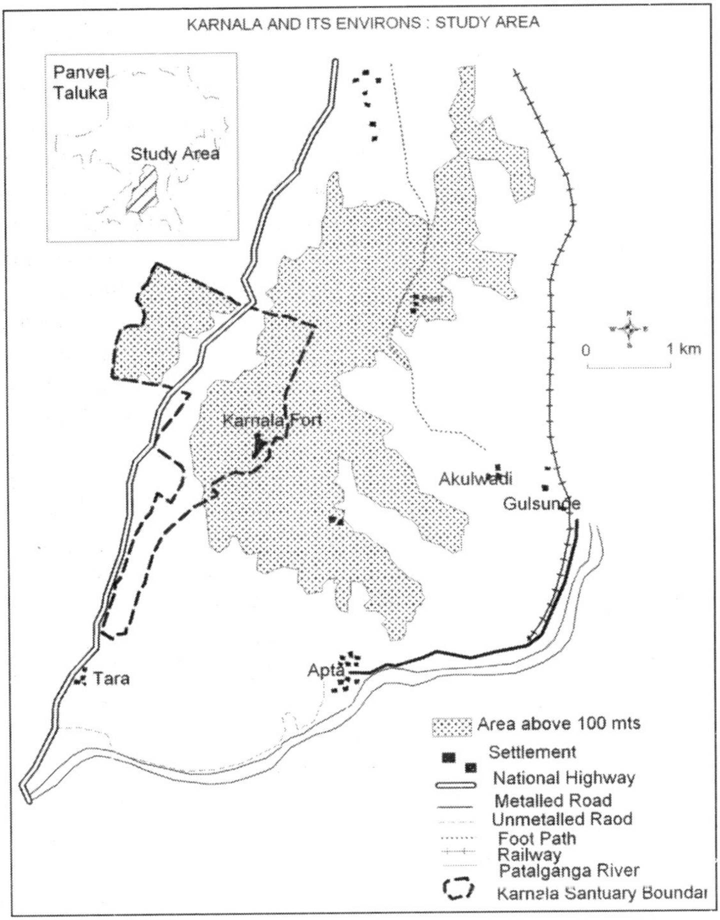

Map 19.3: Karnala and its Environs : Study area

more so as accessibility has increased. Also in the effort to afforest the area, there is a preference for exotic and alien species of the fast-growing variety. It has always been seen that indigenous species are in equilibrium with the prevailing environmental conditions and hence blend with their surroundings. The introduction of non-native plants is a threat to the original flora of the region.

Though a bird sanctuary, the increasing volume of traffic is a menace and birds that used to be sighted in large numbers are fewer and seen more infrequently now. The environment is no longer quiet and undisturbed—a prerequisite for any bird/wildlife sanctuary. The early road graduated to being a highway and there is a demand to increase the width especially in the part that passes through the sanctuary, as 156 people were killed and 250 injured in the year 2009. Social activists and journalists from coastal Raigad, Ratnagiri and Sindhudurg feel that 150 species were earlier residing here, but now the area has been taken over by monkeys and there are hardly any birds. Though the state forest department had recommended allotment of some part of the land from the sanctuary to the Centre, the MoEF and Wildlife Board rejected the proposal.

In our study we tried to bring out the impact of Mumbai-Goa Highway on the biotic community. Through a systematic vegetation analysis we could demonstrate the decline in the quality of vegetation cover as one approaches the road. Since it may not be feasible to shift the highway, we tried to find an area in the vicinity that has diverse vegetation and bird population, where disturbance is minimum, and suggested that a new bird sanctuary may be established there.

Conclusion

There is a need to frame policies and plans judiciously. It is hearty to note that our institutional arrangements are well in place, but their implementation needs to be followed strictly. Thus as far as our natural heritage is concerned, we need to ensure that they are conserved in the best possible way since the environmental quality of any place depends substantially on the indigenous biotic communities, as these have evolved in response to the prevailing conditions and are in an ecologically equilibrated state. However, development cannot be arrested and if a case like the Karnala Bird Sanctuary does occur, where a road graduated to the status of a national highway, it is necessary to identify new areas for conservation purposes.

Impact assessments including Environmental Impact Assessment are a prerequisite for any major development plan. These must be carried out objectively and meticulously by a team of experts which is interdisciplinary in nature.

Any project whether it is for development or conservation has proved to be successful if the local population is involved as stakeholders.

In cities that are struggling to cope with space crunch, where densities of population are extremely high and there is a tremendous stress on the infrastructure, concerted efforts need to be made to preserve the open spaces in and around them. Open spaces and green zones are carbon sinks, regulators of micro-climate and beneficial to the health of the residents. However, it may be difficult to hold back the land 'sharks' and investors who get ample support from the authorities, sometimes flouting norms too. In the larger and long term interests of the citizens and sustainability of the growing megalopolis, the urbanscape needs to have its green and open spaces in adequate proportion. Mumbai suffers from one of the lowest proportions of open spaces per 1000 population in the world (Table 19.3).

Table 19.3: Average Open Space Available in Major Cities of World

City	*Population*	*No. of parks/open spaces within city limits*	*Park/open space (ha) per 1000 residents*
Minneapolis	383000	5694	14.9
Washington DC	572000	7504	13.1
Oakland	399000	3712	9.3
Boston	589000	4865	8.3
Los Angeles	3695000	29801	8.1
Baltimore	651000	5091	7.8
San Francisco	777000	5916	7.6
Philadelphia	1518000	10685	7.0
New York	8008000	49854	6.2
Mumbai	15000000	959	0.03

Source: Harnik, Peter, 2010.

The table reveals the basic flaw in planning or strict implementation of plans as compared to the leading cities of the world. It is almost impossible for any built-up landuse to be reverted to a green area, but at least it is imperative that the ones remaining are preserved. With rising population there is a spurt in construction, and massive development projects are being undertaken across the city and its region. However with awareness about the benefits of open spaces to society among the residents and their growing assertiveness, it is sincerely hoped that plans ensure that there is adequate green cover within a city and in its hinterland and that it is well-preserved and maintained.

REFERENCES

FAO (2005). *State of World's Forests*, Rome.

Forest Survey of India (2009). *India - State of Forest Report*, Dehradun.

Harnik, P. (2010). *Urban Green - Innovative Parks for Resurgent Cities*, U.S.: Island Press.

20
Climate Change, Glacial Recession in Himalayas and Its Impact on Indian Economy

Narender Verma

Introduction

Global climatic change is one of the major problems confronting the world today. Temperature data available since 1000 AD shows that the Twentieth century was unusually warm and the decade of 1990s was the hottest with six warmest years occurring within this decade (Rao *et al.*, 2005). The modern industrial and transport development primarily based on fossil fuels have resulted in increased concentration of Greenhouse Gases in the atmosphere. The concentration of CO_2 and methane has increased by 31 per cent and 149 per cent respectively since the beginning of the industrial revolution in the mid-1700s. These levels are considerably higher than at any time during the last 650,000 years, the period for which reliable data has been extracted from http://en.wikipedia.org/wiki/Ice_core. The burning of fossil fuels has produced approximately three-quarters of the increase in CO_2 over the past 20 years. Most of the rest is due to land-use change, in particular deforestation. Consequently, ithe earth has become warm considerably, particularly after 1750 and over the next century will further warm up as the temperatures are expected to rise between 1-6.3° Celsius. This unprecedented increase in the global temperature has caused adverse impact on the various ecosystems including mountain ecosystem.

Mountains cover about one-fifth of the Earth's land surface, they are an important source of water, energy, minerais, forest and agricultural products as well as area of recreation. Globally the area under mountains is almost 40 million km^2, or some 27 per cent of the total Earth's surface area. Mountain ecosystems support vibrant livelihoods, and include significant watershed resources, biological diversity of unique flora and fauna. They act as a barometer of global climate change. These fragile ecosystems are vulnerable particularly towards the adverse effects of climate change at global level and need specific protection and conservation strategies against the problem (Anthwal *et al.*, 2006). The present paper examines the result of climatic change on the Himalayan ecosystem with primary focus on glacial recession and its impact on the Indian economy.

Climate Change and Glacial Fluctuation

Glacial fluctuation owing to climate change is a natural phenomenon occurring on the earth for the past 5 billion years. There have been repeated large periods of glaciation interspersed by short warm inter-glaciation periods in the past 3 million years. At least 17 major glacial advances have been witnessed in the past 1.6 million years (Goudie, 1983). This cyclical glaciation and inter-glaciation activity is largely attributed to graduai changes in the earth's rotation, tilt, and orbital change around the sun. The recurrence period is estimated to be 100000 years. The most recent, the Last Glacial reached its peak some 20000 to 18000 years ago and came to an end some 10000 years ago (Goudie, 1983). This natural cycle of repeated glaciation seems to have significantly altered due to the human induced greenhouse effect. According to some, it might have begun well before the industrial revolution perhaps 11000 years ago when agricultural revolution occurred that initiated large scale clearance of forests for developing agricultural fields. The Industrial Revolution of the mid-Eighteenth century has dramatically hastened the pace of this change. Prolific emission of greenhouse gases owing to excessive use of fossil fuels has increased the global temperature by 0.6°± 0.2° Celsius since 1900. This increase in the global temperature has caused significant variation in the Glacier Mass Balance, defined as a total loss or gain in the glacial mass at the end of each hydrological year (Paterson, 1998).

Global warming has caused massive retreat of glaciers throughout the world. Many Alpine Glaciers including those of the Himalayas have substantially receded in the past 100 years. According to a study by Dyurgerov and Mier (1997) in which they considered glacier mass balance changes of over 200 mountain glaciers globally, have concluded that reduction in global glacier area amounted to 6000 km and 8000 km over a period of 30 years between 1961-1990. In Montana's Glacier National Park, ice-covered area decreased by 73 per cent (99 km^2 to 27 km^2) from 1850 to 1993. glacierized area in the Alps has decreased by 40 per cent since 1850, with an estimated volume loss of 50 per cent. Spain has 13 glaciers remaining, a decline from 27 glaciers in 1980. Tropical ice caps in the Andes and Africa are disappearing at a similar rate; Mt. Kilimanjaro ice-fields have diminished by 82 per cent by area since 1912, from 12 km^2 to just over 2 km^2 in 2000. Approximately 33 per cent of this retreat has come in the last 20 years (Anthwal *et al.*, 2006). Further various scientific studies suggest that the rate at which the glaciers are presently retreating a quarter of the global mountain glacier mass could disappear by 2050 and up to half by 2100. The mean annual snow cover in the northern hemisphere which was 24.4×10^6 km in the period 1967-1987 has reduced to 23.1×10^6 km since 1988, a reduction by approximately 5 per cent (Robinson and Frei, 1999).

The Himalayas

The Himalayas are among one of the youngest mountains of the world, and the youngest in India. As compared to the Aravalli, Nilgiris, and the Eastern Ghats which are 1500 to 2500 million years old, and the Vindhyachals which are around 1000 million years old the Himalayas date back to only 40-45 million years. They extend for about 2500 km between the Indus and the Brahmaputra in an arc. Their average width ranges from 100 to 400 km along its entire longitudinal extent. The Himalayas extend from the Pamir Knot in the northwest to the valley of the Brahmaputra River in the east. The range passes through Afghanistan, Pakistan, India (Jammu and Kashmir, Himachal Pradesh, Uttarakhand, Sikkim, Assam, Arunachal Pradesh, Manipur, Meghalaya, Mizoram, Nagaland, Tripura and Darjeeling district of West Bengal), China, Nepal, Bhutan, Bangladesh and Myanmar.

The Himalayas—the abode of snow and ice contains over 50 per cent of permanent snow and ice-fields outside the Polar Regions (Table 20.1). This region covers an area of 4.6 million km^2 above 1500 m, 0.56 million km^2 above 5400 m and 3.2 million km^2 above 3000 m (Upadhyay, 1995). They are a large reservoir of glaciers and encompass the world's third largest glacier systems after Antarctica and Greenland occupying about 15 per cent of the mountain terrain (Anthwal *et al.*, 2006). With glacier coverage of approximately 33,000 sq. km, the region is aptly called the "Water Tower of Asia" as it provides around 86,000,000 cubic metres of water annually (Rao *et al.*, 2005).

Table 20.1: Distribution of Permanent Snow and Ice in the Himalayan Region

Sub-region	*Volume (km^3)*	*Surface Area (sm^2)*
Hindukush	930	6,200
Karakoram	2,180	15,670
Himalayas	5,000	43,000
Tibet	4,820	32,150
Total	**12,930**	**97,020**

Source: Bahadur, 1997.

There are 18,065 glaciers with a total area of 34,659.62 km^2 and a total ice volume of 3,734.4796 km^3 (Dahe, 1999) in the entire Himalayan Range. Nanga Parbat, the Nanda Devi group, the Dhaulagiri massif, the Everest-Makalu group, the Kanchenjunga, the Kula Kangri area, and Namche Bazaar are the principal peaks and massifs around which there is major concentration of glaciers in the Himalayas. Regionally the Central Himalayas account for the maximum 52.30 per cent of the total glaciers followed by Western Himalayas 31.26 per cent and Eastern Himalayas 16.44 per cent (Table 20.2).

Glaciers in the Indian Himalayan Region can be mainly classified into the three-river basins of Indus, Ganga and Brahmaputra. The Indus basin has the largest number of glaciers (3,538), followed by the Ganga basin (1,020) and Brahmaputra (662). Siachen 72 km; Gangotri 26 km; Zemu 26 km; Milam 19 km and Kedarnath 14.5 km etc. are the principal glaciers of this region. According to research estimates 17 per cent Himalayas and 37 per cent Karakoram are covered with ice cover. The distribution of glaciers in Jhelum and

Table 20.2: Glacier Resources in the Himalayas

Region	*Number*	*Area/ km²*	*Ice Volume/ km³*
Western Himalayas	5,648	10,284.75	980..7714
Central Himalayas	9,449	20,214.54	2,338.7627
Eastern Himalayas	2,968	4,160.33	414.9455
Total	**18,065**	**34,659.62**	**3,734.4796**

Source: An Overview of Glaciers, Glacier Retreat, and Subsequent Impacts on Nepal, India and China, WWF Nepal Programme, March 2005.

parts of Satluj in the Indus basin and for Bhagirathi, Tista and Arunachal Pradesh in the Ganga-Brahmaputra basin are shown in Tables 20.3 and 20.4.

Table 20.3: Status of the Glacier Inventory of Indus Basin

Basins	*Number of glaciers*	*Glacierised area (km²)*	*Ice volume (km³)*
Jhelum	133	94.0	3.0
Satluj	224	420.0	23.0
Others	3398	33382.0	–
Total	**3755**	**33896.0**	**26.0**

Source: An Overview of Glaciers, Glacier Retreat, and Subsequent Impacts on Nepal, India and China, WWF Nepal Programme, March 2005.

Table 20.4: Status of the Glacier Inventory of Ganga-Brahmaputra Basins

Basins	*Numbers of glaciers*	*Glacierised area (km²)*	*Ice volume (km³)*
Bhagirathi	238	755.0	67.0
Tista	449	706.0	40.0
Brahmaputra	161	223.0	10.0
Others	640	2378.0	–
Total	**1488**	**4062.0**	**117.0**

Source: An Overview of Glaciers, Glacier Retreat, and Subsequent Impacts on Nepal, India and China, WWF Nepal Programme, March 2005.

Statewise Jammu and Kashmir accounts for a maximum 69.90 per cent followed by Uttarakhand 17.80 per cent, Sikkim 8.70 per cent and Arunachal Pradesh 3.20 per cent of the total glaciers (Table 20.5).

Table 20.5: Distribution of Glaciers in India

State	Glaciers	Area (km^2)	Average size (km^2)	Glacier in the state (per cent)
J & K	3136	32000	10.24	69.90
Himachal Pradesh	402	1349	3.35	9.40
Uttarakhand	917	3550	3.87	17.80
Sikkim	450	917	1.50	8.70
Arunachal Pradesh	162	220	1.40	3.20

Source: Geological Survey of India, 1999.

Glacial Fluctuations in the Himalayas

Himalayan Glaciers are constantly retreating since 1850. However, the rate of retreat of these glaciers has significantly accelerated in the past three decades, about 67 per cent of the glaciers are retreating at an alarming rate, a fact confirmed by various scientific studies carried out by various experts and institutions. Global Warming owing to climatic change is one of the principal factors responsible for it. The Himalayan Region including Tibetan Plateau has been consistently warming for the past 100 years. Various studies suggest that warming in the Himalayas has been much greater than the global average of 0.74°C over the last 100 years (IPCC, 2007; Du *et al.*, 2004). The increase in temperature over the Tibetan Plateau during the period from 1955–1996 is about 0.16°C per decade for the annual mean and 0.32°C per decade for the winter mean, and these exceed the increases in the northern hemisphere and for the same latitudinal zone in the same period (Liu & Chen, 2000).

The length of the growing season (daily temperature > 10°C) on the Tibetan Plateau has increased by almost 15 days during the past three decades (Jianchu *et al.*, 2007). Similarly, while the winter mean and maximum temperature in the Karakoram and Hindukush Mountains have shown significant increase, the summer mean and minimum temperature has shown a declining trend. According to Dyurgerov and Meier (2005), many Himalayan glaciers are retreating faster than the world average and are thinning by 0.3-1 m/year. Table 2o.6 shows the rate of retreat of various Himalayan glaciers during various periods.

Among the various Himalayan glaciers Gangotri Glacier is very significant. It is the second largest Indian glacier. Bounded between

Table 20.6: Retreat of Himalayan Glaciers

Sl. No.	*Glacier*	*Period of Measuring*	*Period (year)*	*Recession (metre)*	*Average rate (myr^{-1})*
1.	Milam	1849-1957	108	1350	12.50
2.	Pindari	1845-1966	121	2840	23.40
3.	Gangotri	1935-1996	61	1147	18.80
		1962-1991	29	580	20.0
		1996-1999	3	76	25.33
4.	Tipra Bamak	1960-1986	26	325	12.50
5.	Dokriani	1962-1991	29	480	16.5
		1991-2000	9	161.15	18.0
6.	Chorabari	1962-2003	41	196	4.8
7.	Shankulpa	1881-1957	76	518	6.8
8.	Poting	1906-1957	51	262	5.13
9.	Bara Shigri	1956-1963	7	219	31.28
		1977-1995	18	650	36.11
10.	Chotta Shigri	1987-1989	3	54	18.5
		1986-1995	9	60	6.7
11.	Sonapani	1909-1961	52	899	17.2
12.	Kolai	1912-1961	49	80	16.3
13.	Zemu	1977-1984	7	193	27.5
14.	Arwa Valley	1932-1956	24	198	8.25
15.	Trilokinath	1969-1995	26	400	15.4
16.	Dunagiri	1992-1997	5	15	3.0
17.	Chiba	1961-2000	39	1050	26.9
18.	Meola	1961-2000	39	1350	34.6
19.	Jhulang	1962-2000	38	400	10.5

Source: Anthwal *et al.*, 2006.

30°43′22″ N – 30°55′49″ N (Lat.) and 79°4′41″ E – 79°16′34″ E (Long.), extending in height from 4000 to 7000 m.a.s.l. It is situated in Uttarkashi district of Garhwal Himalaya, Uttarakhand. It is a valley type glacier about 30.20 km long with its width varying from 0.5 to 2.5 km, flowing in NW direction. This glacier comprises three main tributaries, namely Raktvarn (15.90 km), Chaturangi (including Kalandini bamak) (22.45 km) and Kirti (11.05 km) and more than 18 other tributary glaciers that are transverse to subtransverse. The total glacierized area of the catchment is 258.56 sq. km, out of which the Gangotri system comprises 109.03 sq. km, followed by Chaturangi (72.91 sq. km), Raktvarn (45.34 sq. km) and Kirti (31.28 sq. km). The remaining four glaciers contain 29.41 sq. km of glacierized area. The gradient of Gangotri glacier is lowest (0.045), followed by Chaturangi and Raktvarn (0.146 and 0.210 respectively), while Kirti glacier has the highest gradient, viz. 0.317 (Ajay K. Naithani *et al.*, 2001). Apart from being the source of the perennial River Ganga it has a great religious importance for the Indian community. This glacier owing to global warming is retreating at an alarming pace (Fig. 20.1). Observations on the retreat of the Gangotri go back to 1842, and between 1842 and 1935 (Tables 20.6 and 20.7) the snout of the Gangotri glacier was receding at an average rate of 7.3 m a year (Ramchandran, 2008). However, between 1935 and 1996 it retreated by 1,147 m which amounts to an average rate of 19 m a year (Tables 20.5 and 20.6).This implies that the rate of retreat more than doubled during the last century. Over the last 25 years, Gangotri glacier has retreated more than 850 metres, with a recession of 76 metres from 1996 to 1999 alone (Table 20.6). The rate of retreat in the last three decades is more than three times the rate during the earlier 200 years or so during which the glacier retreated by only 2 km. The movement of a glacier leaves behind moraines. The formation and the type of moraines, together with other landforms, provide the key morphological evidence to the previous positions of a glacier, particularly its snout, and provide indications of its advance or retreat. The snout of a glacier is its terminus, the front of the region of melting and evaporation—that usually takes the shape of an ice cave in mountainous terrains like the Himalayas—where the ice-melt turns into a steady flow and the river begins (Ramchandran, 2008). The snout of the Gangotri, the Gaumukh, which appears as an ice-covered cave here, has presently moved approximately 3 km upstream from its position before the retreat of the glacier began around the turn of the 19th century.

An investigation by Anil Kulkarni and others, of 466 Glaciers in Chenab, Parbati and Baspa basins from 1962 using a number of Indian Remote Sensing Satellites, viz. LISS-IV data of IRS-P6 in Parbati basin, and LISS-III data of IRS-1D (Figures 20.2 and 20.3) in Baspa and Chenab basins in Himachal Pradesh shows that the glacial area has reduced from 2077 sq. km in 1962 to 1628 sq. km in 2001-2004, an overall deglaciation of 21 per cent. However, the number of glaciers has increased due to fragmentation. Mean area of glacial extent has reduced from 1.4 to 0.32 sq. km between the 1962 and 2001. In addition, the number of glaciers with higher areal extent has reduced and lower areal extent has increased during the period. Loss in glaciated area for large glaciers was 12 per cent compared to 38 per cent for small glaciers.

Table 20.7: Recession Rate of Gangotri Glacier at Different Time Periods since 1842

Period of observation	*No. of years*	*Ice cave retreat (m/yr)*
1842–1935	93	7.35
1935–1956	21	10.16
July 1956–March 1962	5.6	18.75
March 1962–September 1971	9.5	32.21
1971–1975	4	28.87
1975–1977	2	36.50
1977–1990	13	28.80
1935–1996	61	18.80
1971–2004	33	27.12

Source: Current Science vol. 95 No.1, 10 July 2008, P. 9.

Source: Kargel, NASA, USA.

Fig. 20.1: Retreat of the Gangotri Glacier (Garhwal Himalayas) Snout during the last 220 years

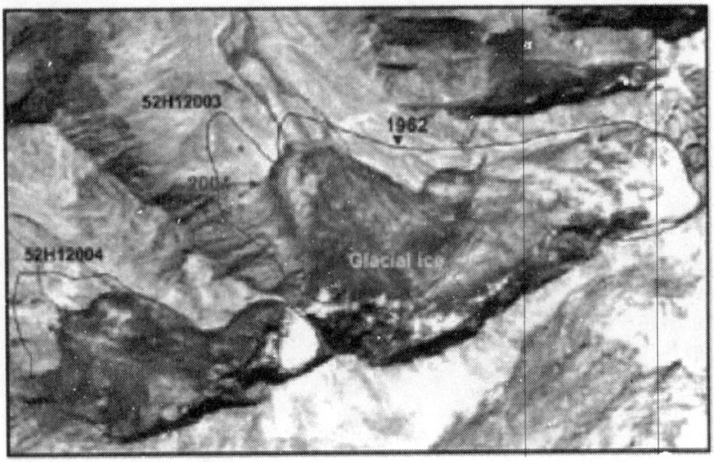

Source: Anil Kulkarni *et al.*, 2007.

Fig. 20.2: Satellite Imagery of Glacier No. 52H12003 and 52H12004 of LISS-IV Sensor showing Glacial Boundary of 1962 and 2004

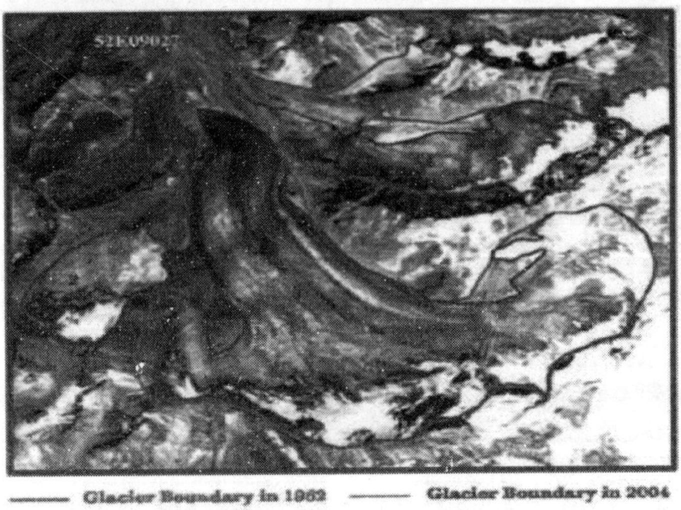

Source: Anil Kulkarni *et al.,* 2007.

Fig. 20.3: Resources at Imagery of LISS-IV Sensor dated 12 September, 2004 of Glacier No. 52E09027. This Glacier Split into Four Glaciers between 1962 and 2004

Impact of Himalayan Glacial Retreat

The rapid rate at which the Himalayan glaciers are receding will have far reaching impact on the ecology, environment, hydrological cycle and the socio-economic condition of the people in the Indian subcontinent as summarized below:

Imbalance in the Volume of Water

The Himalayan region is the source of the nine largest rivers in Asia, the basins of which are home to over 1.3 billion people (Table 20.8).

Table 20.8: Major Rivers of the Himalayan Region: A Statistics

	River		*River Basin*			
	Mean discharge (m^3/s)	*Glacial melt in river flow (%)**	*Area (km^2)*	*Population (in' 000s)*	*Popula-tion density*	*Water availability (m^3/person/ year)*
Indus	5,533	44.8	1,081,718	178,483	165	978
Ganges	18,691	9.1	1,016,124	407,466	401	1,447
Brahmaputra	19,824	12.3	651,335	118,543	182	5,274
Irrawaddy	13,565	Small	413,710	32,683	79	13,089
Salween	1,494	8.8	271,914	5,982	22	7,876
Mekong	11,048	6.6	805,604	57,198	71	6,091
Yangtze	34,000	18.5	1,722,193	368,549	214	2909
Yellow	1,365	1.3	944,970	147,415	156	292
Tarim	146	40.2	1,152,448	8,067	7	571
Total				**1,324,386**		

* Indicative

Source: IUCN/ IWMI, Ramsar Convention and WRI, 2003; Mi & Xie, 2002; Chalise & Khanal, 2001; Merz, 2004.

The most potent impact of glacial recession could be a severe imbalance in the volume of water of major Himalayan rivers. According to a climatic model developed in joint collaboration with the Centre for Ecology and Hydrology funded by the Department for International Development (DFID), UK under the 'Sagarmatha Project' (2004). There will be an increase in river discharge at the beginning causing widespread flooding in the adjacent areas. But after a few decades, this situation will reverse and water levels in these rivers will start declining to a permanent decreased level in a period of 100 years under different climatic scenarios. In the upper Indus, initially the mean flow as compared to baseline will increase between +14 per cent and +90 per cent in mean flows over the first few decades of the 100 year incremental scenario run, followed by –30 per cent to –90 per cent decline in flow by decade 10. For the Ganga, the response of the river at Uttarkashi, flows peak at between +20 per cent and +33 per cent of baseline within the first two decades and then recede to around –50 per cent of baseline by decade 6. In the headwaters of the Brahmaputra, there is a general decrease in decadal mean flows for all temperature scenarios; glaciers are few in this area and flows recede as the permanent snow cover reduces with increasing temperatures.

Availability of Fresh Water

Seventy per cent water is frozen in form of glaciers. The Glacial retreat would initially enhance run off but later the retreating glaciers shall pose a major problem for those dependent on rivers for fresh water supply for daily use, agriculture, industry and others. Glaciers are a continuous source of water for the major perennial rivers of the world. Three of the world's largest rivers, viz. Indus-Ganga and Brahmaputra are fed by the Himalayan Glaciers. According to varying estimates the total amount of water flowing from the Himalayas to the plains to be around 8643 km^3 per year. A reduction in the glacial ice could lead to a substantial decline in the volume of water of these major rivers. River Ganga could transform from perennial stream to a seasonal stream due to lack of water during the dry period during which Gangotri and other tributary glaciers are the primary source of water to this river. It could also increase the conflict for fresh water among the 500 million residents in these three major river basins.

Glacial Lake Outburst Floods

Glacial Lake outburst floods are catastrophic discharges of water resulting from melting of Glaciers. An accelerated retreat of the glaciers in recent times has led to an enlargement of several glacial lakes. As the glacier retreats it leaves a large void behind. Ponds occupy the depression earlier occupied by glacial ice. The moraine walls that act as dam are structurally weak and unstable and undergo constant changes due to slope failures and slumping and there exists the danger of catastrophic failure, causing Glacial Lake Outburst Floods (GLOFs). GLOF is characterized by sudden release of huge amount of lake water that rushes along the stream channel downstream in the form of dangerous flood waves. The propagation of GLOF surges trigger landslides and bank erosion that temporarily block the surge waves and result in a series of surges as the landslide dam breach. There are several such lakes in Uttrakhand Himalayas (127 Lakes) that are vulnerable to GLOFs.

Increased Flooding

Increase in the volume of water on account of glacial melting would increase the water levels in the various Himalayan rivers initially this could lead to unprecedented floods in these rivers causing immense not only to the settlements but also the vast tract of cultivated land in the Indo-Gangetic plains. This could lead to immense loss to life and property as well as farming the main occupation of the millions of residents of these plains.

Damage to Hydro Power Stations

A reduction in water from mountains can severely hamper the hydroelectricity generation prospects for India from the Himalayan Region. Hydropower potential of Himalayan river systems is about 78 per cent of the total Indian hydropower resources. This potential of Indus, Ganga and Brahmaputra river basins is assessed to be 19,988, 10,715 and 34,920 MW at 60 per cent load factor respectively. (Jagdish Bahadur, 1998) India at present is already facing a shortage of power by 10 per cent in the normal period and by 20 per cent in the peak period. Government of India is embarking upon renewable sources of energy with Himalayan region as a potential source for hydropower. However, this could be jeopardized if glacial melting continues unabated at the current pace. Further the increased possibility of Glacial Lake Outburst Floods can cause a severe damage to the various hydroelectric power stations that are already installed or are being installed in the Himalayan Region. This in turn would have an adverse impact on overall development plans of the country.

Impact on Agriculture

Agriculture is the primary occupation of more than 80 per cent of the 500 million residents residing along the Ganga, Indus and Brhamputra Basin. There could be a deleterious impact on agriculture. This would be on account of the imbalance caused in the volumes of water in the river basins of Ganga, Indus and Brhamputra, initially a rise in their water level could cause floods causing substantial damage to the cropped area whereas in later period due to reduction in volume of water, particularly during drier period lesser availability of water for irrigation could affect the irrigated area in the country. According to one estimate the irrigated area could reduce by 37 per cent. This would have a negative impact on the production of food grains in the country. In India, average food consumption at present is 550 g per capita per day. Present annual requirement based on present consumption level (550 g) for the country is about 210 million tonnes (mt), which is almost equal to the current production. However, the country faces major challenges to increase its food production to the tune of 300 mt by 2020 in order to feed its ever-growing population, which is likely to reach 1.30 billion by the year 2020. To meet the demand for food from this increased population, the country's farmers need to produce 50 per cent more grain by 2020 (Mall *et al.*, 2006).

An adverse impact on agriculture could cast its adversity on the industries directly as well as indirectly. The direct impact would be on the agro based industries that directly depend upon agriculture for their raw material whereas indirectly because a substantial proportion of market demand depends upon the health of agriculture, that is responsible for creating demand in the Indian economy.

Conclusion

The overall state of the glacial recession in the Himalayas due to global warming and climatic change is quite worrying. It can have deleterious long term effect on the ecology, environment and the socio-economic condition of the people in the Indian subcontinent. The impacts are likely to be hardest felt at the local level by the most vulnerable in society, particularly the women and children of poor families. Pressures on rural livelihoods may force many landless and poor farmers to seek better employment opportunities in the region's cities, a response, which itself, has numerous social, economic and cultural implications for both the rural and urban areas (Pebley, 1998). Yet, some 25-40 per cent of the urban population in developing countries already live in impoverished slums, with little or no access to water and sanitation (Word Bank, 1997; IPCC, 2001b). Consequently it requires a long term meticulous planning on our part to control the emission of greenhouse gases that are primarily responsible for the present chaotic situation. It is not only developing country like India but all the countries in the world that require to take a drastic step forward to control the existing situation for sustenance of these glaciers in future.

REFERENCES

Anthwal, A., Joshi, V., Sharma, A., Anthwal, S. (2006). 'Retreat of Himalayan glaciers—indicator of climate change', *Nature and Science* 4(4), Marsland Company, Michigan, U.S.A.: 53-59.

Bahadur, J. (1997). 'Himalayan ecohydrology: An emerging subject for restoration of the mountain system'. *Vigyan Prasar*, Department of Science and Technology, New Delhi.

Climate Change (2010). retrieved on 10th December, 2010, from http://en.wikipedia.org/wiki/Climate_change

Frei, A., Robinson, D.A. (1999). 'Northern hemisphere snow extent: regional variability 1972-1994'. *International Journal of Climatology* 19(14):1535–60.

Goudie, A. (1983). *Environmental Change*. Second Edition. Oxford: Clarendon Press: 258.

Kaul, M.K. (1999). *Inventory of Himalayan Glaciers*, Geological Survey of India, 34.

Kulkarni, A.V. (2007). 'Effects of Global Warming on Himalaya Cryosphere', *Jalvigyan Sameeksha,* 22, Published by Indian National Commitee of Hydrology, N.I.H. Roorkee, India.

Kulkarni, A.V. (2007). 'Glacial retreat in Himalaya using Indian Remote Sensing Satellite Data', *Current Science,* 92, No. 1, January, 2007. 69-74

Lemke, P., Ren, J., Alley, R.B., Allison, I., Carrasco, J., Flato, G., Fujii, Y., Kaser, G., Mote, P., Thomas, R.H., and Zhang, T. (2007) "Observations: Changes in Snow, Ice and Frozen Ground". In Solomon, S.D., Qin, M. Manning, Z. Chen, M. Marquis, K.B. Averyt, M. Tignor and H.L. Miller (eds.), *Climate Change 2007: The Physical Science Basis, Contribution of Working Group I to the Fourth Assessment Report of the Intergovernmental Panel on Climate Change*, Cambridge, United Kingdom and New York, NY, USA: Cambridge University Press.

Mall, R.K., Gupta, A., Singh, R., Singh, R.S., Rathore, L.S. (2006). 'Water resources and climatic change: An Indian perspective'. *Current Science*, 90(12), June: 1610-1626.

Naithani, A.K., Nainwal, H.C., Sati, K.K., Prasad, C. (2001). 'Geomorphological evidences of retreat of the Gangotri glacier and its characteristics', *Current Science*, 80 (1), January: 87-94.

Paterson, W.S.B. (1998). *The Physics of Glaciers*, Oxford: Pergamon Press: 318-321

Ramchandran, R. (2001). 'A valley type glacier', retrieved on 12th December, 2010 from www.frontlineonnet.com/fl1807/18070720.htm.

Rao P., Areendran, G. and Sareen, R. (2005). *Potential impacts of climate change in the Uttarakhand Himalayas*. WWF, India.

Retreat of glaciers since 1850 (2010) retrieved on 10th December, 2010 from http://en.wikipedia.org/wiki/Retreat_of_glaciers_since_1850

Thayyen, R.J. (2008). 'Lower recession rate of Gangotri Glacier during 1971-2004'. *Current Science*, 95 (1), July: 9-1.

WWF Report. (2005). *An Overview of Glaciers, Glacier Retreat, and Subsequent Impacts in Nepal, India and China*, ICIMOD, Kathamandu, Nepal,: 29-38

Jianchu, X., Shrestha, A., Vaidhya, R., Erikkson, M. and Hewitt, K. (2007). *The Melting Himalayas, Regional Challenges and Local Impacts of Climate Change on Mountain Ecosystems and Livelihoods*, International Centre for Integrated Mountain Development (ICIMOD) Kathmandu, Nepal.

21

Pattern of Land-Use Changes and Associated Environmental Issues: A Case of Coastal Districts of Kutch, Gujarat

Vishal Gupta and *A.A. Ansari*

Introduction

Land cover refers to the attributes of a part of the earth's land surface and immediate subsurface including biota, soil, topography, surface, groundwater etc. it describes the physical state of the land surface, as in cropland, mountains or forest. The term land cover originally referred to the kind and state of vegetation but has now broadened to include human structure such as buildings or pavement and other aspects of natural environment such as soil type, bio-diversity, surface and ground water.

Landuse refers to the purpose for which humans exploit the land cover. Land-use is the way in which and the purpose for which human beings employ the land and its resources. The common land use includes agriculture, grazing, mineral extraction, building, industrial usage, and recreation.

Land use and land cover play a major role in micro and macro level climatic changes as they lead to significant shifts in the earth atmosphere interactions. The resultant global climate change may in turn force changes in land use and land cover, a cyclic process that may culminate in desertification of land. The social and economical implications of this process have only recently gained recognition.

Environment as a point of concern has been the recent paradigm shift in all the studies, with response to extent and pace of environmental degradation on the globe. Degradation at large is induced in the environment through the alteration of land use practices. The post earthquake period has witnessed significance economic developments in Kutch. The Industrial growth in the region has been not so systematic. The environment is a resource for development if it is utilized skilfully and scientifically on sustainable basis to secure the future need. Successful and sustainable development requires policies that incorporate environmental consideration. Environmental degradation reduces both the quantity and quality of resources, and its impact over human being is spontaneous.

Environmentalists and conservationists have declared Kutch to be one of the most fragile ecosystems in the world. With large tracts of dry land turning waste due to salinity ingress and drought like conditions, Kutch is not in a position to meet the demand of growing population and sudden unprecedented growth in industries. In a four day conference on "Natural Resource Conservation, Use and Sustainability in Drylands", held in December 2006, Indira Highway (Director at Centre for Development Alternatives) said "Kutch is in this vicious circle of droughts which starts with a meteorological cause, followed by hydrological scarcity and then agricultural drought comes, followed by ecological drought. It is a high time that the government plans an integrated approach for sustainable use of natural resources in the region. The neglect has sounded the death knell for the region."

The link between the environment and development involves much more than exploitation of natural resource only, preserving and protecting the ecological equilibrium of our environment is a vital component not only of human development but, also for human survival. At present the Kutch is confronting and combating with multi dimensional environmental degradational problem erupting out of land use changes. The major factor for the same is rapid growth of population particularly along the coastal area and thereby deforestation of mangroves, over exploitation of marine fisheries, deforestation,

Reduction in wet land area, decrease in pastoral land and land degradation due to mining.

Forum for Planned Industrialisation (FPI) has developed a land use plan for Gujarat which recognises that traditional economies are dependent on natural resources. It has identified 13 economic sectors and proposes means for facilitating industrial development without hindering existing livelihoods. It is a fact that a more transparent approach with a good communication line is necessary between the developers, the government and affected communities, so that potential issues are understood, discussed in their infancy and mitigated as well. People's livelihoods are sustainable but they need assistance to improve their settlements and infrastructure. With this background the objective of this paper is to assess the problems associated with coastal and inland areas of Kutch districts because of the effect of human population and their related activities in redefining land use pattern. Apart from it, this paper also attempts to examine the nature of development and its impact on environment in order to put forth certain planning measures.

Study Area

Kutch district stretches roughly from 22°44′ to 24°41′ N latitude and 68°09′ to 71°54′ E longitude. Kutch with an area of 45652 sq.km is the largest district of Gujarat. It is about 23.29 per cent of the total area of the state. Geologically the region is formed of the youngest, upper quaternary formations, and consists of thick sandy and clayey layers of fluvial, fluvio-marine and aeolian origin. The district has an extensive coast line of 322 km, along with the international boundary bordering it from the north and west with Pakistan, on the north east by Rajasthan, on the east by Banaskantha district, on the south east by Rajkot, on the south by Gulf of Kutch and on the west by Arabian Sea. Kutch is a mosaic of landforms in a way that it has peaks, ranges, hills, plateau, alluvial plains, desert plain, grass land, saline encrusted waste land, area with laterite soils etc. The district has a vast reserve of mineral resources such as lignite, bauxite, limestone etc. that fuels the industrial development in the district.

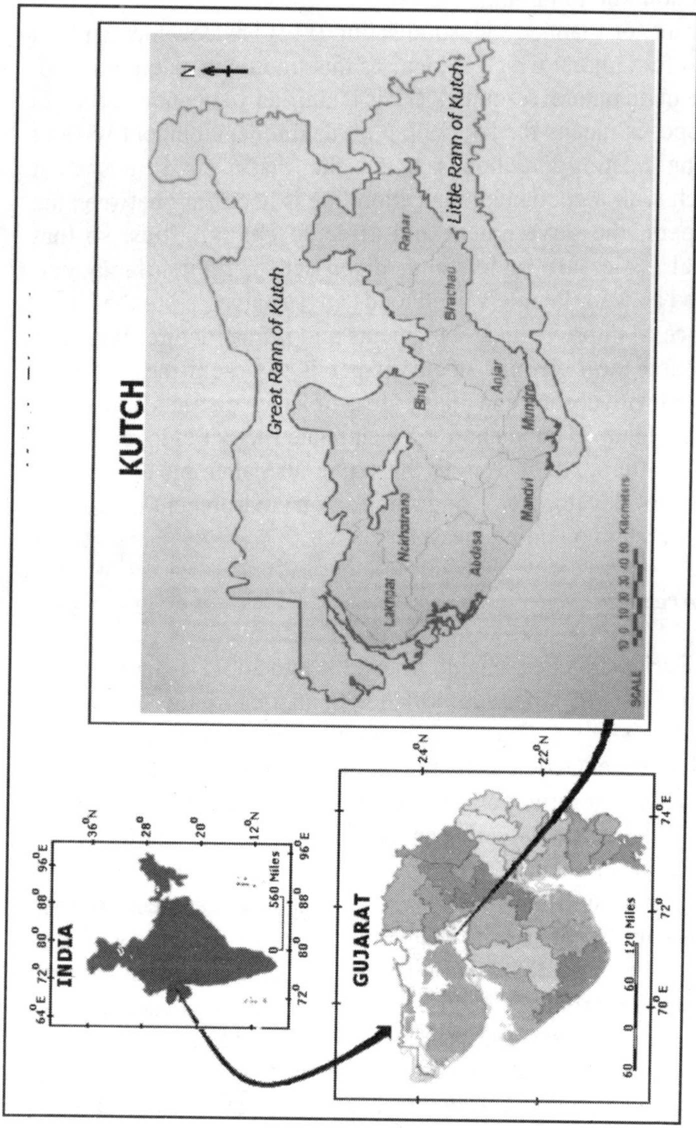

Map 21.1: Kutch District: Location Map

Methodology

This work was made feasible through collection and appraisal of all background information through secondary sources and extensive ground survey. The Survey of India Toposheet on million and quarter scale has also been used to study the physical characteristics of the study area. Statistical information used in this study has been collected from secondary source. Kutch being a restricted area, the accessibility to it has always been limited.

Population

Human beings on earth have been exposed to dual problem, one being the rapid growth of population and other is the pace at which environment is being degraded. According to population census 2001, the district had registered a population 15.83 lakhs which is little above three per cent of the total population of the state. The population of the district has increased almost three times during a period of 100 years that is from 1901 to 2001. On an average every year around 21,000 people are added to the population of the district creating a demand for additional resources. The density of population has increased from 23 persons per square km in 1981 to 28 persons per square km in 1991 and 35 persons per square km in 2001. The density of population is significantly lower because of the vast Ranns which is completely uninhabited. The coastal regions of Gujarat sustain over a million of population being spread over 2000 coastal villages. The Mundra Mandvi area has the highest concentration of rural people in the entire Kutch coast. This rapidly growing population laden with sophisticated technology and industrial development has had significant effects on the ecology and the land use pattern of the coastal and marine environment, in the last decade, the average growth rate of population in the district was 20-22 per cent (1991). Among the talukas, coastal Anjar taluka has recorded the highest growth rate of 38.58 whereas Mundra taluka recorded lowest rate of 8.59. The Decadal growth rate of Lakhpat and Anjar (Coastal Taluka) and Bhuj is greater than the district if taken together. Three talukas, i.e. Bhuj, Anjar (Coastal), Mandvi contribute about 86 per cent urban population of the district. The growing population have always been deterrent to the natural resources and environment. Increase in the population size has created vast gap in between the resources potential and the demand. There has been considerable change in the land use pattern in terms of cultivated area, built up area, forested land etc. of this district. There has been increase of 15 per cent in the cultivated area in the period (1961-1993). All this has increased the pressure on soil. To congregate the demand of increasing population, an intensive ruthless agriculture is being practiced.

Table 21.1: Population Density in Different Talukas of Kutch District

Taluka	*1961*			*1991*		
	Area (sq.km.)	*Population*	*Density*	*Area (sq.km.)*	*Population*	*Density*
Abdasa	2244.8	65118	29.01	2398	86402	36.03
Anjar	1348.3	106189	78.76	1312	265225	202.15
Bhachau	1487.9	68334	45.93	1985	114759	57.81
Bhuj	5525.1	127421	23.06	4457	277215	62.20
Lakhpat	2369.9	18401	7.76	1945	36759	18.90
Mandvi	1352.8	101501	75.03	1406	146034	103.86
Mundra	872.3	54425	62.39	888	68652	77.31
Nakhatrana	1601.6	69632	43.48	1985	116944	58.91
Rapar	2737.6	85419	31.20	3024	150517	49.77
Total	19540.3	696440	35.64	19400	1262507	65.08

Source: District Statistical Abstract, *District Census Handbook*, 1991, Govt. of Gujarat.

Water

Water is the most precious resource of all the resources, especially in the arid and semi-arid environment of Kutch. Out of the 200 million litres per day of the Narmada river water given to Kutch, only 45 million litres per day was allocated for the industry before the 2001-quake. Today, after the promotions and invitations the new industry demand alone stands at 300 million litres per day. The shortfall of 255 million litres per day is being met by a tanker economy based on dangerous ground water suction. Kutch is a traditional dark zone owing to recurrent droughts. Ground water sources are being depleted resulting in lowering down of water table.

Livestock Population

Livestock population have played a significant role in the economy of agrarian society in general and particularly in the economy of arid and semi-arid areas. In the last few decades, there has been tremendous increase in the number and impact of livestock population on the grasslands. The area under grassland, however remains same or is being reduced and degraded eventually under the population (both man and livestock) pressure. Livestock rearing is the chief source of livelihood for the people of Kutch. The population of livestock has increased from 9.4 lakhs in 1962 to 14.1 lakhs in 1992 showing an increase of more than 50 per cent within last 30 years. Because of the excessive livestock rearing, which is the one of the chief source of economy in the region has not only increased the pressure on land but is also destroying the ecology of grassland through relentless grazing. The extent of Banni, Asia's biggest grassland is shrinking because of illegal and over exploitation of the grasses.

Fig. 21.1: Multiple Cropping, Mandvi

Table 21.2: Potential Grazing Areas ('00 Ha.) Available in Kutch

Land Use	*Year*	
	1961–62	1991–92
Forest	1954	2860
Barren and Uncultivable waste	28760	17082
Tree groves	56	–
Cultivable waste	2337	16639
Current fallow	743	1847
Other fallow	1888	235
Pasture	685	701
Total Grazing land	36423	39364
Total Livestock Population ('00)	9401	14126

Source: Process of Desertification in Kutch and Banaskantha Districts of Gujarat, India, GEC.

Reduction of Pasture Land

The long history of industrialisation in Gujarat and the conversion of 'common' land for other purposes caused many pastoralists to lose the grazing land they depended upon. In 1973, the government passed an order allocating 16 ha of *gauchar* (grazing) land per 100 animals per village and prohibited the selling of such land for any other purpose. But nine villages have lost their *gauchar* land to the Mundra SEZ. Land devoted to pasture area has been decreasing while the numbers of livestock have increased 4 fold within a period of about 60 years. Early 1996, the grassland areas (Fig. 21.2) have been reduced from 3.3 million hectares in 1975 to 2.9 million hectares (1995). There have been considerable changes in the Banni Grassland over the period. Earlier the same Banni plain was known for Elephant grass, which now is devoid of simple grasses. Because of the change in the natural ecosystem, trends in arrival of migratory birds have also affected. There is a great concern over the presence of thin layer of salt on the soil of Banni plain. In the maximum part of rann area there is an alternate layer of salt and silt which is the result of ingression of saline water towards the land from sea. The fresh water which moves towards the Banni plain from the northern slope of Bhuj ridges has also been dammed for drinking and agricultural purposes and thereby increasing the salinity of Banni plain.

Fig. 21.2: Banni Grassland

Fig. 21.3: Panoramic view of Rann from Barren Kala Dungar

Deforestation

Gujarat stands far below with its 6.4 per cent of area under forest as compared to the national average of 19.27 per cent. Between 1957 and 1975 about 2.5 million hectares of forest were denotified for agriculture. Further 51913.3 hectares were again denotified to make room for developmental projects. Unauthorized encroachment in the forest for the purpose of agriculture and settlement have been rampant The Kandla port itself has an illegally captured forest area of 18.32 hectares since 1978. Apart from the anthropogenic activities, eco-climatic condition of the area also do not support vegetation, there is more of barren and unculturable land (Fig. 21.3).

Land Degradation due to Mining

The mineral resource and mining have a major role in industrial and economic development. On the other hand, indiscriminate mining of the non-renewable resource not only depletes the natural source, but also disturbed the ecological set up of the region. Mining results into the loss of productive land.

Mining of salt, sands, limestone and lignite coal (Fig. 21.4) has played an important role in exposing this geologically and climatically vulnerable district to other arena of hazards. Mining has not only yielded degradation of soil but also affects the surface and groundwater. Mining of salts in the region has made the area more acidic. The excessive salt content in the water during extraction of salt choke up the mangroves to death. Mining of sand alters the beach morphology directly while its indirect effects are in terms of movement of silt, both onshore and offshore. Stabilization of coastal area witnesses threat from the advancing waves, currents, tides and cyclonic winds etc. as a result of sand mining. The adverse affect of mining of lignite coal in the Punandhro Mines can be well seen in the form of water, air and land degradation for an area of more than 5 km of radius. It is an open cast mine of 1800 hectares and is located near Narayan Sarovar sanctuary, causing repercussions like cutting of trees, soil degradation, vehicle pressure (around 2000 trucks per day), etc. A huge area has been devoted for the same. Pre-mining activities, i.e. removal of soil from the site and further

Fig. 21.4: Lignite Coal Mine, Punandhro

Fig. 21.5: Salt Procurement, Kharaghoda

dumping has completely changed the natural ecosystem. The fly ash can be seen on the ground, leaves of the trees, stagnant water etc. or anything that comes in the way. Release of fly ash in the environment results in considerable change to tree and shrubs species, diversity of the surrounding areas are likely to affect the habitat and threaten species like the chikara, wolf and caracae in and around project area. The excavated/mining area and filled up area, both are devoid of any vegetative cover. The mined area seems to be turned upside down, entire filled up area has been extensively rilled and gullied. Bentonite mining has also adversely affecting top soil in Mandvi taluka. Limestone mining by the cement industry is creating its own hazards.

Development of Industries

Before 2001, salt, cattle and traditional craft constituted Kutch's economy. Now it is one of the fastest growing economic and industrial hub of Gujarat. Its geographical location is very favourable for Transportation and Trade. It has two major ports, Kandla and Mundra which are closest to Gulf countries and Europe by sea route. Road qualities are also one of the best in India. Kutch is rich in minerals, with very large reserve of Lignite, Gypsum and lot of other minerals. Kutch has got tax break for Industries for 15 years after the major earthquake in January 26, 2001. And now houses Asia's largest Cement Plan, Sanghi Cement. Kandla—India's largest port by cargo handling. Mundra is one of the most sophisticated high-tech ultra modern port. Mandvi has one of the largest windmill farms concentration. Kutch region produces 70 per cent salt of India. The Belts From Towns (1) Shamkhiyali-Bhachau-Gandhidham-Kandla, (2) Gandhidham-Mundra, (3) Gandhidham-Adipur-Anjar-Bhuj and (4) Bhachau-Bhuj, houses hundreds of Industries and is known as Silver corridor of India. The major Industries in Kutch are Import-Export, Transportation and Cargo Handling, Cement, Mining, Ports, Salt, Tourism, Agriculture, Animal Husbandry, Timber, Real Estate, Edible Oil (More than 50 per cent of India's edible oil is processed here) and lot more.

The environment is in the state of equilibrium till it is unhindered by the anthropogenic activities, but the fact is that the coastal environment has been degraded as a result of the growth of Industries. Industrial discharge has been hazardous to the fishes and detrimental to the growth of coral reefs. Estuaries environment is being destroyed through dumping from ship, Pipe line discharge, industrial and municipal waste, sewage sludge and dredged material containing heavy metals, toxic non-metals, chlorinated hydrocarbons, petroleum hydrocarbons oil spillage (burnt engine oil/crude oil/vessel to vessel transport). Several reefs have been gradually degraded due to lifting of sand and corals for the cement industry. However, these have been checked by declaring the area as national parks and sanctuary. Construction of national highway to provide basic infrastructure to the industries is another indirect effect of industrialization.

The Gujarat government declared Kutch as an "Imported Timber Conversion Zone (ITCZ)", freeing the region from the 2002 Supreme Court ban on wood-based industries to put brakes of forest depletion. According to the Kutch Timber Association Kutch is about to become Asia's largest centre for wood-based industries, this would translate into more than 100 plywood units, another 200 saw mills, and 50,000 additional jobs in talukas of Gandhidham, Anjar, Mundra and Bhachau.

Mangroves

Mangroves are the natural barriers against cyclones, Tsunamis and Salinity ingress besides providing livelihood for hundreds of fishermen on the coast. As per the report (2001) of marine and water resources division the land cover of Mangroves have been reduced significantly. Clearing of mangroves of Gulf of Kutch for developmental purpose will have a severe effect and is not too far. The Bocha island area which was supposed to be left undisturbed is also being cleared by excavators and bulldozers. The growths of Industries and salt pans have made a large encroachment on the area, which were earlier covered by mangroves. The bunds which, are made around the mangroves to trap the saline water from the sea had has an adverse effect on mangroves and gradually the salt chokes them to death. Also, use of mangroves in the form of kitchen fuel wood and charcoal (Fig. 21.6) is widely practiced.

Fig. 21.6: Charcoal made out of Mangrove Timbers

The projects like Sanghi and Adani have contributed to large scale destruction of patches of mangroves and have disturbed the only means of livelihood of local people. However, these industries may become boon provided it adopts the sustainable approach. In between 1975 and 1982 the Gulf of Kutch has lost 72.5 per cent of mangrove area. Mundra region has also lost its 260 hectare of mangroves within a period of 11 years, i.e. 1988-1999. Effects of destruction of Mangroves is manifold such a destablishing of the coastal area, which will again have several indirect effects on human beings, loss of Biodiversity, deposition of sediments on coral through back tidal impact, local weather phenomenon etc. Till a decade ago there were a variety of mangroves which attracted more than 100 species of fishes to the shores. But now only three species are found and they are Avacenia marina, Rhizophora and Ceriops.

Over Exploitation of Marine Fishes and Corals

The mangroves shelter juvenile fish and microorganisms that form the basis of the food chain, loss of mangroves means a reduction in fish-catch. There is reportedly a 30 per cent reduction in the fish catch already. Deepening of shipping channels to allow access to cargo vessels with a draught of up to 17.5 metres destroys the benthic environment that many commercial fish species depend upon. Increased shipping traffic destroys fishing nets as well. Increasing pressure of population in the coastal area has lead to the over exploitation of marine fishes. In between 1998-99, there was a significant fall is the catch, i.e. around 5.61 lakh tonnes to that of an annual figure of 7.02 lakh tonnes. Fishes of export quality has gone down drastically in the recent decades and which may have serious repercussions. Fishing activity is developed through out the coastline of Kutch, however fish catch are significant at Mandvi, Mundra, Navlakhi and Jakhayu. Millions of tonnes of crude oil enter the Gulf of Kutch and the Mundra port has plans to develop berths for handling chemicals. Deep within the gulf, the waters around the Mundra port experience a slow rate of water recycling and therefore, any pollution leak from industry would take a long time to be flushed out, stagnating, suffocating and killing the marine life. The coastal areas have also been exploited by the growing population in terms of extraction of corals; development of ports and other infrastructures has exploited the coastal areas. All these have led to severe siltation on the off shore reefs, also the effluent discharge from the industries and the houses has increased the pressure on ecosystem. Latest available record says that the coral reef area has decreased by 43 per cent since 1975 (Kohli, 2002). Corals, a most fragile ecosystem if conserved properly can be of great utility for mankind. Coral increased the shore stability and also offers protection of natural disasters. Coral like paracyanthus indices has been declared endangered and many like Acropra species is probably extinct.

Reduction of Wetland

Kutch is rich in wetlands (A treasure house of biodiversity) of international significance. Migratory birds from Central Asia and Russia visit these places during winter but in recent times, however, urbanization, agriculture development, industrialization, mining activity and other anthropogenic interference have adverse effect on wetland. The decrease in wetland areas in Kutch, a cause for concern, shrunk from 1.5 million hectare (1971) to 1.07 million hectare (1992). Narayan Sarovar, low-lying areas of Banni Plain, Dhaulavira are the important sites where the water foul visits regularly.

Sea Ingress

Removal of mangrove trees along the coast has led to the sea-ingress in the land area in the immediate vicinity. The trees not only prevented sea-ingress, they allowed the coastal water to remain pristine allowing unique sea life to survive near the shore. Because of the sea ingress several fishes have simply vanished. Also, large area of grasslands including useful medicinal plants such as *Piloo,* which is also a source of fodder for the cattle and goats next to mangroves have been destroyed because of sea ingress suffered by various areas. The marginal sites which has front with the Ranns and the adjoining sea are very much affected by the sea ingress. North of Lakhpat, Jakhayu Port, Navlakhi, the western part of Little Rann of Kutch and the west of Khavda hills, all are affected by the ingress of sea.

Major Land Use and Environmental Changes Identified in the Study Area:

- Change of hydrology and the flow of sediments.
- Desertification and dune encroachment along the coastal areas.
- Mangroves destruction through extensive wood utility.
- Ecological habitat destruction due to coastal development.
- Construction of aquaculture ponds in the coastal area.
- Damage of coral reefs through unsustainable fishing practice.
- Removal of corals for the tourist trade.
- Construction of coastal highway right on beach and dumping of rubbles and rubbish materials seaward.
- Physical damage caused by tourists diverse in shallow coastal waters.
- Over-exploitation of coastal resources.
- Increased coastal and marine problems.
- Unplanned urban development, disposal of raw materials and untreated Industries wastes.

Driving Forces for Land Use Changes

- Rapid growth of population.
- Poverty among the coastal communities.
- Lack of alternative livelihood, gradual decline in fish catch because of over exploitation.
- Profit making large enterprises which mercilessly exploit the resources, playing with the interest of local communities.
- Lack of awareness about management and resource sustainability among the peoples.
- Lack of serious efforts by government in implementing conservational programmes.

Measures Required to Prevent Further Damages

- Development should be in consonance with the environment.
- All projects must follow Environment Impact Assessment.
- Industrial Development Corporation must see that all company must follow the norms pertaining to environmental issues.
- Reforestation should be initiated in all those areas where mangroves have depleted. And in future, efforts should be made to prevent such degradation.
- To ensure rational development of the region, Coastal Zone Management should be implemented. Environmentally effective coastal management depends upon accurate and comprehensive scientific data, based on which policy can be made. The basic problem lies in the non-availability of geographic data of coastal area.
- Survey on environmental and social impact due to industrialization should be mandatory before giving clearance to the industries.
- Integrated Coastal Zone Management should be implemented. It is relatively a recent concept, which aims to utilize the natural resources in perpetuity, by periodically taking stock of the environmental response to use and harvest pressure and incorporate such indices in planning and implementing economic development.

- Mundra mangrove reserve forest and the Jakhau area in the Gulf of Kutch have been identified as specific mangrove sites by the Gujarat Ecology Commission, and is in the process of restoration.

The coastal zone is a geographical zone unlike true ocean or land yet lying in between the regime of ocean and land. To meet our future demand our next goal is to exploit the littoral resources such as coral reefs, mangroves, fisheries, fauna, beaches etc.

Integrated programmes to manage land and coastal water resource, i.e. Coastal Zone management programmes has really helped a lot in uplifting the economic and social prosperity of the coastal communities. Improved fisheries productivity, increased tourist flows and revenue, pollution free coastal zone and warning of natural calamities and protection of lives and property from sea storm are among the practical benefit of coastal zone management. Conservation of the resources of Coastal zone often clashes with coastal area development and land use changes.

Conclusion

At present the Kutch district, particularly the coastal area is exposed to intense and sustained environmental pressures from a range of driving forces which have been increasing in their intensity over many decades. The problems along the coastal area is increasing exponentially because of rapid growth of population and the rate and extent of land use change which crop up through transforming it to special economic zone. Land use changes in Kutch is largely regulated by an integrated complex of socio-economic and bio-climate interrelationships. The resources of coastal area are being utilized recklessly by the ever-increasing coastal population and in a state of destruction because of over exploitation if it is not checked in time.

REFERENCES

Government of Gujarat (1991). *District Statistical Abstract, District Census Handbook* 1991, Government of Gujarat.

Gujarat Ecology Commission and Gujarat Institute of Desert Ecology, April 2000, *Ranns and Desertifications (SEAP, Gujarat), Phase I Report*, Vadodara, Gujarat.

Kohli, S. (2002). "Landuse and environmental change due to recent development in coastal region of Kutch district." (unpublished) Ph.D. thesis, Deptt. of Geography, M.S. University of Baroda.

Process of Desertification in Kutch and Banaskantha Districts of Gujarat, India, GEC.

22

Arsenic Pollution and Its Sustainable Eco-remediation from Aquatic Ecosystems: An Overview

Prabhat Kumar Rai

Introduction

Population development and environment are inextricably linked with each other. Generally, there are two school of thoughts, one advocating development in view of expanding population, while, other emphasizing the issue of environment protection and sustainable development. In order to calm down the heat generated from this debate, a middle path making an equitable balance between development and environment is the need of the hour.

The larger problems of environmental degradation caused by excessive and uncontrolled use of water and chemical substances pose a big challenge to researchers as well as policy makers (Rai, 2011). In 2001, for example, more than six billion pounds of the 522 chemicals listed on the Toxics Release Inventory (2001) were released both on- and off-site in the U.S. (see Kirchhoff, 2005). There are over 70,000 chemicals in use in the world, and over 4,000,000 on the American Chemical Society's computer list of chemicals (Rai, 2008a; Rai, 2009a; Rai, 2010c; Rai, 2011). Although chemicals (inorganic and organic) provide society with a wide range of benefits, particularly increased agricultural and industrial productivity, however, they have the potential to cause considerable health and environmental problems throughout their life cycle, from production through to disposal (Rai, 2011). The rapid pace of industrialization and urbanization has given birth to the problem of water pollution of the aquatic ecosystems (Rai *et al.,* 2007; Rai and Tripathi, 2007b; Rai, 2007a; Rai, 2007b; Rai, 2008a; Rai, 2008b; Rai, 2008c; Rai, 2008e; Rai, 2009; Rai, 2009a; Rai, 2009b; Rai, 2009c; Rai 2010a; Rai, 2010b; Rai, 2010c; Rai, 2011).

Arsenic in drinking water sources is a cause for concern in many parts of the world (Table 22.2). However, in Indian subcontinent, large-scale arsenic contamination in the Gangetic Delta region in Bangladesh and India have generated significant amount of media attention (Chakraborti, 2008). In Bangladesh...groundwater concentrations of arsenic reach 2 mg l^{-1}; the World Health Organization's provisional guideline value for drinking water is only 0.01 mg l^{-1} (Ruiz and Romero, 2002). In this part of the world alone, more than 100 million people are presently at risk of developing arsenic related health hazards (Mudhoo, 2011). In Bangladesh arsenic contamination is possibly the largest mass poisoning case in the world now. The Green Revolution has been identified theoretically to be the recent cause of the problem which has involved large-scale unplanned withdrawal of groundwater, gradually denuding the arsenic deposited under the fertile delta of Bangladesh millions of years ago by the rivers from the Himalayas or some other source (Mudhoo, 2011).

For many years, the greatest use of arsenic was for poisons, and in particular, insecticides. Lead arsenate, $Pb_3(AsO_4)_2$ was a common insecticide. Arsenic has numerous applications as a semiconductor and other electronic applications as indium arsenide, silicon arsenide, and tin arsenide. Besides several industrial uses of arsenic derivatives, arsenic is finding increased uses as a doping agent in solid-state devices such as transistors and an essential component in cancer therapy. Gallium arsenide is used as a laser material to convert electricity directly into coherent light. Arsenic is used in bronzing and for hardening and improving the sphericity of shot (Mudhoo, 2011).

Despite some uses, arsenic poses serious threat to living beings particularly humans. Arsenic causes genetic toxicity, developmental and reproductive toxicity (Chakraborti *et al.*, 2003), immunotoxicity, biochemical and cellular toxicity (Santra *et al.*, 2000), and chronic toxicity (Mudhoo, 2011). The mechanisms involved in arsenic toxicity can be both cytotoxic and genotoxic. Cytotoxic mechanisms result in cell death and examples include the inactivation of cell enzymes, and allosteric inhibition (Liu *et al.*, 2001; Mudhoo, 2011). Arsenic appears to induce predominantly large mutations resulting in nonviable cells, such as those observed using a human-hamster hybrid cell mutagenic assay, where arsenite induced the deletion of millions of base pairs at the CD59 locus (Mudhoo, 2011). Many heavy metals have affinity for sulfydryl bonds and can alter protein structure, leading to disruptions of metabolic processes (Gochfeld, 1997; Mudhoo, 2011). The binding with sulfydryl groups by arsenite compounds has the potential to influence a wide range of metabolic activities including cellular glucose uptake, gluconeogenesis, fatty acid oxidation, and production of glutathione (Jones, 2007; Mudhoo, 2011). Rossman (2003) mentioned the mechanism of arsenic carcinogenesis and presented the arsenic toxicity mechanism in humans. In humans, arsenic compounds are metabolized by methylation followed by excretion in the urine (Rossman, 2003).

There are several global historical landmarks and stories related with the misuse of arsenic. Among those it is worth to mention the story of "Death by wall paper" related with French Emperor Napoleon Bonaparte. After winning the War of Waterloo, the British force reportedly used arsenic to kill French Emperor Napoleon Bonaparte (Mudhoo, 2011). Due to its use by the ruling class to murder one another and its potency and discreetness, arsenic has been called the Poison of Kings and the King of Poisons (Mudhoo, 2011). After his death on May 5, 1821 (Hindmarsh and Corso, 1998), it was announced that Napoleon died due to stomach cancer. But many years after his death, the hair of Napoleon, reserved at a museum, was tested and it was proved that he was killed by arsenic poisoning. Thus, in 1960, activation analyses at the Harwell Nuclear Research Laboratory of the University of Glasgow, London, of authenticated hairs of Napoleon Bonaparte taken immediately after his death confirmed Napoleon's chronic arsenic poisoning on the island of St. Helena. The arsenic concentration in Napoleon's hair was 13 times more than the normal level and it was considered to be the reason of his death (Mudhoo, 2011).

Arsenic (atomic number 33) is ubiquitous and ranks 20th in natural abundance, comprising about 0.00005 per cent of the earth's crust, 14th in the seawater, and 12th in the human body (Mandal and Suzuki, 2002; Mohan and Pittman Jr., 2007). Arsenic is a naturally occurring element that can be found in atmospheric, marine, mining, freshwater, groundwater and terrestrial environments (Mandal and Suzuki, 2002; Mudhoo, 2011). Its concentration in most rocks ranges from 0.5 to 2.5 mg kg^{-1}, though higher concentrations are found in finer grained argillaceous sediments and phosphorites (Pendias and Pendias, 1989; Mandal and Suzuki, 2002; Mohan and Pittman Jr., 2007). It is a silver-grey brittle crystalline solid with atomic weight 74.9; specific gravity 5.73, melting point 817°C (at 28 atm), boiling point 613°C and vapour pressure 1mm Hg at 372°C. Since its separation in 1250 A.D. by Albertus Magnus (Mandal and Suzuki, 2002; Mohan and Pittman Jr., 2007), this element has been a prime cause of water quality perturbations. The WHO provisional guideline of 10 ppb (0.01 mg L^{-1}) has been adopted as the drinking water standard. However, many countries have retained the earlier WHO guideline of 50 ppb (0.05 mg L^{-1}) as their standard or as an interim target including Bangladesh and China. In 2001, US-Environmental Protection Agency (UEPA) published a new10 ppb (0.01 mg L^{-1}) standard for arsenic in drinking water, requiring public water supplies to reduce arsenic from 50 ppb (0.05 mg L^{-1}). The flow of arsenic in global cycle is shown in Fig. 22.1.

Problem of Arsenic (As) Pollution in Aquatic Ecosystems

Arsenic (As; a semi-metal or metalloid) is one of the most hazardous contaminants that may be present in the aquatic environment. It derives its origin from both natural and anthropogenic sources. As contamination in aquatic ecosystem pose serious threat to aquatic biodiversity and drinking of contaminated ground water pose severe health hazards in humans. As hazards, lead, mercury, arsenic, and cadmium are ranked first, second, third, and sixth hazards respectively on the list of the US Agency for Toxic Substances and Disease Registry (ATSDR), which lists all hazards present in toxic waste sites

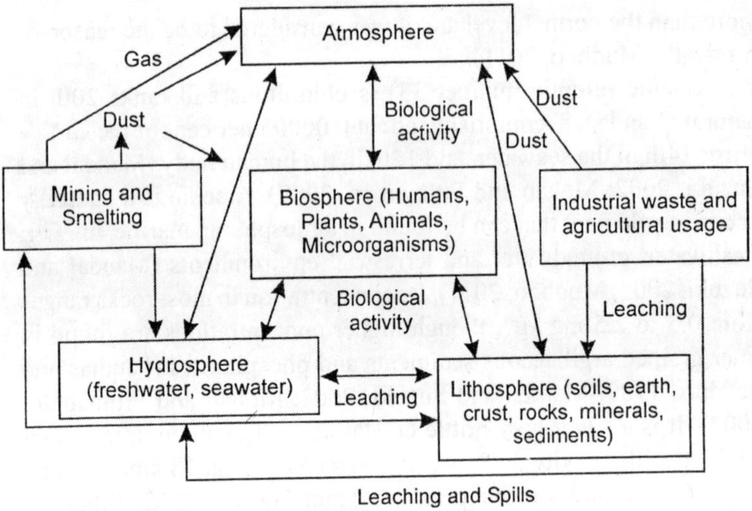

Fig. 22.1: Flow of Arsenic in the Global Cycle (Adapted from Mudhoo, 2011)

according to their prevalence and the severity of their toxicity. Arsenic is particularly deleterious in aquatic ecosystems due to the process of bio-methylation, leading to formation of methyl arsine, methylarsonic acid etc. (Fig. 22.2).

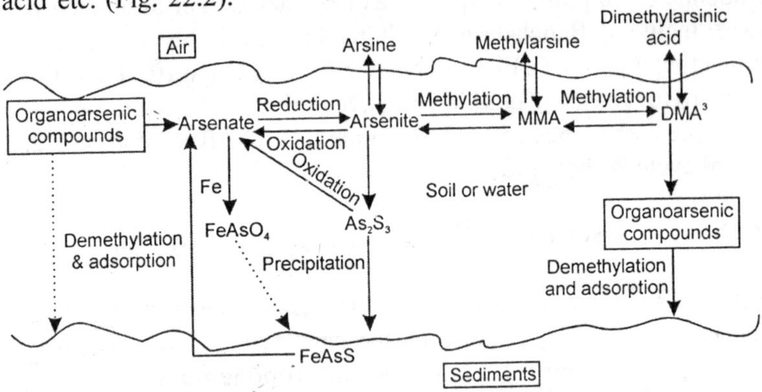

Source: Roy and Saha, 2002, p. 38; Jones, 2007, p. 2; Mudhoo, 2011, p. 435.

Fig. 22.2: Chemical Conversions Associated with Arsenic in Environment

As far as dispersion or transport of arsenic in natural waters is concerned there may be several modes as mentioned below (Mudhoo, 2011):

1. release through rock weathering,
2. transport and deposition by water and ice,
3. release of arsenic to the ground and surface water due to increasing *p*H,
4. accumulation in humic soil horizons,
5. remobilization by reduction in water-saturated soils and stagnant ground waters,
6. remobilization by using P-rich fertilizers or dung, and
7. oxidation, precipitation, and dilution in surface waters.

Recently, in developing countries major problem have arisen such as ground water contamination with Arsenic in Bangladesh (Alam *et al.,* 2003) and its simultaneous bio-concentration inside vegetables (Queirlo *et al.,* 2005). Global reports indicated that among 21 countries in different parts of the world affected by groundwater arsenic contamination, the largest population at risk is in Bangladesh followed by West Bengal in India (Mohan and Pittman Jr., 2007). Ground water contamination in the eastern part of Uttar Pradesh, India should be monitored through extensive networking projects since several districts, e.g. Ballia is severely affected. Human health effects of arsenic include cardiovascular problems, skin cancer and other skin effects, peripheral neuropathy and kidney damage (WHO, 1997).

At the current time, Arsenic poses a serious threat as mentioned in introduction. Therefore, their removal is an essential prerequisite, prior to disposal into the aquatic ecosystems.

Literature survey revealed scarce work on phyto-accumulation of Arsenic, a metalloid which has drawn recent environmental concern. Therefore, more concentrated works are recommended in this regard. Rahman *et al.* (2007) showed that the *Spirodela polyrhiza* L. accumulates arsenic by physico-chemical adsorption and via the phosphate uptake pathway when arsenate was added to the solutions. Their results indicated that *S. polyrhiza* L. would be a good arsenic phytofiltrator.

High Arsenic concentrations have been reported recently from the USA, China, Chile, Bangladesh, Taiwan, Mexico, Argentina, Poland, Canada, Hungary, Japan and India (Mohan and Pittman Jr., 2007). Arsenic (As; a semi-metal/metalloid) pollution in ground water is much prevalent in developing countries like India and Bangladesh (Rai, 2009a) causing skin lesions culminating in to cancer.

Desirables of Arsenic (As) Treatment Technologies

The chemicals management methodology for metals with reference to As should be:

- Compatible;
- Cost effective;
- Flexible enough to handle fluctuations in the quality and quantity of effluent feed;
- Reliable by operating continuously;
- Robust enough to minimize supervision and maintenance;
- Selective enough to remove only the contaminant metals under consideration; and
- Simple enough to minimize automation and the need for skilled operators (Eccles, 1999).

Rarely, any chemical technology is able to address aforesaid measures in totality. In general, conventional technologies, e.g. chemical precipitation, ion exchange, sedimentation, micro-filtration and reverse osmosis are not eco-friendly and concomitantly contaminate the aquatic bodies (Rai, 2009a). Table 22.1 provides an overview of the treatability of physico-chemical treatments for industrial effluent with their advantages and disadvantages.

Low-Cost Adsorbents for Arsenic Removal

Use of several low cost adsorbents were attempted for the removal of Arsenic as conventional chemical methods like commercial and synthetic activated carbons, air oxidation, chemical oxidation, alum coagulation, iron coagulation, lime softening, activated alumina, iron coated sand, ion-exchange resin, nanofiltration, reverse osmosis, electrodialysis, foam flotation and many more were experimented (McLellan and Rock, 1986; Viraragharan and Ayyaswami, 1987; Couillard, 1992, Couillard, 1994; Allen and Brown, 1995; Nasir *et al.,* 1998; Brown *et al.,* 2000; Altundogan *et al.,* 2000; Gupta and

Table 22.1: Summary of the Treatability of Physico-Chemical Treatments for Industrial Effluent (Type of Treatment, Advantages, Disadvantages and References) [Modified after Rai, 2009a; Rai, 2009c]

Sl. No.	*Conventional methods*	*Advantages*	*Disadvantages*
1.	Chemical precipitation	Low capital cost, simple operation	Sludge generation, extra operational cost for sludge disposal
2.	Coagulation–flocculation	Shorter time to settle out suspended solids, improved sludge settling	Sludge production, extra operational cost for sludge disposal
3.	Dissolved air flotation	Low cost, shorter hydraulic retention time	Subsequent treatments are required to improve the removal efficiency of heavy metal
4.	Ion exchange	No sludge generation, less time consuming	Not all ion exchange resin is suitable for metal removal, high capital cost
5.	Ultrafiltration	Smaller space requirement	High operational cost, prone to membrane fouling
6.	Nanofiltration	Lower pressure than RO (7–30 bar)	Costly, prone to membrane fouling
7.	Reverse osmosis	High rejection rate, able to withstand high temperature	High energy consumption due to high pressure required (20–100 bar), susceptible to membrane fouling

Ali, 2002; Fan *et al.,* 2004; Mohan and Chander, 2006a; Mohan and Chander, 2006b; Sneddon *et al.,* 2005; Mohan *et al.,* 2007).

Immobilization and adsorption of Arsenic on rice husk (Nasir *et al.,* 1998), lignite (Allen and Brown, 1995; Mohan and Chander, 2006a; Mohan and Chander, 2006b; Sneddon *et al.,* 2005), peat (McLellan and Rock, 1986; Viraragharan and Ayyaswami, 1987; Couillard, 1992, Couillard, 1994; Brown *et al.,* 2000), hydroxylapatite (Sneddon *et al.,* 2005), char (Fan *et al.,* 2004; Mohan *et al.,* 2007) and red mud (Altundogan *et al.,* 2000) were relevant researches pertaining to development of low cost adsorbents for Arsenic. Also aforesaid adsorbents attempted to remove the adverse impact of harmful carbon (Gupta and Ali, 2002). However, these adsorbents were unable to fulfil all the above-mentioned criteria of good adsorbents; therefore, the scientific quest was towards the eco-remediation of Arsenic through phytoremediation technology.

Wetland plants can absorb pollutants in their tissue and provide a surface and an environment for microorganisms to grow (Rai, 2009a). These plants stimulate both aerobic decomposition of organic matter and the growth of nitrifying bacteria (Rai, 2008a). In aquatic ecosystems of tropics, bio-agents, specifically, wetland plants are preferred in heavy metals management from aquatic ecosystems due to low cost, frequent abundance in aquatic ecosystems and easy handling (Rai, 2008a; Rai, 2009a).

Eco-Remediation of Arsenic (As) Management: Phytoremediation

As discussed earlier in this paper, problem of Arsenic pollution of groundwater is particularly frequent in developing countries, there is an integrated effort going on for its phytoremediation. Table 22.2 lists the countries facing the problem of Arsenic with their permissible limit.

Table 22.2: Countries Affected by Arsenic Contamination and Maximum with Permissible Limits for Drinking Water

Country	*Maximum Permissible Limits (µg/L)*	*Country*	*Maximum Permissible Limits (µg/L)*
Argentina	50	India	10
Taiwan	10	Mexico	50
Bangladesh	50	USA	10
Vietnam	10	Nepal	50
China	50	New Zealand	10
Chile	50		

Source: Mohan and Pittman Jr., 2007, p. 53.

Initially, some low-cost adsorbents listed for As treatment including treated slags, carbons developed from agricultural waste (char carbons and coconut husk carbons), biosorbents (immobilized biomass, orange juice residue), goethite and some commercial adsorbents, which include resins, gels, silica, treated silica, however, immobilized biomass adsorbents offered outstanding performances (Mohan and Pittman Jr., 2007). Therefore, phytoremediation technology may be a potent alternative for As removal in aquatic ecosystems. Pros and cons involved with phytoremediation technology are mentioned in Table 22.3.

Table 22.3: Advantages and Limitations of Phytoremediation Technology

Advantages of phytoremediation technology	*Limitations of phytoremediation technology*
Due to wide abundance and diverse habit (submerged, free floating, rooted etc.) wetland plants are effective in microbes assisted heavy metals removal from surface water as well as sediments.	Paucity of work on removal of metals from sediments
In situ treatment of heavy metals is possible	Most of the phytoremediation works are confined in labs and have not been scaled up at industrial level in case of shallow ground water, soils, and sediments
Solar driven, thus does not require any chemicals or growth medium for biosystems	Mass transfer limitations associated as with other biotreatments
Costs 10–20 per cent of conventional/ chemical treatments	Relatively slower than conventional/ chemical treatments. However, bacterial pretreatment and chemical chelates (EDTA) enhance removal of heavy metals.
High public acceptance	Toxicity and bioavailability of biodegradation products is yet to be explored
Fewer side-effects in the form of air and water emissions	Contaminants may be mobilized into the ground water
Conserves natural resources	Influenced by soil and climate conditions of the site under rehabilitarion

Source: Rai, 2008, p. 133; Rai, 2009, p. 697.

A common feature of most types of phytoremediation is arsenic tolerance and accumulation, and both tolerance and accumulation involve arsenic compartmentation and translocation (Zhu and Rosen,

2009). Arsenic accumulation in plants is modulated by a network of functional genes, their expression patterns, and temporal/spatial coordination (Zhu and Rosen, 2009).

Further, Mohan and Pittman Jr. (2007) demonstrated through literature that the chemical treatments for Arsenic were not cost-effective. Robinson *et al.* (2006) sampled 28 species of aquatic macrophytes and 11 terrestrial species from ambient waters, sediments, and soils for arsenic tolerance. Huang *et al.* (2006) investigated the bioavailability, soil-to-plant transfer, and associated health risks of arsenic in soils collected from paddy rice fields and vegetable fields in suburban areas of some major cities of Fujian Province, China. Liu *et al.* (2008) investigated the effects of arsenic stress on growth, nutrition and arsenic uptake, and speciation in shoots and roots of winter wheat (*Triticum aestivum* L.) and recorded more accumulation in roots. Mokgalaka-Matlala *et al.* (2008) assessed the effects of As^{3+} and As^{5+} on the growth of roots, stems, and leaves and the uptake of arsenic, micro- and macro-nutrients, and total amylolytic activity and to elucidate the phytotoxicity of arsenic to the mesquite plant (*Prosopis juliflora* × *P. velutina*).

Pteris vittata (Chinese brake fern) is a well-known As hyperaccumulator (Ma *et al.*, 2001; Yu *et al.*, 2011). The plant possesses an exceptional ability to take up, translocate and tolerate As (Zhao *et al.*, 2009). When grown in As-contaminated soils, *P. vittata* accumulates As in fronds often more than 10 times the concentration in the soil (Caille *et al.*, 2004; Kertulis-Tartar *et al.*, 2006; Salido *et al.*, 2003; Shelmerdine *et al.*, 2009; Tu and Ma, 2002; Yu *et al.*, 2011).

Visoottiviseth *et al.* (2002) investigated that among 36 plant species, only two species of ferns (*Pityrogramma calomelanos* and *Pteris vittata*), a herb (*Mimosa pudica*), and a shrub (*Melastoma malabrathricum*), seemed suitable for phytoremediation.

Rahman *et al.* (2007) showed that the *Spirodela polyrhiza* L. accumulates Arsenic by physico-chemical adsorption and via the phosphate uptake pathway when arsenate was added to the solutions. Their results indicated that *S. polyrhiza* L. would be a good Arsenic phytofiltrator. Appreciable Arsenic accumulation was recorded in duckweed, *Spirodela polyrhiza* (Rahman *et al.*, 2007) and *Lemna gibba* (Mkandawire and Dudel, 2005), in comparison to traditional chemical technologies. The potential of *Garcinia cambogia*, an indigenous plant found in many parts of India was assessed to remove trivalent Arsenic from aquatics and its excellent performance in batch culture revealed that fresh and immobilized *G. cambogia* biomass could lead to the development of a viable and cost-effective technology for Arsenic removal in groundwater (Kamala *et al.*, 2005). Alvarado *et al.* (2008) monitored the removal of arsenic by the Water Hyacinth (*Eichhornia crassipes*) and Lesser Duckweed (*Lemna minor*). A multidisciplinary research effort that integrates the work of natural sciences, environmental engineers and policy makers is essential for greater success of green technologies as a potent tool of Arsenic management.

Conclusion

Multifaceted issues or paradigm of sustainable development should be appropriately addressed in the discipline of environmental management (Rai, 2011). Arsenic is introduced into the environment through both geological and anthropogenic processes and is considered as a global contaminant. Arsenic is among the top carcinogens, and arsenic elevation in soil (and thus food) and drinking water has been reported to affect millions of people around the globe (Zhu & Rosen, 2009). In concert with examining the feasibility of chemicals management technology at regional scale, sustainable development requires a framework for integrating environmental policies and development strategies in a global context (Rai, 2011). The implementation of green chemistry, the design of chemical products and processes that reduce or eliminate the use and generation of hazardous substances, is extremely essential in order to promote sustainability (Kirchhoff, 2005). The collaborative efforts of academia, industry, and government are needed to advance sustainability through the adoption of green chemistry in pollutant's amelioration (Kirchhoff, 2005).

Eccles (1999) through cost benefit analysis proved that biological processes for heavy metals removal are cheaper, when compared to the conventional technologies. The expansion of this research work has promoted the eco-friendly as well as cost-effective and hence sustainable technology in the form of phytoremediation (Rai &

Tripathi, 2007a; Rai, 2008a; Rai, 2009a). The use of plants for remediation of metals offers an attractive alternative, because it is solar driven and can be carried out *in situ*, minimizing cost and human exposure (Rai, 2009a). Therefore, phytoremediation technology promotes the use of plants for environmental cleanup.

Unfortunately, in literatures there is paucity of empirical data on comparative cost-effectiveness of conventional and phytoremediation technology, in relation to Arsenic, due to certain constraints (Mohan and Pittman Jr., 2007). Pertaining to Arsenic remediation, the cost of individual adsorbent depends on local availability, processing required and treatment conditions which is quite non-uniform in developing countries like India (Mohan and Pittman Jr., 2007). Costs will vary when the adsorbents are made for developed countries, developing countries or underdeveloped countries. Further, it is impractical/not feasible to measure cost-effectiveness by selecting numerous chemicals and biological materials together at a time.

In nutshell, phytoremediation of arsenic is an eco-friendly and cost-effective approach. However, there exists a paucity of empirical data on the green technology for arsenic removal. More attention of phytotechnologist is required in this area for mass scale removal of this deleterious semi-metal.

Future Prospects in Arsenic Phytoremediation

Moreover, development of genetically engineered arsenic hyper accumulators is an area which is still underdeveloped and may be future prospect (Zhu and Rosen, 2009, p. 220). Genetic modification of fast-growing and high-biomass crop plants to accumulate substantial arsenic in above ground tissues for phytoextraction and/or volatilization, or to sequester arsenic in roots for phytotstabilization should be investigated. Recently, Lin Nie *et al.* (2002) have shown the first clear evidence of arsenic phytoremediation using transgenic canola (*Brassica napa*) plants. These plants express the *Enterobacter cloacae* UW4 1-aminocyclopropane-1-carboxylate (ACC) deaminase (Ruiz and Romero, 2002).

X-ray absorption spectroscopy (XAS) has been used to study the uptake and co-ordination chemistry of elements such as arsenic and chromium in different plant species including *Pteris vittata* and

Prosopis spp., *Brassica juncea* (Indian mustard) and *Pisum sativum* (Castillo-Michel *et al.,* 2007), however, more works should include this instrumental advancement in order to understand the physiology and biochemistry of arsenic phytoremediation in totality.

REFERENCES

Alam, M.G.M., Snow, E.T., Tanaka, A. (2003). Arsenic and heavy metal contamination of vegetables grown in Samta village, Bangladesh, *Sci. Total Environ.,* 308, 83-96.

Allen, S.J., and Brown, P. (1995). Isotherm analysis for single component and multicomponent metal sorption onto lignite, *J. Chem. Tech. Biotechnol*, 62: 17-24.

Altundogan, H.S., Altundogan, S., Tumen, F. and Bildik, M. (2000). Arsenic removal from aqueous solutions by adsorption on red mud. *Waste Manage.* 20 (8): 761-767.

Alvarado, S., Gu′edez, M., Lu′e-Mer′u, M.P., Nelson, G., Alvaro, A., Jes′us, A.C. and Gyula, Z. (2008). Arsenic removal from waters by bioremediation with the aquatic plants Water Hyacinth (*Eichhornia crassipes*) and Lesser Duckweed (*Lemna minor*). *Bioresource Technology,* 99: 8436-8440.

Brown, P.A., Gill, S.A. and Allen, S.J. (2000). Metal removal from wastewater using peat, *Water Res*, 34: 3907-3916.

Caille, N., Swanwick, S., Zhao, F.J., McGrath, S.P. (2004). Arsenic Hyperaccumulation by *Pteris vittata* from arsenic contaminated soils and the effect of liming and phosphate fertilisation. *Environmental Pollution*, 132: 113-120.

Castillo-Michel *et al.* (2007). Use of X-ray absorption spectroscopy and biochemical techniques to characterize arsenic uptake and reduction in pea (*Pisum sativum*) plants. *Plant Physiology and Biochemistry*, 45: 457-463.

Chakraborti, D. (2008). *Groundwater arsenic contamination in India, its health effects and an approach for mitigation (1988-2008)*. Paper presented at the International Convention on Water resources Development and Management at Birla Institute of Technology and Science, Pilani, Rajasthan, India, October.

Chakraborti, D., Mukherjee, S.C., Pati, S., Sengupta, M.K., Rahman, M.M., Chowdhury (...) Chakraborti, D. (2003). Arsenic groundwater contamination in Middle Ganga Plain, Bihar, India: A future danger? *Environmental Health Perspectives,* 111: 1194-1201.

Couillard, D. (1992). Appropriate wastewater management technologies using peat. *J. Environ. Syst.*, 21: 1-19.

Couillard, D., (1994). The use of peat in wastewater treatment. *Water Res.* 28:1261-1274.

Eccles, H. (1999). Treatment of metal-contaminated wastes: Why select a biological process? *Trends Biotechnol.,* 17: 462-465.

Fan, M., Marshall, W., Daugaard, D., Brown, R.C. (2004). Steam activation of chars produced from oat hulls and corn stover. *Bioresour. Technol.* 93 (1), 103-107.

Gochfeld, M. (1997). Factors influencing susceptibility to metals. *Environmental Health Perspectives* 105: 817-833.

Gupta, V.K., Ali, I. (2002). Adsorbents for water treatment: low cost alternatives to carbon, in: A.T. Hubbard (Ed.*), Encyclopedia of Surface and Colloid Science*, Marcel Dekker, pp. 136-166.

Jones, F., T. (2007). A broad view of arsenic. *Poult Sci* 86: 2-14.

Kamala, C.T., Chub, K.H., Charya, N.S., Pandey, P.K., Rameshd, S.L., Sastrye, A.R.K., Sekhar, K.C. (2005). Removal of arsenic (III) from aqueous solutions using fresh and immobilized plant biomass. *Water Research,* 39: 2815-2826.

Kertulis-Tartar, G.M., Ma, L.Q., Tu, C., Chirenje, T. (2006). Phytoremediation of an arsenic-contaminated site using *Pteris vitrata* L.: a two-year study. *International Journal of Phytoremediation* 8: 311-322.

Kirchhoff, M.M. (2005). Promoting sustainability through green chemistry. *Resources, Conservation and Recycling,* 44: 237-243.

Liu, Q., Hu, C., Tan, Q., Sun, X., Su, J., and Liang, Y. (2008). Effects of As on As uptake, speciation, and nutrient uptake by winter wheat (*Triticum aestivum* L.) under hydroponic conditions. *Journal of Environmental Sciences* 20: 326-331.

Liu, S.X., Athar, M., Lippai, I., Waldren, C.A., and Hei, T.K. (2001). Induction of oxyradicals by arsenic: Implications for mechanisms of genotoxicity. *Proceedings of the National Academy of Sciences of the United States of America* 98: 1643-1648.

Ma, L.Q., Komar, K.M., Tu, C., Zhang, W.H., Cai, Y., Kennelley, E.D. (2001). A fern that hyperaccumulates arsenic. *Nature* 409: 579.

Mandal, B.K., Suzuki, K.T. (2002). Arsenic round the world: a review. *Talanta,* 58: 201-235.

McLellan, J.K., Rock, C.A. (1986). The application of peat in environmental pollution control: a review. *Int. Peat. J.* 1: 1-14.

Mkandawire, M., Dudel, E.G. (2005). Accumulation of arsenic in *Lemna gibba* L. (duckweed) in tailing waters of two abandoned uranium mining sites in Saxony, Germany. *Sci. Total Environ,* 336: 81-89.

Mohan, D., Chander, S. (2006a). Removal and recovery of metal ions from acid mine drainage using lignite—a low cost sorbent. *J. Hazard. Mater.*, 137: 1545-1553.

Mohan, D., Chander, S. (2006b). Single, binary and multicomponent sorption of iron and manganese on lignite, *J. Colloid Interf. Sci.* 299 (1):76-87.

Mohan, D., Pittman Jr., C.U., Bricka, M., Smith, F., Yancey, B., Mohammad, J., Steele, P.H., Alexandre-Franco, M.F., Serrano, V.G., Gong, H. (2007). Sorption of arsenic, cadmium, and lead by chars produced from fast pyrolysis of wood and bark during bio-oil production. *J. Colloid Interf. Sci.*, 310(1): 57-73.

Mohana, D., Pittman Jr, C.U. (2007). Arsenic removal from water/waste water using adsorbents—A critical review. *Journal of Hazardous Materials* 142: 1-53.

Mokgalaka-Matlala, N.S., Flores-Tavizon, E., Castillo-Michel, H., Peralta-Videa, J.R., and Gardea-Torresdey, J.L. (2008). Toxicity of arsenic (III) and (V) on plant growth, element uptake, and total amylolytic activity of mesquite (*Prosopis Juliflora* × *P. Velutina*). *International Journal of Phytoremediation* 10: 47-60.

Mudhoo, Ackmez, Sharma, Sanjay Kumar, Garg, Vinod Kumar and Tseng, Chin-Hsiao (2011). Arsenic: An Overview of Applications, Health, and Environmental Concerns and Removal Processes, *Critical Reviews in Environmental Science and Technology* 41: 5, 435-519.

Nasir, K., Shujaat, A., Aqidat, K., Jamil, A., (1998). Immobilization of arsenic on rice husk, *Adsorpt. Sci. Technol.* 16 (8): 655-666.

Nie, L. *et al.* (2002). Phytoremediation of arsenate contaminated soil by transgenic canola and the plant growth-promoting bacterium Enterobacter cloacae CAL2. *Plant Physiol. Biochem.* 40: 355-361.

Pendias, H., Pendias, K. (1989). *Trace elements in soil and plants*. CRC Press, Florida.

Queirlo, F., Stegen, S, Restovic M., Paz, M., Ostapczuk, P. Schwuger M.J., Munoz L. (2000). Total arsenic, lead, and cadmium levels in vegetables cultivated at the Andean villages of Northern Chile. *Sci. Total Environ,* 255: 75-84.

Rahman, M.A., Hasegawa, H., Ueda, K., Maki, T., Okumura, C., Rahman, M.M. (2007). Arsenic accumulation in duckweed (*Spirodela polyrhiza* L.): A good option for phytoremediation. *Chemosphere,* 69: 493-499.

Rai, P.K., Tripathi, B.D. (2007b). Microbial contamination in vegetables due to irrigation with partially treated municipal wastewater in a tropical city. *International Journal of Environmental Health Research*, 17 (5): 389-395.

Rai, P.K., Mishra, A., and Tripathi, B.D. (2010). Heavy metals and microbial pollution of river Ganga: A case study on water quality at Varanasi. *Aquatic Ecosystem Health and Management,* 13(4): 352-361.

Rai, P.K. (2008e). Ecological investigation on heavy metal pollution of G.B. Pant Sagar and its phytoremediation. Ph.D. Dissertation, Banaras Hindu University, Varanasi, India, p. 156.

Rai, P.K. (2010a). Microcosm investigation on Phytoremediation of Cr using *Azolla pinnata. International Journal of Phytoremediation,* 12: 96-104.

Rai, P.K. (2011). An eco-sustainable green approach for heavy metals management: two case studies of developing industrial region. *Environ Monit Assess* DOI 10.1007/s10661-011-1978-x; Published online 5 April, 2011.

Rai, P.K. (2007a). Phytoremediation of Pb and Ni from industrial effluents using *Lemna minor:* an eco-sustainable approach. *Bull. Biosci.,* 5(1): 67-73.

Rai, P.K. (2007b). Wastewater management through biomass of *Azolla pinnata*: an eco-sustainable approach. *Ambio,* 36(5): 426-428.

Rai, P.K. (2008a). Heavy-metal pollution in aquatic ecosystems and its phytoremediation using wetland plants: An eco-sustainable approach. *Inter. J. Phytoremediation,* 10(2): 133-160.

Rai, P.K. (2008b). Mercury pollution from chlor-alkali industry in a tropical lake and its biomagnification in aquatic biota: Link between chemical pollution, biomarkers and human health concern. *Human and Ecological Risk Assessment: An International Journal*, 14: 1318-1329.

Rai, P.K. (2008c). Phytoremediation of Hg and Cd from industrial effluents using an aquatic free floating macrophyte *Azolla pinnata. International J. Phytoremediation,* 10(5), 430-439.

Rai, P.K. (2009). Heavy metals in water, sediments and wetland plants in an aquatic ecosystem of tropical industrial region, India. *Environ. Mon. Assess.* 158: 433-457.

Rai, P.K. (2009a). Heavy metal phytoremediation from aquatic ecosystems with special reference to macrophytes. *Critical Rev. Environ. Sci. Technol.,* 39(9): 697-753.

Rai, P.K. (2009b) (Last update 2010). Phytoremediation of heavy metals in a tropical impoundment of industrial region. *Environmental Monitoring and Assessment*, 165: 529-537.

Rai, P.K. (2009c) (Last update 2010). Seasonal monitoring of heavy metals and physico-chemical characteristics in a lentic ecosystem of sub-tropical industrial region, India. *Environmental Monitoring and Assessment,* 165: 407-433.

Rai, P.K. (2010b). Heavy Metal Pollution in Lentic ecosystem of Sub-tropical Industrial region and its Phytoremediation. *International Journal of Phytoremediation* 12(3): 226-242.

Rai, P.K., Sharma, A.P., Tripathi, B.D. (2007). Urban environment status in Singrauli Industrial region and its eco-sustainable management: A case study on heavy metal pollution, In: *Urban Planing and Environment, Strategies and Challenges*, (Lakshmi, Vyas, Ed.). pp. 213-217. McMillan Advanced Research Series.

Rai, P.K., Tripathi, B.D. (2007a). Heavy metals removal using nuisance blue green alga *Microcystis* in continuous culture experiment. *Environmental Sciences* 4(1): 53-59.

Rai, P.K., Tripathi, B.D. (2008). Heavy metals in industrial wastewater, soil and vegetables in Lohta village, India. *Toxicological and Environmental Chemistry,* 90(2): 247-257.

Rai, P.K., Tripathi, B.D. (2009). Comparative assessment of *Azolla pinnata* and *Vallisneria spiralis* in Hg removal from G.B. Pant Sagar of Singrauli Industrial region, India. *Environmental Monitoring and Assessment,* 148: 75-84.

Robinson, B., Kim, N., Marchetti, M., Moni, C., Schroeter, L., Van Den Dijssel, C., Milne, G., and Clothier, B. (2006). Arsenic hyperaccumulation by aquatic macrophytes in the Taupo Volcanic Zone, New Zealand. *Environmental and Experimental Botany*, 58: 206-215.

Rossman, T.G. (2003). Mechanism of arsenic carcinogenesis: An integrated approach. *Mutation Research* 533: 37-65.

Roy, P., and Saha, A. (2002). Metabolism and toxicity of arsenic: A human carcinogen. *Present Science* 2: 38-45.

Ruiz, J.M., Romero, L. (2002). First evidence of arsenic phytoremediation. *Trends in Plant Science* 7(9): 384.

Salido, A.L., Hasty, K.L., Lim, J.M., Butcher, D.J. (2003). Phytoremediation of arsenic and lead in contaminated soil using Chinese brake ferns (*Pteris vittata*) and Indian mustard (*Brassica juncea*). *International Journal of Phytoremediation* 5: 89-103.

Santra, A., Maiti, A., Das, S., Lahiri, S., Charkaborty, S.K., and Mazumder, D.N. (2000). Hepatic damage caused by chronic arsenic toxicity in experimental animals. *Journal of Toxicology: Clinical Toxicology* 8: 395-405.

Shelmerdine, P.A., Black, C.R., McGrath, S.P., Young, S.D. (2009). Modelling phytoremediation by the hyperaccumulating fern, *Pteris vittata*, of soils historically contaminated with arsenic. *Environmental Pollution* 157: 1589-1705.

Sneddon, R., Garelick, H., Valsami-Jones, E. (2005). An investigation into arsenic (V) removal from aqueous solutions by hydroxylapatite and bonechar, *Mineral. Mag.* 69 (5): 769-780.

Tu, C., Ma, L.Q. (2002). Effects of arsenic concentrations and forms on arsenic uptake by the hyperaccumulator ladder brake. *Journal of Environmental Quality* 31: 641-647.

USEPA, Federal Register, 66 (14) (2001) 6976-7066.

Viraragharan, T., Ayyaswami, A. (1987). Use of peat in water pollution control: a review. *Can. J. Civil Eng.* 14: 230-232.

Visoottiviseth, P., Francesconi, K., Sridokchan, W. (2002). The potential of Thai indigenous plant species for the phytoremediation of arsenic contaminated land. *Environmental Pollution* 118: 453-461.

WHO (1997), *Health and environment in sustainable development*. Geneva: WHO.

Ye, W., Khan, M.A., McGrath, S.P., Zhao, F.J. (2011). Phytoremediation of arsenic contaminated paddy soils with *Pteris vittata* markedly reduces arsenic uptake by rice. Environmental Pollution Article in Press, doi:10.1016/j.envpol.2011.07.024.

Zhao, F.J., Ma, J.F., Meharg, A.A., McGrath, S.P. (2009). Arsenic uptake and metabolism in plants. *New Phytologist* 181: 777-794.

Zhu, Y., Rosen, B.P. (2009). Perspectives for genetic engineering for the phytoremediation of arsenic-contaminated environments: from imagination to reality? *Current Opinion in Biotechnology*, 20:220-224.

23

Diversity of Weed Species and Soil Seed Bank in Indian Dryland and Irrigated Agroecosystems

Rajani Srivastava

Introduction

In recent year ecological problems caused by intensification of land use practices have become a topical issue (Tilaman *et al.*, 2002; Roschewitz *et al.*, 2005). Intensive agriculture is based on crop monocultures, high inputs of mineral fertilizers and pesticides to the farming system, and intensive tillage of soil. Application of such methods found to have a negative impact on species richness or agroecosystem diversity. Various diversity indices serve as useful tools to evaluate the effect of different management practices on floristic diversity (Waldhardt *et al.*, 2003). While concern about the loss of species diversity has mainly been focused on the natural habitat, the diversity of weed species and soil seed bank occupying agroecosystem has received much less attention. In Indian tropical region, few workers have paid attention on weed flora and soil seed bank of agroecosystem (Saxena & Ramakrishnan, 1984; Tripathi & Yadav, 1987; Misra *et al.*, 1995). Except few (Srivastava & Singh, 2005) scant information available on comparative evaluation of weed and soil seed bank diversity in different cropping systems. In India, out of total arable land about 65 per cent land functions under dryland conditions (rainfed only, no irrigation resulting in moisture deficit), and remaining 35 per cent arable land is under irrigation (rainfall augmented with irrigation). Two agroecosystems represent widely differing agroecological conditions (in terms of water management, tillage and fertilizer), showing different soil properties (Srivastava & Singh, 2002), which may affect the composition and dynamics of soil seed bank.

The present study, carried out in Indian dry tropics, was focused on comparative analysis of two contrasting agroecosystems (dryland and irrigated), both located in close vicinity with the following objectives:

(i) to evaluate the differences in species composition and species diversity between two agroecosystems, and
(ii) to compare diversity and composition of soil seed bank in dryland and irrigated rice based agroecosystems.

Methods

This two-year (November 1998-February 2001) study was carried out in the dıyland and irrigated Farm of the Institute of Agricultural Sciences located within the campus of the Banaras Hindu University, Varanasi (25° 18′ N lat. and 83° 1′ E long., 76 m above mean sea level). Both of these Farms are established before more than 50 years.

This region has a tropical sub-humid climate, characterized by strong seasonality with respect to temperature and rainfall. The long-term average annual rainfall is about 1100 mm.

The soil of the study sites belongs to the order Inceptisols, and sub-order orchrepts, showing thin, pale brown surface horizon, and neutral to slightly alkaline reaction. The top 10 cm soil in dryland and irrigated agroecosystems, is neutral in reaction, with 0.72 per cent and 0.81 per cent organic C and 0.07 per cent and 0.08 per cent total N, respectively (Srivastava & Singh, 2002).

Three permanent plots (10 m × 9 m each) were established at each study site in November 1998. Six sub-plots (2 sub-plots per plot, each 50 cm × 50 cm) were permanently demarcated at each site for weed analysis. Crops grown were rice (*Oryza sativa* var. NDR 118) in rainy season alternated in winter with lentil (*Lens esculentus* var. Pant 639) in the dryland; rice (*Oryza sativa* var. HUR 36) and wheat (*Triticum aestivum* var. HUW 234) alternated in the irrigated agroecosystem.

During winter and rainy crop periods at monthly interval composition and density of weeds in agroecosystems were recorded, beginning 30 days after crop transplanting/sowing. In this study species composition and density of weeds recorded during mid-cropping season (60 days after crop transplanting/sowing) are presented.

For soil seed bank estimation, within each sub-plot, soil was excavated from a 20 cm × 20 cm area in three successive 10 cm depths, twice during the annual cycle, once after the rainy season harvest (November-December) and again after the winter harvest (April-May). Depth samples of soil (6 sub-plots × 3 depths per site; each corresponding to 20 x 20 cm, 10 cm depth) were air dried and weighed. The seed losses due to germination during the air-drying period were also quantified. The size of seed bank was estimated in depth samples by seedlings emergence method. All eight counts were made at 15 days intervals during both cropping seasons.

The following species diversity coefficients were calculated from the weed density values:

Shannon-Wiener index, $H' = -\Sigma pi \log pi$ (Shannon and Weaver 1963), Simpson index, $D = 1-\Sigma pi^2$ (modified by Berger & Parker, 1970), Species richness, $d = (S-1)/\ln N$ (Margalef, 1958). where S = total number of species in the sample, N = total number of individuals of all species, pi = proportion of all individuals which belong to species I (number of individuals of each species i/N). The similarity (IS_{sp}) between sites A and B was calculated (using maximum number of species occurring during the two cropping seasons) as follows:

$$IS_{sp} = \sum \frac{(Mw/Mg)}{a+b+c} \times \frac{100 \times Mc}{Ma+Mb+Mc}$$

(Mueller-Dombois & Ellenberg, 1974), where Mw = the smaller quantity of a species common to stands A and B, Mg = the greater quantity of a species common to stands A and B, a = number of species occurring in stand A only, b = number of species occurring in stand B only, c = number of species common to both stands A and B, Mc = sum of quantities of species common to stands A and B, Ma = the sum of quantities of species unique to stand A, Mb = the sum of quantities of species unique to stand B.

Results

Species Composition and Density

Comparable number of weed species (12-14), mostly annuals, was recorded in the two agroecosystems during both cropping seasons (Appendix Table 23.1). Weed species composition differed with agroecosystems and with seasons. Both dryland and irrigated agroecosystems had only 25 per cent and 38 per cent similarity with each other during winter and rainy cropping seasons, respectively (Appendix Table 23.1). During mid-season (after 60 days of crop sowing/transplanting), *Anagallis arvensis* and *Chenopodium album* were the dominant forbs in dryland agroecosystem, contributing 89 per cent shoots during winter season. In the irrigated agroecosystem, *Chenopodium album* and *Melilotus indica* were the dominant forbs (contributing 83 per cent forb density); *Phalaris minor* was the dominant grass (78 per cent grasses/sedges shoots).

Soil Seed Bank

Both post-rainy-season and post-winter-season soil seed banks in two agroecosystems showed distinct seasonal variations in number of seedlings emerging in rainy and winter seasons (Appendix Table 23.2). It is evident that the soil seed banks were stabilized in both agroecosystems. Total number of seedlings emerging from post-rainy-season seed bank in different soil depths ranged 816-17481 seedlings m^{-2} in the dryland agroecosystem and 1728-7168 seedlings m^{-2} in irrigated agroecosystem. Seedlings emerging from the post-winter-season seedbank (1253-17979 and 1849-6083 m^{-2} in dryland and irrigated agroecosystems), closely matched the number of seedlings from post-rainy-season seed bank. In both seasonal phases of the seed bank the maximum seedling emergence (83-98 per cent of the annual total) occurred during the succeeding rainy season and only a small fraction appeared in the winter season. The rainy season began about 7 months after the post-rainy-season soil collection, and about 3 months after the post-winter-season collection.

In general, the seed banks of the two agroecosystems were differentiated with respect to the proportion of broad leaf weeds and grasses/sedges (Appendix Table 23.2). The post-winter-season and post-rainy-season phases of the seed banks, however, showed broad similarity in both agroecosystems. In the dryland agroecosystem the share of broad-leaved weeds (e.g. *Ammannia baccifera, Anagallis arvensis, Chenopodium album* etc.) in total seedling emergence at 0-10 cm soil depth was greater (58-61 per cent of total) compared to the same depth in the irrigated agroecosystem (35-42 per cent). Grasses/sedges showed the reverse trend, their share being 34-35 per cent in the dryland agroecosystem, compared to 54-59 per cent seedling emergence in the irrigated agroecosystem. As the soil depth increased, seedling emergence of broad leaved weeds decreased in both agroecosystems. The share of grasses/sedges in total seedling emergence increased with the increase in soil depth. Thus, at 0-10 cm soil depth broad leaved weeds were the major contributor in total seedling emergence in dryland agroecosystem whereas in the irrigated agroecosystem grasses/sedges were the main contributor.

In dryland agroecosystem, *Anagallis arvensis* and *Chenopodium album* were the major species during winter season contributing 87-91 per cent of the total winter seedling emergence from 0-20 cm soil depth (Appendix Table 23.2). Major proportion of seedlings in the irrigated agroecosystem was shared by *Chenopodium album* and *Melilotus indica,* totalling to 65-78 per cent of the total seedlings in the three depths. During the rainy season *Ammannia baccifera, Echinochloa colona* and *Cyperus rotundus* were the major seedling species and these three species together contributed 71-81 per cent of the total seedling emergence. Corresponding contribution of these three species in the irrigated agroecosystem was 83-95 per cent. Compared to the dryland agroecosystem, the density of *Echinochloa colona* seedlings was approximately doubled in the irrigated agroecosystem and the density of *Ammannia baccifera* was reduced to about half.

Diversity of Weed and Soil Seed Bank

Variations in richness, evenness and diversity of species and soil seed bank are shown in Appendix Table 23.3. The irrigated agroecosystem exhibited higher above ground weed species diversity (*H'*) than dryland agroecosystem in the rainy season ($p < 0.01$); during the winter difference was smaller ($p < 0.1$). While diversity (*H'*) varied significantly ($p < 0.001$) between cropping seasons in the irrigated agroecosystem, the seasonal variation in dryland was small ($p < 0.10$). The seasonal trends of species diversity and species evenness were broadly similar in both agroecosystem (data not shown, as reported by Srivastava & Singh, 2005).

In case of soil seed bank diversity, with the increase of depths diversity increases in dryland agroecosystem whereas in irrigated agroecosystem reverse trend have seen. Mean diversity values are slightly greater in irrigated agroecosystem in post rainy season (1.90 and 0.79 c.a. 1.82 and 0.73) as well as post winter soil seed bank (1.66 and 0.74 c.a. 1.53 and 0.68) compared to dryland agroecosystem. Species diversity and species richness in both agroecosystem soil seed bank was greater in soil collected after rainy season whereas it was lower in soil that collected after winter season.

Discussion

Diversity indices measured in this study were within the range of those reported for various cropping systems in diverse geographical areas (Clements *et al.*, 1994; Derksen *et al.*, 1995). Values reported in the literature for Shannon's H′ for weed communities generally are <2.0. Low plant diversity appears to be typical of arable land and intensively managed grassland (Wilson *et al.*, 2003). In the presently studied dryland and irrigated agroecosystems diversity fall within these values. Changes in above ground weed diversity between presently studied agroecosystems also reflected in soil seed bank. After both cropping seasons, upto 30 cm soil depth, soil seed bank diversity was greater in irrigated agroecosystem compared to dryland, this is due to distinctly more diverse seed bank in upper soil layer (0-10 cm) in irrigated agroecosystem. Srivastava and Singh (2006) reported that in high input irrigated agroecosystem (in terms of nitrogen and water addition), relatively fewer terrestrial and higher aquatic weeds in irrigated agroecosystem, as well adjusted to grow in flooded condition in rainy season and moist condition during winter to make efficient use of resources. Diverse above ground weed community shed their seeds and diverse germinating soil seed bank obtained from upper soil layer in this system. Whereas in low input dryland agroecosystem (less nitrogen and no additional water) encountered by more adverse condition and reserve more seeds in lower (10-20 and 20-30 cm) soil layers. This may be adaptive mechanism of storage of weed diversity in dryland agroecosystem.

Diversity-dominance relations closely display the information summarized in species diversity and evenness measure (Wilson, 1991). Although the dryland and irrigated agroecosystems differ markedly with respect to species composition, both systems at all soil depths show broadly similar log normal diversity-dominance curves. These curves suggest the dominance by few (2-3) species having high abundance and larger number of species having low abundance in these systems. Appearance of the majority of low abundance species in winter in dryland agroecosystem and both seasons in irrigated agroecosystem indicates seasonal differences in the availability of microsites supporting these species. The seasonal differences were less marked in case of germinating soil seed bank in both agroecosystems up to 30 cm soil depth.

Above ground weed species abundance and their depthwise distribution of soil seed bank vary between two agroecosystems. This is evident from low similarity index in both cropping seasons. It has been reported that variations in species abundance and density of weed flora between agroecosystem is due to varying cropping practices like increased use of fertilizers, irrigation and hydrology (Moody, 1996; Kent *et al.*, 2001; Johnson & Kent, 2002). Soil seed bank density declined in irrigated agroecosystem in both post rainy and post winter seasons soil seed bank (compared to dryland). This be related to lowered weed seed production, particularly of the annual weeds. Annual weeds, most of them broad leaved type, generally show greater density and higher seed output per individual in the dryland agroecosystem. Greater seed production in the dryland agroecosystem seems to be a survival strategy of weeds under unfavourable, prolonged moisture deficit condition (Sen, 2000). The suppression or elimination of several broad leaved weed species due to change from dryland to irrigated condition leads to near total depletion of seeds of these species from the soil seed bank. In the total soil seed bank density major contribution by *Ammannia baccifera* in dryland agroecosystem and *Ammannia baccifera* and *Melilotus indica* in irrigated agroecosystem of broad leaves weeds, shows its wide tolerance range. *Echinochloa colona* and *Cyperus* spp., both abundant in presently studied soil seed bank of agroecosystems, have been recorded among the most important weeds across the world (Holm *et al.*, 1977). *Echinochloa colona* is a major weed of rice under both flooded and upland conditions (Shad & Siddiqui, 1996); its close resemblance with the rice seeds and seedlings results in getting it transplanted with rice. Few species resist control measures in agroecosystems and become dominant due to adaptation to the cropping system (Bühler *et al.*, 1997). The conversion of dryland to irrigated condition also affects the relative abundance of several seed bank species. Build up of very large seed bank of *Echinochloa colona* and *Cyperus rotundus* in the irrigated agroecosystem indicates that in future these species may become predominant weeds with the spread of irrigation. *Echinochloa colona* has been reported to succeed over the previously dominant weeds in Asia where direct seeding has been replaced by the rice transplanting (as in the present irrigated agroecosystem) (Ho & Itoh, 1991). On the other hand, considerable reduction in the seed bank of *Ammannia baccifera* in the irrigated agroecosystem shows the possibility for disappearance of this species in future.

In above ground weed flora, there was distinctly higher density of forbs than grasses/sedges during winter season and reverse dominance in rainy season in both agroecosystems. It has been reported that monocotyledonous grasses/sedges (C_4 type) respond more rapidly than broad leaved forbs (C_3 type) to warm and moist condition whereas during winter season grasses/sedges likely to decrease (Shad & Siddiqui, 1996; Srivastava & Singh, 2006). During both seasons, some weeds are abundant in both agroecosystem due to their wider tolerance to varying moisture condition, but some are dominant to either drier soil dryland condition (e.g. *Anagallis arvensis*) or in wet irrigated condition (*Melilotus indica, Phalaris minor*). In both agroecosystem, the decline of soil seed bank density with increasing soil depths in both seasons soil seed bank is obviously related to less downward movement of seeds and creation of anaerobic condition in soil, simultaneously related to decrease seed density.

Conclusion

To sum up, this investigation documents the patterns of diversity and abundance of weed species seed production and their soil seed bank associated with two rice-based agroecosystems. Conversion of dryland to irrigated agroecosystem involves elimination and recruitment of weed species, resulting in distinctly changed weed community showing increased weeds and soil seed bank diversity in irrigated condition. Although water management suppresses the density of several weeds, it leads to dominance of some potentially noxious weeds (e.g. *Phalaris minor, Melilotus indica*). Approximately double soil seed bank size and greater seed production and higher diversity at lower depth may show an adaptive mechanism of storage of weed diversity in dryland agroecosystem.

Acknowledgements

We thank the Head and the Programme Co-ordinator, Centre of Advanced Study in the Department of Botany for providing laboratory and library facilities. Thanks are due to various authorities of Institute of Agricultural Sciences, BHU, for providing long term irrigated and dryland sites and other field facilities.

APPENDIX

Table 23.1: Mid Season Weed Composition and Density in Dryland and Irrigated Agroecosystems. Values in Parenthesis Show the Number of Annual Species

Species	*Winter season*		*Rainy season*	
	Dryland	*Irrigated*	*Dryland*	*Irrigated*
Total species	14 (11)	12 (10)	12 (8)	13 (10)
Common species	10		7	
Similarity index	38%		25%	
Exclusive species	4	2	5	6
Broad-leaved weeds (no. of individual m^{-2})				
Alternanthera sessilis (L.) R. Br. ex. Dc	5		4	
Graphalium indicum anct.	1			
Solanum nigrum L.	1			
Anagallis arvensis L.	153	16		
Chenopodium album L.	106	74		
Lathyrus aphaca L.	4	2		
Lathyrus sativa L.	7	3		
Polygonum plebejum R. Br.	6	3		
Rumex crispus L.	7	4		
Melilotus alba Desr.		7		
Melilotus indica All.		93		
Lindernia ciliata (colsm.) Pannell			8	
Lindernia crustaceae (L.) F.V. Muell			3	
Sphaeranthus indicus L.			3	
Ammannia baccifera L.			46	33
Corchorus acutangulus Lamk.			16	16
Eclipta alba (L.) Hassk				1
Ipomoea aquatica Forsk				10
Ludwigia parviflora Roxb.				5
Polygonum hydropiper L.				14
Sub-total	**290**	**202**	**80**	**79**

(*Contd.*)

(*Contd.*)

Species	Winter season		Rainy season	
	Dryland	Irrigated	Dryland	Irrigated
Grasses/sedges (no. of individual m^{-2})				
Sporobolus diander (Retz.) P. Beauv.	12			
Phalaris minor L.	13	104		
Cynodon dactylon (L.) Pers.	8	10	9	
Dichanthiun annulatum (Forsk.) stapf.	8	11	8	4
Cyperus spp L.*	12	9	52	48
Echinochloa colona (L.) Link			133	98
Cyanotis axillaries D. Don			3	2
Commelina benghalensis L.				2
*Marsilea minuta L.***				2
Sub-total	**53**	**134**	**205**	**156**
Total (broad-leaved + grasses/sedges)	**343**	**336**	**285**	**235**

* Sum of *Cyperus iria* L. and *Cyperus rotundus L.*;

** Pteridophyte.

Table 23.2: Seed Bank (no. × 10^3 m^{-2}) Estimated by Seedling Emergence Method and its per cent Species Composition in Soil Collected from Three Depths in Two Agroecosystems

Seasons/species		*Agroecosystems (soil depth, cm)*					
		Dryland			*Irrigated*		
		0-10	10-20	20-30	0-10	10-20	20-30
Post–rainy–season seed bank		17481	3034	816	7168	3443	1728
Broad leaved weeds (%)							
Melilotus indica	W	–	–	–	5	5	1
Chenopodium album	W	1	6	–	5	5	1
Anagallis arvensis	W	2	6	1	1	2	0.4
Corchorus acutangulus	R	3	5	11	4	1	2
Lindernia spp.*	R	3	5	6	1	2	1
Ammannia baccifera	R	49	31	22	19	25	3
Grasses/Sedges (%)							
Cyperus iria	W+R	2	5	14	7	3	68
Echinochloa colona	R	15	12	8	26	37	23
Cyperus rotundus	W+R	17	25	23	26	16	1
**Other species		8	6	5	7	4	1
Post–winter–season seed bank		17979	5104	1253	6083	3285	1849
Broad leaved weeds (%)							
Melilotus indica	W	–	–	–	7	4	2
Anagallis arvensis	W	0.3	0.1	–	–	–	–
Chenopodium album	W	1	0.1	–	3	1	–
Corchorus acutangulus	R	3	3	6	5	2	3
Lindernia spp.*	R	4	7	3	1	1	0.4
Ammannia baccifera	R	53	45	41	26	28	29
Grasses/Sedges (%)							
Cyperus iria	W+R	0.3	3	9	4	2	1
Cyperus rotundus	W+R	16	17	26	20	18	6
Echinochloa colona	R	18	21	9	30	43	57
**Other species		4	4	6	4	1	1

* *Lindernia ciliata, Lindernia crustaceae.*

** *Alternanthera sessilis, Commelina benghalensis, Cyanotis auxillaris, Cyperus globosus, Dichanthium annulatum, Ipomoea aquatica, Lathyrus aphaca, Lathyrus sativa, Ludwigia parviflora, Phalaris minor, Sporobolus diander.*

Table 23.3: Species Diversity (Shannon-Weiner: H′; Simpson: D) and Species Richness (d) of Soil Seed Bank at Three Depths in Two Agroecosystems. Values in Parenthesis show Range of Respective Indices of above ground Weeds

Depth (cm)	Species diversity				Species richness	
	H′		D		d	
	Dryland	Irrigated	Dryland	Irrigated	Dryland	Irrigated
			Post rainy season soil seed bank			
0-10	1.70	2.01	0.70	0.82	1.64	1.80
10-20	1.99	1.83	0.81	0.77	1.62	1.97
20-30	2.09	0.99	0.85	0.49	1.79	1.21
Total (0-30)	1.80	1.90	0.73	0.79	1.71	1.91
	(1.51-1.75)	(1.63-1.81)	(0.66-0.73)	(0.77-0.79)	(1.96-2.26)	(1.26-1.93)
			Post winter season soil seed bank			
0-10	1.47	1.82	0.65	0.79	1.74	1.49
10-20	1.58	1.49	0.72	0.71	1.41	1.36
20-30	1.64	1.16	0.74	0.58	1.12	1.20
Total (0-30)	1.53	1.66	0.68	0.74	1.68	1.50
	(1.71-1.86)	(1.33-2.07)	(0.72-0.85)	(0.70-0.81)	(2.05-2.15)	(1.48-1.93)

REFERENCES

Berger, W.H., and Parker, F.L. (1970). Diversity of planktonic forminifera in deep-sea sediments. *Science,* 168, 1345-1347.

Buhler, D.D., Hartzler, R.G., and Forcella, F. (1997). Implications of weed seed bank dynamics to weed management. *Weed Science,* 45, 329-336.

Clements, D.R., Weise, S.F., and Swanton, C.J. (1994). Integrated weed management and weed species diversity. *Phytoprotection,* 75, 1-18.

Derksen, D.A., Thomas, A.G., Lafond, G.P., Loppky, H.A., and Swanton, C.J. (1995). Impact of post emergence herbicides on weed community diversity within conservational tillage systems. *Weed Research,* 35, 311-320.

Ho, N.K. and Itoh, K. (1991). Changes in weed flora and their distribution in the Muda area. Paper Presented at the 8th MADA/TARC Quarterly Meeting, 3 November 1991, Alor Setar Malaysia.

Holm, L.G., Plucknett, D.L., Pancho J.V. and Herberger, J.P. (1977). *The World's Worst Weeds: Distribution and Biology.* East West Centre, Honolulu, Hawaii, USA: University Press of Hawaii.

Johnson, D.E., and Kent, R.J. (2002). The impact of cropping on weed species composition in rice after fallow across a hydrological gradient in West Africa. *Weed Research,* 42, 89-99.

Kent, R., Johnson, D.E., and Becker, M. (2001). The influence of cropping system on weed communities of rice in Cote d'Ivoire, West Africa. *Agriculture Ecosystem and Environment,* 87, 299-307.

Margalef, D.R. (1958). Information theory in ecology. *Gen Systematics,* 3, 36-71.

Misra, J., Pandey, H.N., Tripathi, R.S., and Sahoo, U.K. (1995). Dynamics of buried seed populations and seedling cohorts of two dominant weeds in a hill agroecosystems of the humid subtropics of India. *Research Population and Ecology,* 37(1), 1-7.

Moody, K. (1996). Weed community dynamics in rice fields. In: Naylor, R. (ed.), *Herbicides in Asian Rice: Transitions in Weed Management.* Palo Alto California: Institute for International.

Mueller-Dombois, D., and Ellenberg, H. (1974). *Aims and Methods of Vegetation Ecology.* New York: John Wiley and Sons.

Roschewitz, I., Gabriel, D., Tscharntke, T., and Thies, T. (2005). The effects of the landscape complexity on arable weed species diversity in organic and conventional farming. *Journal of Applied Ecology,* 42, 873-882.

Saxena, K.G., and Ramakrishnan, P.S. (1984). Herbaceous vegetation development and weed potential in slash and burn agriculture (Jhum) in North-Eastern India. *Weed Research,* 24, 135-142.

Sen, D.N. (2000). Weeds in rainfed cropping in Indian desert. In P.K. Jha, S.B. Karmacharya, S.R. Baral, and P. Lacoul (eds.) *Environment and Agriculture*, Nepal: Ecological Society, 223-228.

Shad, R.A., and Siddiqui, S.U. (1996). Problems associated with *Phalaris minor* and other grass weeds in India and Pakistan. *Experimental Agriculture,* 32, 151-160.

Shannon, C.E., and Weaver, W. (1963). *The Mathematical Theory of Communication.* University of Illinois Press, Urbana.

Srivastava, R., and Singh, K.P. (2005). Species diversity in dryland and irrigated agroecosystems and marginal grassland ecosystem in dry tropics. *Community Ecology,* 6(2), 131-141.

Srivastava, R., and Singh, K.P. (2006). Weed populations and soil seed bank in dryland and irrigated agroecosystem. In H.N. Pandey and S.K. Barik (eds.), *Ecology, Diversity and Conservation of Plants and Ecosystems in India*, Vol. 1, 110-128.

Srivastava, R., and Singh, K.P. (2002). Variations in soil organic C and total N storage due to cultivation practices in the Gangetic Plain, India. *International Journal of Ecology and Environmental Sciences,* 28, 193-199.

Tilaman, D., Cassmon, K.G., Matson, P.A., Naylor, R., and Polaskey, S. (2002). Agricultural sustainability and intensive production practices. *Nature,* 418, 671-677.

Tripathi, R.S., and Yadav, A.S. (1987). Population dynamics of *Eupatorium adenophorum* Spreng. and *Eupatorium riparium* Regel in relation to burning. *Weed Research,* 27, 229-236.

Waldhardt, R., Simmering, D., and Albrecht, H. (2003). Floristic diversity at the habitat scale in agricultural landscapes of central Europe—summary, conclusions and perspectives. *Agriculture Ecosystem and Environment,* 98, 79-85.

Wilson, J.B. (1991). Methods for fitting dominance—diversity curves. *Journal of Vegetation Sciences,* 2, 35-46.

Part V
Policies and Planning

24

Data for Development Policy, Planning and Evaluation Sources, Nature and Limitations

Aalok Ranjan Chaurasia

Introduction

Sustainable development is all about improving the lives of the people without compromising the quality of life of future generations. This implies that development policies and programmes must be oriented towards the development and welfare needs of the people while preserving the environment. In other words, development policy and planning must be based on the hard evidence about the people, how they live and how their lives are changing. Development policy and planning must also take into account how the population stock—the size and structure of the population—and the life style pattern are impacting upon the environment. Development data provide the hard evidence about the people and their life styles and, therefore, are critical for the success of all development policies and programmes. Development data may be likened to money in an economy and blood in the system of human being. However, the current understanding about development data is at best fragmentary and awareness about sources, nature and limitations of different types of development data remains poor. At the same time, lack of high quality development data and access to existing data are fundamental challenges to the use of development data for evidence-based planning and evolution of sound development and welfare policies, evaluate development programmes and lead development activities that are directed towards improving the quality of life of the people.

Evidence-based development policy and planning may be contrasted with the opinion-based development policy and planning. Fig. 24.1 describes how development data can contribute to evidence-based development policy and planning and how lack of the development data or the inability to access the development data or limited capacity to use the available data can shift the emphasis from evidence-based development policy and planning to opinion-based policy and planning. Since, the opinion-based development policy and planning is not usually based on an objective assessment of the ground situation and actual development and welfare needs of the people, it generally turns out to be normative in scope and has limited impact in terms of the improvement in the quality of life of the people.

Fig. 24.1 describes how lack of development data or unawareness about development data or non-availability of and limited access to the development data as well as poor to very poor capacity of analysing and using the available data leads to opinion-based development policy and planning. The figure also shows how lack of development data or the inability of using the available data may lead to a subjective evaluation of the impact of development programmes and activities in terms of improvements in the quality of life of the people and in preserving the environment. The figure clearly emphasises the need of establishing institutional frameworks and mechanisms that may ensure easy access to and easy availability of development data of different nature at different tiers of development administration system. At the same time, there is also a pressing need to increase the awareness and facilitate easy access and uninterrupted availability of different types of development data and to build up the capacity to use the development data for development policy and planning and for objectively assessing the impact of development programmes and activities.

Development data can broadly be classified into three categories. The first category of development data are related to the people, beneficiaries of all development activities. Since improvement in the quality of life of the people is the ultimate objective of all development

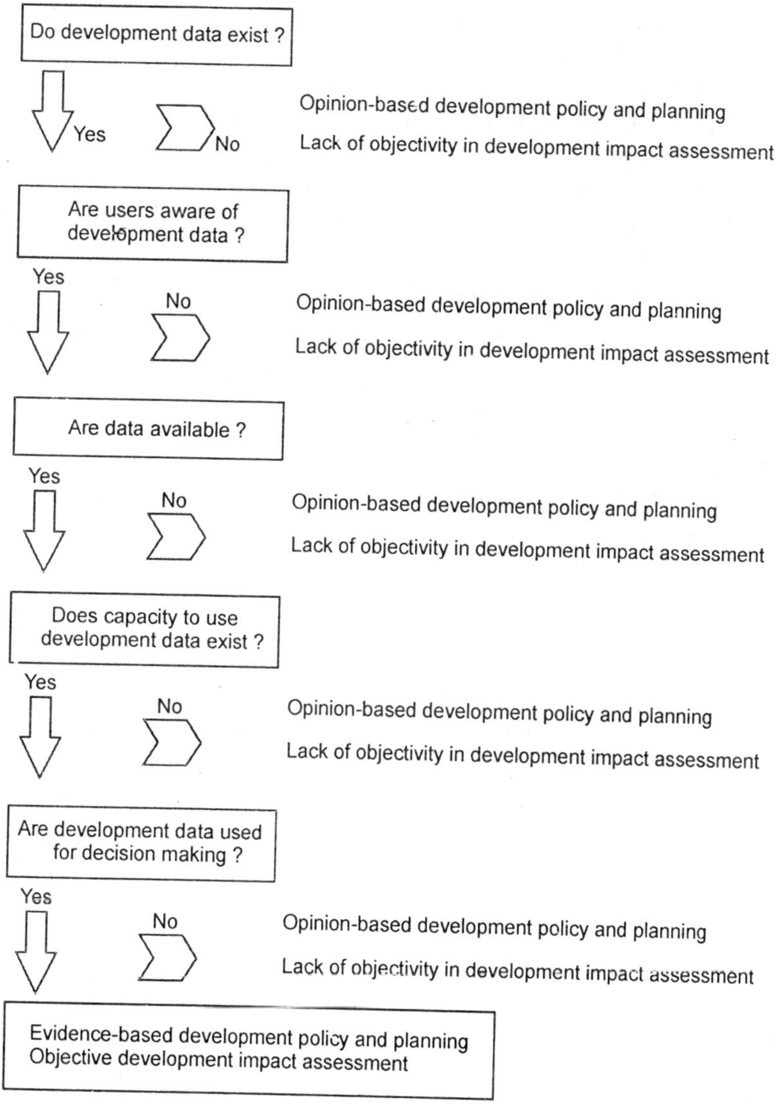

Fig. 24.1: Development Data and Development Policy Planning and Evaluation

activities, data related to population stock—size, structure, quality, etc.—have important implications for development policy and planning.

The second category of development data are related to social and economic development processes. This may include resources needs for development programmes and activities and their quality, data related to development programmes implementation and information related to outputs of development programmes and activities that impact upon the quality of life of the people as well as on the environment.

The third category of development data are related to environmental conditions that impinge upon the quality of life of the people and that have potential of endangering the very existence of life on the planet Earth. It is important that links are established between the date related to people, data related to social and economic development processes and data related to the environment so as to ensure that the process of social and economic development does contribute to improving the quality of life of the people without having a negative impact on the environment.

The most commonly used framework for development is the results framework. This framework breaks down development activities according to the logical components of the implementation processes and for each component, objectively verifiable indicators are defined for planning, monitoring and evaluating development programmes and activities. While developing or selecting indicators, it must be recognised that basic orientation of development processes varies by administrative and managerial hierarchy. Basic orientation of social and economic development policy and planning at national and state level is radically different from that at the grassroots level. At the grassroots level, the development process must focus on development and welfare needs effectiveness and the capacity efficiency of the development administration system. By contrast, at national and state levels, the primary concern of development is on the realised efficiency. At the grassroots level, development indicators should be simple, easy to measure, devoid of technical jargons and capable of straightforward interpretation. More complex development indicators may be retained at higher levels of development administration and management where necessary technical capacity exists or may be created. This implies that development data needs vary by the hierarchy of the development administration system, although it is possible to establish a link between development data needs at different tiers of development administration. This is an important consideration in discussing the availability of development data at different tiers of development administration.

The purpose of the present paper is to present an overview of sources of the three categories of development data described above and to discuss the nature, scope and limitations of the available development data in the context of evidence-based development policy and planning, evaluating development programmes and leading development activities. The need for discussing the availability, scope and limitations of development data has become all the more in recent years because of the decentralisation of the development planning process and the demand for quantifying the progress in terms of improving the quality of life of the people. Success of the decentralised approach to development planning depends crucially on the availability and use of local level development data for planning for meeting the development and welfare needs of the people.

Development Data Related to People

Data related to the people is perhaps the most important development data which reflect the quality of life of the people. For example, the most widely used indicator of measuring the quality of life of the people is the expectation of life at birth which indicates the average number of years, a person is likely to survive from birth under the prevailing mortality conditions. Mortality, in turn is shaped by a host of social, economic, cultural, environmental and technology related factors.

There are a number of sources which provided data related to the people in India. These sources are described, in some detail, in the following pages.

Population Census

The most widely known of source of data related to people in India is the decennial population census. India has the unbroken series of decennial population census now spanning more than a century. Among the developing countries of the world, the record of India in conducting census of the population at every 10 year is beyond parallel. The unbroken series of decennial population census in India provide an extraordinary valuable storehouse of data related to the people. The synergistic possibilities of analysis of the quality of life of the people inherent in successive census descriptions of the people of the country are truly remarkable. In the hands of skilled analyst, data available through the population census can be transformed into reliable estimates of population characteristics reflecting the size and structure of the population and quality of life of the people.

The history of first synchronous population census in India dates back to 1981, although, the earliest literature suggests that some kind of population count was maintained in India during the *Vedic* period. The celebrated *Arthashastra* by Kautilya written in the 3rd Century BC prescribed the collection of population statistics as a measure of state policy for taxation. It contained a detailed description of methods of conducting population, economic and agricultural census. During the regime of the Mughal king Akbar, the administrative report *Ain-e-Akbari* included comprehensive data pertaining to population, industry, wealth and many other characteristics that reflected the quality of life of the people at that time. A systematic and modern population census, in its present form, was however conducted non-synchronously between 1865 and 1872 in different parts of the country. This effort culminating in 1872 which is often labelled as the first population census of India. Since 1881, population census in India have been undertaken uninterruptedly once every ten years. The latest population census, conducted in 2011 was fifteenth in the series. Population census in India collects a range of data related to the quality of life of the people at the individual level as well as at the household level. A description of the data related to the people collected at the 2001 population census is summarised in Box 24.1.

Population census is the only source of data which covers all people, all households and all habitations of the country. However, prior to the 2001 population census, the analysis of the information collected at the census was confined to only 20 per cent of the households and, on the basis of this analysis, estimates were prepared for the entire country. At the 2001 population census, for the first time, efforts were made to analyse the information collected from all households and individuals.

Box 24.1: Data available through population census

Data Related to Households

- Use of the census houses
- Condition of census houses used as residence
- Predominant material of the roof, wall and floor of the census houses
- Type of structure of census houses
- Number of dwelling rooms
- Ownership status of the house
- Number of married couples and whether they have independent sleeping rooms
- Source of drinking water (e.g., Tap; Hand pump; Tube well; Well; Tank; Pond, River; Canal; Spring; others) and its location
- Source of lighting (e.g., Electricity; Kerosene; Solar energy; Other oils, Any other; No lighting)
- Availability of bathroom, type of latrine and type of drainage for waste water
- Availability of separate kitchen and type of fuel used for cooking (e.g., Firewood; Crop residue; Cow dung cake; Coal, Lignite, Charcoal; Kerosene; LPG; Electricity; Bio-gas; others)
- Availing of banking services and availability of the specified assets (e.g., Radio, Transistor; Television; Bicycle; Motor Cycle, Moped; Car, Jeep, Van; None of these

Data Related to People

- Relation to head of the household
- Sex
- Age at last birthday
- Current marital status
- Age at marriage
- Religion
- Name of Scheduled Caste/Scheduled Tribe
- Mother tongue
- Other languages known
- Highest educational level attained
- Attending educational institution
- Disability status
- Worker/non-worker
- Main Worker/Marginal Worker

(*Contd.*)

(*Contd.*)

- Economic activity of the main or marginal worker
- Non-economic activity of marginal worker and non-worker
- Marginal worker or non-worker-seeking/available for work
- Distance and mode of travel to place of work
- Place of Birth
- Place of last residence
- Duration of residence at place of enumeration
- Reason for migration
- Number of children ever born (For ever married women only)
- Number of children surviving (For currently married women only)
- Area of land cultivated (For cultivation households only)
- Tenancy status

Source: Registrar General and Census Commissioner of India.

Civil Registration

Another source of data related to people are the data about births and deaths available through the civil registration system. The registration of births and deaths in India was started on voluntary basis and there was no uniformity in statistical returns resulting in both under-registration and incomplete coverage vital events. In order to unify the civil registration activities, the Registration of Births and Deaths Act, 1969 was enacted under which registration of all births and deaths is mandatory. However, despite the enactment of the Act way back in 1969, the completeness of registration of births and deaths in several States and Union Territories in the country still remains far from satisfactory. In the developed countries, information available through the civil registration system is a very valuable source of development data related to the people. In India, on the contrary, information available through the civil registration system is rarely used in development research and development policy and planning because of a host of factors, the most important of which is the incompleteness of the reporting of vital events.

Sample Registration System

The Sample Registration System was introduced in India on a pilot basis in the year 1964-65 on a pilot basis to generate reliable and continuous data related to key demographic indicators at national and sub-national (State) levels. In 1969-70, the system was launched throughout the country. The system has since been providing data on vital events on a regular basis. The Sample Registration System is the only source of demographic data in India which provides data related to key demographic indicators on an annual basis.

The Sample Registration System is based on a dual record system. Field investigations under the system comprise continuous enumeration of births and deaths in a sample of villages/urban blocks by a resident part time enumerator, and an independent six monthly retrospective survey by a full time supervisor. Data obtained through these two sources are matched. The unmatched and partially matched events are re-verified in the field to get an unduplicated count of correct events. The advantage of this procedure, in addition to elimination of errors of duplication, is that it leads to a quantitative assessment of the sources of distortion in the two sets of records making it a self-evaluating technique. At present, the system is operational in 7,597 sample units (4,433 rural and 3,164 urban) spread across all States and Union territories of the country and covers about 1.5 million households and 7.10 million population. Estimates available through the Sample Registration System are known to be associated with random errors of unknown origin. In order to eliminate these errors, the practice is to use three-year moving average instead of annual estimates available through the system. The system is designed to provide development data related to people at the national and state level only, although, there has been some attempts to analyse the information available through the system at the regional level in each state. The system does not provide estimates of key demographic indicators at the district and below district level. However, the system does provide estimates of key demographic indicators separately for different population groups including rural and urban population as male and female population and, therefore, is of little use in decentralised development planning as well as monitoring and impact assessment of development activities.

National Family Health Survey

The National Family Health Survey was instituted in 1992-93. It is a large-scale, multi-round survey which is based on statistically representative sample of households throughout India. Three rounds of the survey have been conducted since 1992-93. The survey provides national and state level information related to fertility, infant and child mortality, the practice of family planning, maternal and child health, reproductive health, nutrition, anaemia, utilization and quality of health and family planning services. Each round of the National Family Health Survey has two specific goals: (a) to provide essential data on health and family welfare status of the people that may be used for health and family welfare policy and planning, and (b) to provide information on emerging yet important health and family welfare issues. The third round of the survey, conducted in 2005-06 has also collected information related to men and unmarried women, and on such issues as HIV/AIDS related behaviour, attitudes towards family life education for girls and boys, use of Integrated Child Development Scheme, men's involvement in maternal care, health insurance, etc. Moreover, estimates of HIV prevalence have also been provided by the survey on the basis of the blood sample collected in every state of India. Estimates of HIV prevalence are available for India and for Andhra Pradesh, Karnataka, Maharashtra, Manipur, Tamil Nadu and Uttar Pradesh.

District Level Household Survey

The district level household survey was commissioned in 1998-99 by the Government of India, Ministry of Health and Family Welfare to provide information about family planning, maternal and child health, reproductive health of ever married women and adolescent girls, utilisation of maternal and child health care services at the district level. This is the only survey which provides development data related to people at the district level. The survey also provides information related to new-born care, post-natal care and coverage of beneficiaries under different government schemes such as *Janani Suraksha Yojana*. The third round of the survey was conducted in 2007-08 which also provided information of the facilities available in the public health institutions—Sub-health Centre, Primary Health Centre, Community Health Centre and District Hospital accessible to the sampled villages. This is the only country-wide survey that provides development data of the people at the district level. The survey, however, does not provide development data for different social groups separately because the sample covered in one district is very small.

Medical Certification of Causes of Death

The scheme of medical certification of the cause of death was launched by the Government of India in selected states during the Third Five-year Plan as part of the civil registration system. The scheme provides cause-specific mortality profiles and is a key indicator for analysing the health trends of the population in a scientific manner The information available through the scheme is of considerable use to public health planners, administrators, medical professionals and research workers, and for assessing the effectiveness of public health programme. Medical certification of the cause of death under the civil registration system has got statutory backing under the Registration of Births and Deaths Act, 1969. Ideally, the scheme is to be implemented in all hospitals of the country so that all deaths can be certified medically about the cause of death. However, by 1999, only about 15 per cent of the estimated deaths in India could be medically certified about the cause of death.

Development Data Related to Social and Economic Conditions

Unlike data related to the people, there are not many sources of data related to social and economic conditions in India. In general, one relies upon data and statistics provided by the government and non-government agencies involved in the implementation of various social and economic development programmes and activities. These official data and statistics about social and economic conditions, however, is not a reliable source for assessing social and economic conditions because they are associated with a number of errors. In fact, a comprehensive development management information system does not exists in the country at any tier of development administration that can generate data related to social and economic conditions for development planning, monitoring and evaluation as well as for development research.

National Sample Survey

The National Sample Survcy was initiated in 1950 as a nation-wide, large-scale survey conducted in successive rounds to fill up data gaps for development policy and planning. Each round of the survey covers a specific subject of contemporary interest. Like the Sample Registration System, the National Sample Survey also provides estimates up to the state level only, although, information available through the survey has also been used for below state level analyses also. Perhaps, the most encouraging aspect of the National Sample Survey is that the unit level data are available either on payment or free of cost after entering into a memorandum of understanding with the National Sample Survey Organisation.

Till date 65 rounds of the national sample survey have been conducted throughout the country. These surveys have provided reliable data on various aspects of social and economic perspectives in India and in its constituent states. Some of the key areas that have been covered under these surveys include:

- Consumer expenditure
- Employment and unemployment
- Land and livestock holdings
- Housing conditions
- Village facilities
- Urban slums
- Disability

It has also been conceived that the constituent states of India will also evolve mechanism on the lines of national sample survey to organise state specific sample survey to general data related to social and economic conditions on a regular basis. However, in most of the states, no such mechanism has been evolved. There exists a very substantial capacity gap at the state level to organise such sample survey which is both complex and multi-dimensional in scope and coverage. As such, assessment of social and economic conditions below the state level in India is a mere guess. Some very limited information and hard data are available through the population census but this information is grossly inadequate. Village level data about social and economic conditions in India is largely non-existent other than that generated through population census. Since, population census is conducted at an interval of 10 years, the social and economic data generated through population census is or very limited use for development policy and planning.

Development Data Related to Environment

Issues related to environment, as it is related to population and development, can be grouped into two broad categories—local level environmental issues and global environmental concerns. In any case, all the environmental concerns of population and development may be conceptualised in the framework of natural resources use and associated wastes generated which, in turn, depend upon the type of technology used. The most commonly used framework linking environmental concerns with population, development and technology is the IPAT framework. This framework conceptualises the environmental impact (I) in terms of population (P), affluence or level of development (A), and the state of technology (T). The advantage of this simple model is that it can be applied in the context of global environmental concerns as well as in the context of local environmental issues. Using the IPAT framework, the environmental impact of human activity may be explored in two ways—in terms of resource used and in terms of wastes generated as the result of resource use. One way of exploring the impact of human activity on resource use is the analysis of the trends in energy use. In this context, the IPAT framework may be translated into

$$E = P^*g^*(e/g) = P^*g^*I$$

where E denotes total energy use, g denotes the gross national product per capita or per capita income and is a measure of affluence and e denotes the per capita energy consumption so the ratio i=e/g is the energy intensity which reflects the structure of the economy and the efficiency of energy use and, therefore, is a measure of what is known as the technology intensify. For example, a service-oriented economy is generally argued to be less energy intensive than an industry-oriented economy which is in the early stages of industrialisation and which requires large amount of energy for its social and economic production system. Similarly, the impact of the human activity on the environment may also be assessed and analysed in terms of the wastes generated. Since fossil fuel is the primary source of energy and CO_2 is released as the result of burning fossil fuels, the IPAT framework may be translated into what is known as the Kaya identity:

$$C = P*g*i*k$$

where C is the total CO_2 emission and k is the carbon intensity. It is the ratio of total CO_2 emissions per unit of energy use. It reflects the fuel mix of the energy system, in particular, the share of the energy from renewable sources and energy from non-renewable resources; and, among the non-renewable resources, the share of highly carbon intensive sources such as coal and less carbon intensive sources such as natural gas.

The above simple yet straightforward approach of analysing population, development and environment relationship has also been criticised for a number of perceived flaws. The primary argument is that any analysis of population, development and environment relationship must take into consideration the social, cultural and political contexts of human activity. It is also argued that P, A and T variables may at best be viewed as proximate variables of the environmental impact. They, themselves, are determined by a host of indirect but more fundamental factors associated with human activity which are not accounted by the simple IPAT model. In order to circumvent many of the perceived limitations of the conventional IPAT model, a stochastic version of the simple PAT model has also been developed and applied. Nonetheless, frequent availability of information related to population size and growth as well as trends in affluence measured in terms of income per capita has prompted widespread use of the simple IPAT approach. Moreover, the simple IPAT model illustrates an important consequence of the multiplicative relationship between the driving forces—each variables amplifies changes in other variables in the model. As a result, a small change in technology may have only a small absolute effect on energy use and CO_2 emissions in a country with a small population and low levels of income, while the same change would have a much greater effect in a populous, affluent country.

Country level estimates of energy use and CO_2 emissions have been prepared by a number of international organisations including World Bank, Earth Trends and Carbon Dioxide Information Analysis Centre of the United States, Department of Energy. However, such information is not available for the Indian states. The environment database maintained by the Environment Information Centre set up by the Government of India, Ministry of Environment and Forests does not contain state specific information about energy use and CO_2 emissions. The primary objective of the Centre is to act as a one-stop source for quality environmental data on India. It acts as a repository of Indian environmental data and provides site specific information in a timely manner to facilitate environment related planning and decision-making. Unfortunately, the data available through the Centre is not freely available. This is a major hindrance against the use of the data.

A major challenge to analysing the environmental impact of population and development at the local level is deciding about the indicators of environmental impact that may be causally linked to population and development scenario. These indicators need to be developed in the locally specific context and, therefore, may vary from one place to another. This variation in the local context of population, development and environment relationship inhibits evolution of a standard framework of analysis of population, development and environment interaction as has been evolved at the national and global level.

Conclusion

India needs a comprehensive data system that can meet data demand for analysing population, development and environment interactions. Such a system does not exists at present, although efforts have been initiated in this direction through setting up of the Environment Information Centre. This data system needs to be evolved in the context of data needs for population, development and environment policy, planning, monitoring and evaluation. There are four major aspects in this context:

1. The data type and its associated information system;

2. The aggregation level of the data;
3. The measurement of the data; and
4. The usage of data.

Major data characteristics of the proposed population, development and environment data system are shown in Fig. 24.2.

	Population, development environment data system				
Data type system	Population Data Bank	Develop-ment Indicators Data Bank	Quality of Life Indicators Data Bank	Develop-ment Programme Indicators Data Bank	Environ-ment Indicators Data Bank
Aggregation level	Country State District Sub-district Panchayat Village	Country State District	Country State District Sub-district Panchayat Village	Country State District Sub-district Panchayat Village	Country State District Sub-district Panchayat Village
Measure-ment type	Quantitative	Quantitative	Quantitative Qualitative	Quantitative Qualitative	Quantitative Qualitative
Usage	Policy and programme planning Advocacy and legitimation	Policy and programme planning Advocacy and legitimation	Monitoring and evaluation Advocacy and legitimation	Monitoring and evaluation Advocacy and legitimation	Impact assessment Advocacy and legitimation

Fig. 24.2: Basic Characteristics of Population, Development, Environment Data System

The population, development, environment data system conceptualised above may be viewed as an 'institutional mechanism' or, more specifically, as a 'tool' to support development policy and planning and to facilitate evidence-based monitoring and evaluation of development programmes and activities in the context of improvements in the quality of life and impact on the environment. It is also expected to help streamline and coordinate regular compilation and generation of relevant information to analyse population, development, interactions and monitor progress towards sustainable development.

25
India's Population Policy: Operational Framework for Implementation

Arun K. Singh

Introduction

India, the second most populous country in the world, has roughly around seventeen per cent of the world's population living on less than three per cent of the land area. At the onset of the twentieth century India's population was about 238 million, by 2001 it crossed the billion mark and as per 2011 census provisional data it is 1.21 billion. On an average the country added 16 million people annually in the1980s and 18 million annually in the 1990s until the present (Planning Commission, 2007). As compared with the global demographic figures, the world population has increased by threefold during the last century (from 2 billion to 6 billion) whereas India's population has increased by nearly five times till 2011. Although, the crown of the world's most populous country is on China's head for decades, India is all set to surpass it by 2030 when its population will be more than 1.53 billion people. In 1950, shortly after India gained independence, India's population was 68 per cent that of China. Now it is 90 per cent. China's population currently increases by a little less than 7 million per year while India's population grows by about 18 million. As per UN estimates India's population will likely pass China's shortly after 2020. Surprisingly since 1971 there has been a continuous decrease in India's decadal population growth rate, but because of our huge demographic base the increase in absolute number is significantly large as more than 50 per cent of India's current population is below the age of 25 and over 65 per cent below the age of 35 (Haub & Sharma, 2011). An analysis of birth rates and death rates pattern of the last hundred years (1901-2001) further reveals that the country is in the middle stage of demographic transition with fertility and mortality both started declining throughout the country because of improvements in the health sectors and general thought process of the people. Consequently the pace and magnitude of this decline varies over the space and there is a clear cut north-south divide even in the growth pattern. This becomes more clear if we take into account the case of Bihar and eastern Uttar Pradesh (BIMARU states) on the one hand and Kerala (developed state) on the other.

Broadly speaking a population policy aims to modify the demographic behaviour of a given population in a desired direction which could be possible through regulating *reproductive policies, health policies, migration and urbanization policies* or *family and welfare policies which* intends to enhance the well-being of individuals and families. It includes the measures and programmes designed to contribute to the achievement of economic, social, demographic, political and other collective goals through affecting critical

Table 25.1: India: Population Size and Growth (1901-2011)

Census Year	*Population In thousand*	*Growth over decade*		*Annual growth rate*
		In thousand	*In per cent*	
1901	238,396			
1911	252,093	1,3697	5.7	0.6
1921	251,321	-772	-0.3	0
1931	278,977	27,656	11.0	1.0
1941	318,661	39,683	14.2	1.3
1951	361,088	42,428	13.3	1.3
1961	439,235	78,147	21.6	2.0
1971	548,160	108,925	24.8	2.2
1981	683,329	135,169	24.7	2.2
1991	846,421	163,092	23.9	2.2
2001	1,028,737	182,316	21.5	2.0
2011	1,210,193	181456	17.6	1.6

Source: Based on Haub & Sharma, 2011.

demographic variables, namely size and growth of population, its geographical distribution and its demographic characteristics (UNESCO, 1980). Population policy of any country can no longer confine itself to the reduction of birth rate alone. It must be a comprehensive strategy for overall improvement of social, economic and demographic fabric of a nation. The official realization that population is not merely about numbers but about the health and quality of life of people in general and women and children in particular is a way forward in a country like India where "woman dies approximately every five minutes while giving birth or because of birth related complications; millions of women 'agree' to discontinue their studies and get married, 'consent' to have sex, 'decide' to become pregnant, 'choose' to continue in abusive relationship, 'demand' sex selective abortions, 'prefer' to stay at home rather than work or *vice versa*, 'volunteer' to participate in clinical trials for hazardous contraceptives and 'prefer' to neglect their own health and nutrition" (Sen, 2010). Such preferred choices are not exclusively associated with poverty, lack of education rather they are inevitable outcomes of a natural order of hierarchy closely woven into the fabric of social institutions. So there is a need for a better and more widespread understanding that the number of children desired by any couple depends on a large and complex interrelated number of socio-economic and cultural factors, and that any policy action seeking to control population must seriously take all these variables into consideration.

Pre-Independent India and Issues Related with Population Policy

India has the distinction of being the first developing country to adopt a well defined population policy favouring family planning in the early fifties of the last century. Today it has become an integral and accepted part of overall socio-cultural and economic development policy and programmes. India's present population policy and debates on population size has its roots in the pre-independence era when a large number of social reformers, thinkers and voluntary organizations tried to influence and persuade the Government to adopt a policy of population control as a measure to protect the health of the mother.

In fact one of the first writings on *'the Population Problem in India'* was by P.K. Wattal in 1916 when he talked about the consequences of unbridled population growth and its impact on development. R.N. Tagore (1928) expressing his views on population issues said *"Birth control movement is a great movement because it will save women from enforced and undesired maternity"*. R.D. Karve (1925) was the first individual to open a family planning clinic which culminated into his resignation from teaching job. In 1932 Government of Madras opened birth control clinic in the Presidency and in the same year All India Women's Conference at Lucknow passed a resolution recommending that "men and women should be instructed in methods of birth control in recognized clinics". National Planning Committee (1935) strongly supported family planning and came up with the following recommendations:

- size of the Indian population is the concern for economic planning;
- family planning and limited number of children is essential for family happiness;
- increase in marriage age and discouragement of polygamy are desirable....;
- an eugenic programme should include the sterilisation of persons suffering from transmissible diseases of a serious nature such as insanity and epilepsy; and
- maintenance of vital statistics and the carrying out of periodic demographic surveys are necessary; and appropriate machinery devised for the purpose.

In 1936 Dr. A.P. Pillai conducted a course on family planning and in 1940 P.N. Sapru moved a resolution in the Council of States for the establishment of birth control clinics in the country. Support for birth control was evident when Bengal Famine Inquiry Committee (1945) in its report contained a chapter on the potential dangers to the economy and life of people arising out of rapid population growth, especially a population living in abject poverty and deprived of the bare necessities of life. Similarly, Bhore Committee (1943) recommended a suitable health infrastructure for the country and stressed the need for a national programme of family planning for improving the health status of population. Looking at the developments made in the pre-independence period Gunnar Myrdal (1968) rightly remarks, *"Intellectual ground work has been laid when the colonial era came to an end, a rational policy could be pursued in the era of independence."* The two reports (Bengal Famine Inquiry Committee and Bhore Committee) paved the way for the Government of India to adopt a population policy to promote reduction in the number of children born to Indian couples. India's population policy which is an outcome of debate and discourse initiated in pre-independence India is based on certain guiding principles and philosophy:

- The community must be prepared to feel the need for the services in order that those may be accepted when provided.
- Parents alone must decide the number and spacing of the children.
- People should be approached through the media or trusted leaders.
- The services should be made available to the people as near to their homes as possible.
- The services will have greater relevance and effectiveness if they are made an integral part of medical and public health services, especially of the mother and child health programmes. (Quoted from Agrawala, 1985)

Population Policy Related Issues in Independent India

Setting up of Family Planning Association in 1949 was the first effort in independent India which aimed to advice the nation on the population related issues. In April 1950, Government of India appointed a Population Policy Committee under the Chairmanship of the then Minister of Health and upon its recommendations; a Family Planning Cell was created in the office of the Director General of Health Services. The First Five Year Plan document (1952) highlighted the urgency of family planning and stressed the population stabilization at a level consistent with the requirement of national economy which could be achieved through reduction in the birth rate. As an important component of its developmental strategy it sought to reduce the birth rates to the extent necessary to stabilize the population at a level consistent with the requirements of the national economy. The Government initially adopted *clinic approach* and the focus was on to present the accurate picture of the factors contributing to the rapid population increase, discover and disseminate the suitable techniques of family planning and make advice on family planning. This approach was based on western model, i.e. *Planned Parenthood Organization* (PPO) in which target group was supposed to approach the centre for obtaining services. This resulted in the increase of number of clinics from 147 in 1950 to 4165 in 1961. But in a developing country like India, its success was under question because of existing socio-cultural ethos and value system. What to talk about going to health care centre of the young couple and discussing the matter related to their family planning, they were not socially permitted to meet each other in the daytime. Consequently the 1961 census showed a continued rise in the population growth rate and an increase in the fertility levels. The government came forward with clear cut demographic goal in 1962, i.e. to reach a crude birth rate of 25 by 1972 and the clinic approach was replaced by *extension approach* aiming at bringing the message and services to people in far flung areas through a network of primary health centres. Thus creation of social climate in favour of small family norms, provision of readily accessible services, adoption of effective family planning methods by all eligible couples, stimulating social changes such as increasing marriage age, education and employment of women, accelerating over all development and continue research and evaluation became the guiding principles of population policy during the Third Plan (1961-66). But unless the children are provided with preventive and curative medical services, parents will not go for small family norm. In this context Mrs. Gandhi (1980) categorically remarked *"So, they must be ensured better health, education, and of course, even employment later on for sustainability of the society. It is part of national planning as a whole"*. In 1974 India came up with a slogan "Development is the best contraceptive" in the UN World Population Conference at Bucharest. This statement was interpreted and misinterpreted leading India to defend and change its stand in the words "family planning is not going to work unless we improve the standard of living of those who are below poverty line." Soon it was realized that the demographic situation is grim and the country cannot wait for the economic development leading to population control and there is a need for 'frontal attack on the problems of population'.

Emergency and Aftermath

Up to 1976 official policy towards population was a part of socio-economic development and for the first time, a National Population Policy was formulated and adopted by the Parliament in April 1976 which inspired the state governments to 'pass suitable legislation to make family planning compulsory for citizens' and to stop child bearing after three children, if the 'state so desires' (Srinivasan, 2006). The strategy during this period can be termed as *Coercion* and the number of sterilizations done in India during April 1976 to March 1977 was 8.26 million, more than the total number done in the previous five years and more than the number done in any other country in the world until that time. If the effectiveness of the Population Policy (1976) is measured by the number of sterilizations performed then it could be held as a success (Gwatkin, 1979). However, from a rights based perspective, if its effectiveness is measured by deaths, violence, or rights compromised in an attempt to achieve goals of the Population Policy, initiatives were unsuccessful. The fear psychosis culminated into debacle of the regime in 1977 and it is believed that the mandate was the first in the world in which fertility behaviour was a deciding factor (Cassen, 1978). The new government adopted new population policy which was basically a damage control device and rejected any type of coercion and compulsory sterilization. Thus the period after 1977 can be termed as *Recoil and Recovery* phase for the family planning programme. The change of government again in January 1980 marked a turning point and helped to restore the family planning programme garbed as family welfare programme. The emphasis on sterilizations continued though on a voluntary nature. While facing a TV interview with Chhabra (1981) the then Prime Minister remarked:

> "I agree with the view that human rights aspect is very important, especially of course for the woman because she has the entire burden of child bearing and child rearing and yet her opinion is hardly ever asked for; nor is there much consideration

> for her welfare. And, unfortunately the child is one who suffers from this. Today birth should be a matter of choice of both parents so that they feel responsible for the child and give it its due rights." She further added "because we do not believe in coercion: we think that there should be persuasion. But this must be done on a massive educational scale and unless everybody is in favour, it will take a very long time and may defeat the whole purpose of it. I think we have to involve not just the department that is in charge of this, but the whole population, specially schools, the whole youth movement, women's organisations, other institutions which are concerned with any type of social welfare. I mean everybody gets into it to create the right atmosphere. One part which is very important is that we must ensure better health, because if there is any chance of children dying more, then people will not be willing to risk that. So, they must be ensured better health, education, and of course, even employment later on. It is part of national planning as a whole."

Above comments of Mrs. Gandhi reflects her changed ideas, philosophy and vision related with population policy which stressed elements of human right, reproductive right and welfare of the children. Commenting on the view expressed above Bose remarks "In India poverty and population make a vicious circle: the birth rate is high because people are poor, and people are poor because the birth rate is high. This is simplistic Economics." *This seems to be a chicken-egg story.* Today the concept of development is far less concerned with per capita income and more concerned with equity, equality, empowerment and quality of life issues. Bose further adds "India will succeed in implementing its population policy if it injects into the programme the health and education consciousness of Kerala, the people's involvement as in Gujarat, the contribution of the organised sector as in Maharashtra and the rural prosperity and dynamism of Punjab. But at the implementation level the time-bound, target-oriented approach was revived and efforts to encourage the use of reversible methods were initiated in the comeback Congress regime in 1980. Incentive payments were vigorously promoted, leading to the violation of women's rights. Among this entire discussion one has to look at the basic principle and philosophy of India's population policy that *'Parents alone should decide the number and spacing of the children'* not the state as it happened to be. The small family norm is thus visualised as a part of the development process, regardless of the economic exercises on the complex relationship between population growth and economic development.

Democratic Decentralization and Population Policy

The global movement in the 1980s and early 1990s witnessed a paradigm shift. Policymakers, planners, government and members of the civil society started arguing for the need to understand social and human development that was centred on the basic needs of people, to ensure lives free from deprivation. There was advocacy for basic issues like population and health including reproductive right, maternal mortality, education and gender equity. The primary health care including family planning, primary education and provision of certain basic amenities to the people such as drinking water and roads became the responsibility of the panchayats and there was provision of reservation of one-third of the seats in panchayats for women members. Thus at the grassroots level gender issues and women empowerment became the reality under the provisions of 73rd CAA. The International Conference on Population and Development (ICPD), held in Cairo in 1994, happens to be a landmark where it was explicitly agreed that the advancement and protection of women's human rights should be central to government efforts to address population and development issues. The ICPD Programme of Action focused attention on gender equality, equity, and women's empowerment. Further, it declared that *"the promotion of the responsible exercise of these rights for all people should be the fundamental basis for government- and community-supported policies and programmes in the area of reproductive health, including family planning."* It was at this point that globally as well as locally there was an increasing use of concepts reproductive and sexual rights as well as larger human rights framework. In response to these developments, there was growing recognition within the Government of India to rethink the family planning programme and the government appointed an expert committee (1993) led by M.S. Swaminathan which in its report (1994) defied the target based approach, asserted a policy that was pro-poor, pro-nature and pro-women and argued that population stabilization would be reality through wide ranging changes in social and economic development. The Maternal and Child Health approach was broadened to include reproductive health, child health, neo-natal health and reproductive right.

In 1996, the government took the radical decision of abolishing method-specific contraceptive targets and replacing it with Target-free Approach which intended to reverse the top-down, target-centred population policies. The basic tenets of the approach were:

- to provide services according to client needs,
- emphasizing quality of care as an equal part of any health programme, and
- providing a wider choice of safe contraceptive methods and strengthening reproductive health services.

The new approach represents a dramatic change in the culture of the Family Welfare Programme and attention has been focused on gender concerns. The Reproductive and Child Health Programme seeks to address gender issues impinging on women's health by improving quality of care, including promoting better interaction between providers and clients; increasing the availability of female health care providers at the primary health care level; addressing neglected concerns of women such as reproductive tract infections; addressing the needs of neglected population sub-groups such as adolescents; organising gender sensitisation training for stakeholders; encouraging male involvement in reproductive health; and facilitating women's and men's participation in programme monitoring through client feedback (World Bank, 1997).

Table 25.2: Features of the Target and Target Free Approach

Issue	*Target approach*	*Target free approach*
Driving force	Targets	Community needs
Orientation	Provider	Client
Concern	Target Achievement	Quality of care
Goal	Demographic impact	Reproductive health status
Approach	Top down	Bottom up

Source: Bhan & Panda, 2010.

National Population Policy (2000)

Population policy of any country can no longer confine itself to the reduction of birth rate alone. It must be all comprehensive strategy for overall improvement of social, economic and demographic fabric of a nation. Thus the strategy for decreasing the rate of population growth, limiting our number in the young and reproductive age groups, enhancing our employment potential, diversifying economic base, improving level of literacy and education and balancing sex ratio should constitute the elements of any population policy. Unfortunately, since independence India has been over emphasizing the role of positive measures (contraception, sterilisation and abortion) at the name of population policy to keep the number under control without paying much attention to the dynamic interaction between the demographic structure and social and economic development and implication of such interaction for limiting the rate of population growth (Chandna, 2008). In this context the new National Population Policy 2000 *(NPP 2K)* is a comprehensive document, which addresses the population problems taking into account the related issues like child survival, maternal health, women empowerment and employment and contraception.

> "In the new millennium, nations are judged by the well-being of their people; by level of health, nutrition and education; by the civil and political liberties enjoyed by their citizens; by the protection guaranteed to children and by provisions made for the vulnerable and the disadvantaged. The vast number of people in India can be its greatest asset if they are provided with the means to lead healthy and economically productive lives" (National Population Policy 2000).

The spirit of India's NPP2K is guided by important considerations of human development, equity, reproductive and human right. This makes a significant shift from earlier population policy which was primarily demographically-driven and target-oriented. The policy has well spelt immediate objectives (to address unmet need for contraception, health care infrastructure and health personnel and to provide integrated service delivery for basic reproductive health and child care to all), midterm goals (to bring total fertility rate to the replacement level) and long term goals (to achieve population stabilization by 2045, at a level consistent with the requirement of sustainable economic growth, social development and environmental protection).

To achieve these goals the NPP2K proposes a multipronged strategy. The operational strategies and strategic themes as outlined in the NPP2K emphasize decentralized programme implementation based on an assessment of community needs and local contexts. The operational strategies outlined in the policy envision a decentralized, integrated approach to health service delivery implemented by local institutions like the Panchayati Raj Institutions (PRI) and village self help groups. The policy also lays emphasis on information, education and communication services, i.e. using public media and other tools to increase knowledge and awareness of the need for family planning and the options available to clients. For the first time in the history of India's population policy, the national rural health mission (NRHM) has incorporated a framework after wider consultation of with diverse stakeholders including civil society groups. In addition a community monitoring group has been formed to monitor activities at village, block and district levels. In the following segment of this paper an operational framework has been suggested for grassroot level implementation.

Operational Framework for Policy Implementation

Operational Framework refers to a schematic diagram (Fig. 25.3) indicating the range of activities in a socio-organizational set up. Usually two approaches are adopted in the population policy implementation context. The first one is the target based which was in practice before 1996. In this approach targets were fixed in terms of number of sterilisations to be performed and accordingly incentives and disincentives were given. The second approach is beneficiary based where the focus is on meeting the basic health and family welfare needs of the people which is determined by the prevailing social and cultural environment as well as social and economic development and host of other factors which are beyond the scope of traditional population stabilization programme. The beneficiary based operational framework for the population policy implementation is driven by the simple concept that the implementation process must strive for meeting the health and family welfare needs of all people in a cost effective, cost efficient and operationally feasible manner. This approach has been advocated and adopted in India since 1996 but its success requires a radical change in approach and attitude of the people at grassroots level. A diagrammatic model presented here follow the beneficiary based approach and starts with the married woman being in the focus as she enters the socially recognized, sexually active reproductive life and then moves to different stages of production in course of giving birth to a child. The operation framework recognizes two sets of interventions; the first meets the specific family welfare needs and the second contributes towards creating enabling environment. In both cases the key issues are accessibility, availability and acceptability of the services by the people. The operational framework identifies that for the five

Sl. No.	*Target Groups*	*Interventions*	*Concerning deptts.*
1.	Currently married but not in the family way	Registration of all the married women. Launching a service package which ensures that conception does not take place before 18.	Deptt. of Economics and Statistics Deptt. of Women and Child with the support of Family Welfare Deptt.
2.	In the family way	Comprehensive maternity care package according to the need.	Deptt. of Women and Child with the support of Family Welfare Deptt.
3.	In the process of delivering a child	Safe delivery package (attended by properly trained persons and efficient referral network) to minimize the risk of death due to delivery related complications.	Deptt. of Women and Child with the support of Family Welfare Deptt.
4.	Having less than two children	A service package that ensures a birth interval of at least three years. Postnatal package suited to specific need of women.	Deptt. of Women and Child with the support of Family Welfare Deptt. and Health Deptt.

(*Fig. 25.1 Contd.*)

post emergency period Recoil and Recovery approach appeared as a damage control device. Mrs. Gandhi in her come back regime (1980) emphasized changed ideas, philosophy and vision related with population policy which stressed elements of human right, reproductive right and welfare of the children. She also advocated for participation of different stakeholders like civil society, local democratic institutions or government departments (women and child development, education, health and family welfare etc.). Commenting on the view Bose in an interview with Chhabra (1981) remarked that development is not concerned with increase in income rather it is concerned more with equity, equality, empowerment and quality of life. This could be visualized only when there is a convergence of models applied in different parts of the country. But at the implementation level time-bound, target-oriented approach dominated the scene as it happened in pre-emergency period. This truly reminds us the statement of Pandit Jawaharlal Nehru while addressing the youth in university convocation "thought *without action is abortion as action without thought is folly*". The commitment shown by Mrs. Gandhi coupled with global movement in the 1980s and 1990s in the form of democratic decentralization and Cairo Conference became explicit and proved to be the foundation stone for the latest population policy and the people started advocating for basic issues like population and health including reproductive right, maternal mortality, education and gender equity. The Government of India appointed an expert committee which in its report (1994) defied the target based approach, asserted a policy that was pro poor, pro- nature and pro-women and argued that population stabilization would be reality through wide ranging changes in social and economic development. In this context the new National Population Policy 2000 *(NPP2K)* is a comprehensive document, which addresses the population problems taking into account the related issues like child survival, maternal health, women empowerment and employment and contraception. The policy has well spelt immediate objectives (to address unmet need for contraception, health care infrastructure and health personnel and to provide integrated service delivery for basic reproductive health and child care to all), midterm goals (to bring total fertility rate to the replacement level) and long term goals (to achieve population stabilization by 2045, at a level consistent with the requirement of sustainable economic growth, social development and environmental protection). To achieve these goals the NPP2K proposes a multi pronged strategy. The operational framework discussed in the text reveals how enabling environment could be created in which different stakeholders would play their respective role to achieve the objectives at the grassroots level to fulfil the statement of E.F. Schumacher 'think globally and act locally'.

REFERENCES

Agarwala, S.N. (1985). *India's Population Problems*. New Delhi: Tata McGraw-Hill.

Bhan, G. and Panda, P. (2010). Revisiting India's population policies. In A.K. Shiv Kumar, P. Panda and R.R. Ved (eds.) *Handbook of Population and Development*, 57-71. New Delhi: Oxford University Press.

Cassen, R.H. (1978). *India: Population, Economy, Society*. New York: Holmes and Meier Publishers.

Chandna, R.C. (2008). *Geography of Population: Concepts, Determinants and Pattern*. Ludhiana: Kalyani Publishers.

Chhabra, R. (1981). Population policy in India: two comments. *Population and Development Review*, 7(1), 61-64.

Gwatkin, D.R. (1979). Political will and family planning: the implications of India's emergency experience. *Population and Development Review*, 5(1), 29-59.

Haub, C. and Sharma, O.P. (2006). India's population reality: Reconciling change and tradition. *Population Bulletin*, 61(3).

Haub, C. and Sharma, O.P. (2011). India releases latest census results showing population catching to China. Retrieved on 14 Jan. 2011 from www.prb.org

Myrdal, G. (1968). *Asian Drama: An Inquiry into the Poverty of Nations*. London: Penguin Press.

Planning Commission, Government of India (2007). *Report of the Working Group on Population Stabilisation for 11th Five Year Plan (2007-12)*. Retrieved on 14 Jan. 2012 from www.nic.in.

Ranjan, A. and Dass, I. (2005). *Decentralised Planning for Population Stabilisation*. Bhopal: Population Resource Centre, Madhya Pradesh.

Sen, Kalyani-Menon (2010). Women in India's national population policy. In A.K. Shiv Kumar, P. Panda and R.R.Ved (eds.), *Handbook of Population and Development*, New Delhi: Oxford University Press, 160-68.

Srinivasan, K. (2006). Population policies and family planning programmes in India: A review and recommendations, *IIPS Newsletter.* January 2006.

Tagore, R.N. (1928). *Family Planning News*, 1(12).

World Bank (1997). *World Development Report, 1997*. Retrieved on 21 Aug., 2011, from www-wds.worldbank.org.

26

Irrigation Management and Inclusive Growth: AKRSP (India)'s Intervention in Tribal Gujarat

Jyotirmayee Acharya

Introduction

In the new millennium, the importance of water as a gendered resource has emerged as never before. Currently, integrated water resources management is being promoted by several international organizations, national governments and civil societies, Gender and Water Alliance and World Water Forum and so on. The national consultative meeting held in July 2007 on 'towards conflict resolution on water conflicts in India' lead by Society for Promoting Participative Ecosystem Management concerns on the growing recognition that managing water as much as land and biomass is going to be a critical challenge[i] for future economic growth and agricultural sustainability. The debate in the field of social exclusion while represents an important loss of capability in its own right as well as being a cause of other deprivations which further remove the individual or group from access to the good life or the means to acquire it (Sen, 2000). Mainstreaming gender concerns in the small irrigation management schemes relevant to this paper is that giving effective operational value to the policy initiatives/ changes would enable the equal participation, representation and rights of women in the rural livelihood and thereby, improving the effectiveness and sustainability of those strategies. The World Bank defines Participatory Irrigation Management (PIM) as *'the involvement of irrigation users in all aspects of irrigation management, and at all levels'*. This implies that water users participation in planning, design, construction, operation and maintenance also, include their participation in financing, deciding on rules and norms, monitoring and evaluation of irrigation systems.

Practitioners in the water sector (Shah *et al.*, 2006; Parthasarathy, 2006; Oza, 2007; SEWA, 2003) on PIM in Gujarat[2] and elsewhere observed that the farmers with small size of landholdings have derived highest benefit in crop productivity when they have played sound role in collective decision making on the aspects of their livelihoods through the PIM association. Problem encountered in the state irrigation department are inadequate operation and maintenance, inequitable distribution of water, no incentives for saving water and low recovery of water rates. The problem is compounded by lack of any effective institutional check on the management, opportunity cost of women and marketing, lack of engagement of the NGOs working with the communities. Government's institutional reform issues are now giving importance to the PIM, even more importance than price and trade policy reforms (National Water Policy, 2002; Gujarat Water Users Participation Management Act, 2007). However, the ability of the local communities and governments to deal with water management is constrained by the absence of gender sensitive analysis, gender disaggregated data and reliable tools, information networks and the professional and institutional capacities. Nevertheless, while implementation of PIM policy found all very important but by themselves were not able to influencing goals, strategies, resource allocation, and concrete outcomes.

Shah (1993) pointed out that the market in groundwater in Gujarat is highly competitive, efficient and individualized, bearing all the marks of what economists define as 'economic rationality'. The politics of water scarcity and price fixing in the community cut across the issues of caste, class, gender, leadership and so on (Dubash, 2002; Hardiman, 1998). Social exclusion is a process and a situation that prevents individuals or groups in such cases from full participation in asserting their rights.

Cultural politics of water offers the analytical framework within which questions of power and gender inequality, conflicts and compromises, which shape the waterscape and agricultural productivities need to be addressed. Agarwal (1994) has argued that land defines the social status and political power in the village. Yet in South Asia approximately 3 to 10 per cent of rural women own land which they cultivate (International Fund for Agricultural Development, 2002). The exclusion of women from the intra-household resources and assets constrict their productive performances limits their participations in the land-based livelihoods and entrepreneurships. Feminist studies in India (Agarwal, 1994; Kelkar, 2005) address the question whether women's lack of ownership and control of agricultural land affects productivity. The taskforce on Technological Empowerment of Women in Agriculture of National Commission for Women in India emphasises that lack of title to land makes it difficult for women farmers to access institutional credit (Swaminathan, 2005). However, agricultural policy-making including irrigation agencies or land reforms and practitioners those seek to reduce poverty and promote agricultural growth have hardly designed and implemented the gender equity considerations and recognition that farmwomen's control over land and on-farm irrigation management in the households and community institutions.

Small Irrigation Schemes and Tribal Water User Association

Marginalised communities and tribal particularly in the dry and semi-arid region those who lack effective livelihood choices from agriculture and allied activities often accept entrenched poverty, deprivation and exclusion as a way of life (Sen, 2000, Shah & Yagnik, 2007)[3]. While food processing, cotton manufacturing and chemical industries attract male wage-workers and contractual-labourers from the rural and urban-fringe women play a crucial role in the household economy, agricultural production and water related management activities. Previously used to hill slope cultivation have now shifted to subsistence agriculture, fodder farming and pockets of commercial farming since last four generations. Irrigation taps water from canal and other local reservoirs, but mismanagement still deprives tail-enders. Tube/bore well excavating ground water is an alternative for the farmers in the areas without canal system. However, agricultural productivity remains largely rain-fed owing partly to inclining and undulating land with run-off of top soil and canal mismanagement practices and partly to lack of effective management of land and monsoon rainfall harvesting techniques, financial services and access to agricultural inputs and market.

Study found that large commercial farmers and cash-crop growers of Gujarat are bigger investor and have greater control. In fact, the concept of a water user association (WUA) formed around a well was introduced by AKRSP (India) specifically to integrate the needs of these marginal and small farmer including women (Agarwal *et al.*, 1999; Koopen *et al.*, 2001). AKRSP (India) began field operation in 1985 and has since become one of the larger grassroots NGO (working with 84,229 households of 800 villages of 32 blocks, see APR, 2007) and popular for Sustainable Community-based Approaches for Livelihood Enhancement (SCALE, 2002-2011) in the environmentally challenged and economically vulnerable regions of Gujarat and Madhya Pradesh. While working on PIM, watershed projects, drinking water and Roof Rain Water Harvesting Structure and technologies the role of AKRSP (India) for rapport building through Participatory Rural Appraisal, village micro-plan and organizing community institutions, capacity building of the government officials and members of WUAs including land holding operations till they acquire competence[4] to become self-managers in Gujarat and Madhya Pradesh cannot be ignored. Coastal Salinity Prevention Cell as an interface institution developed with the collaboration of NGOs, experts, the Government and Sir Ratan Tata Trust, has started implementing PIM in the command areas and developed a detail data base of 1200 salinity affected villages across the Verawal, Kutch and Junagadh districts of Gujarat. The water quality testing laboratory at Surendranagar district was given ISO certificate in 2007.

To evaluate the status of canal network of Sardar-Sarovar Narmada Nigam Limited in South Gujarat a joint inspection by farmers of 66 villages and the irrigation officials was carried out in 2007. Furthermore, the Meghal River Basin treatment for water rejuvenation and irrigation through 89 village-institutions of 52 villages around at Junagadh district is another pioneering collaborative effort. Efforts are being taken to integrate women's interest, tribal people's contribution in terms of labour, cash and commitments to manage the water structure are crucial as investment motivates self and collective participation.

Given the performances of AKRSP (India) on integrating the interest of the small and marginal landholder into large and small scale PIMs, Tribal-Sub-Plan (TSP) of Government of Gujarat supported small water irrigation schemes such as Group-Well (GW) since 2001. In the year 2006, Japanese Fund for Poverty Alleviation also sponsored 30 integrated GW package for 300 users divided among 30 villages. The expected outcome is to reduce the out-migration, increase food security and income from Rs. 3,000 to Rs. 18,000 per year among the poor tribal.

The post-development era perception about water management and agricultural productivity is thus characterised not so much by a 'retreat of the state', as by radical restructuring of relations between the discursive and institutional regimes of the state and civil society. Difficult for NGOs to compete with the current subsidy policy of government though an innovative component of this is that the enabling environment by AKRSP (India) made utilize irrigation infrastructure in an effective manner. Regarding the decentralize governance through PIM process initiated a partnership between government (TSP), NGOs, donors (Asian Development Bank), and water users especially SHG women.

Given the structural condition the labour and capital embodied in agricultural landscape would not only be building and cultivating physical structure, viz. group-well, check-dams, and crops—but also, how services and collective efforts have creating, sustaining and altering relations between women and men over time. With the objective of centre-staging gender equality the strategies are to assess the effectiveness of farmwomen's participation in the Irrigation schemes two questions were addressed in this study. What contribution does the small water harvesting irrigation schemes of AKRSP (India)'s intervention make to bring about change in the trend of women's participation in agricultural practices and reduction in poverty? How the current modes of women's participation in the irrigation management under AKRSP (India)'s interventions have brought change in their lives?

Feminist studies in India[5] and elsewhere pointed that unequal gender relations need to be addressed both as a cause and as a factor in the intensification of poverty (Acharya, 2003). The definition on "gender mainstreaming" as a major strategy for the promotion of gender equality was provided by the United Nations (1997). As a concern of the international women's movement for alternative strategies to move gender related issues out of the periphery and "mainstream"[6] into development decision-making. In the years following the Beijing Conference in 1995, gender mainstreaming has been referred to as a comprehensive strategy that involves the integration of women-oriented programmes and gender issues into development institutions. Gender mainstreaming process would provide an understanding of why promotion of gender equality[7] is important for overcoming the costs of women's marginalisation (Acharya, 2003, Kelkar, 2005). Enhancing women's agency in the land and water management interventions requires action to ensure access to finance including microfinance services, building of women's organisations, effective participation in water user associations, community forum and local governance, land rights, and strategies to stage women's performances that contribute transforming gender relations in community and public ideologies (Mukhopadhayay, 2004). Thus, it is imperative to understand the ways in which links between sites of memories come into play with these epistemic structures. This paper attempts to chart this ground, narrating new forms of engagement that blur the boundaries between gender, water and agricultural productivity.

Personal Narratives Methodology

By invoking the historical exploitation and exclusion of the disempowered, the genre of post-colonial and post-modern feminist writings constantly point towards the possible advantages of rethinking ethics, gender policy and cultural politics in the post-colonial discourse (Spivak, 1988; Mohanty *et al.*, 1991; Acharya, 2003). They argue that quantitative method, like post-colonial socio-economic reform, has been co-opted by patriarchal capitalist societies to make the situation in which women live invisible (Agarwal, 2003; Mies, 1998; Kelkar, 2005; Kabeer, 2005). The trajectory of ground water management by women in this study demands a closer scrutiny of the transformatory potential of SHGs[8] and its efforts to access to credit and the ways they inculcate the ethos of political decentralization in local panchayat (Agarwal & Sivaramakrishna, 2001; Acharya, 2003). Personal narratives methodology (Berg and Acharya, 2004) within a feminist framework is used to assess farmwomen's experiences and perceptions about the impact of PIM project of AKRSP (India) enhancing collective actions and enabling the process of inclusion and women's capabilities. The significance of women managing group-well, often narrated orally, to shed light on their perspectives on new challenges to what overtake as "farmwomen".

This study draws cases of women leaders and focused group discussions on their participation in the water harvesting projects especially in the group-well schemes from three Mahila Vikas Mandals located in Badtal, Sathvav and Bhatkahi villages under Mahila Jagruti Sangathan (Women's Federation), at Mandvi Block of Netrang District (Table 26.1).

As a staff of the Research and Monitoring unit of AKRSP (India), I had several open ended conversations with various stakeholders (community representatives including men who were supportive or learnt lessons from the gender inclusive practices), AKRSP (India) personnel and government officials over a period of time. However, experience of women is imperative at this juncture to understand the connection between gender equality and resource management patterns and sites of memories come into play with these epistemic structures and in turn changing gendered ideologies. A rigorous gender analysis of individual and community based management would give an idea about the significance of women's participation in decision making. The outcome of the gender analysis would be of great relevance in centre-staging women's concerns in designing of livelihood interventions for agriculture and irrigation purpose. This investment in the discursive and symbolic practices of imagining landscapes, through narratives and cases, may seem subtle but it yields real effects and possibilities (Mohanty, 1991; Acharya, 2003).

The data was collected during November 2006 to February 2007 at various times. Individual interviews and group conversations with the SHG members of Mandavi and Sagbara women's federations at different localities: federation office, home veranda and group-well sites at agriculture-field. Members are farmers, agricultural labourers

Table 26.1: Overview of Women's Participation in the Water Harvesting Projects, Mandvi

Type of project	*No. of project*	*Managed by women*	*Major performances*
Lift Irrigation	1		Women joined as a nominal member. Gradually they have taken various responsibilities for implementation, such as Extension Volunteer and Credit Service Agent, Committee member etc.
Pump Irrigation	11	1	Women manage diesel pump for pumping well water
Group Well scheme	41	13	The MVM has been actively participating in all phases of project activities, especially for planning, implementation and asset management.
Check Dam	5	1	Asses creation and management
		3	Involved indirectly in all phases of project cycle
Group Well work ongoing		2	Their performances rated as good and rate chart was mandated by MVM
Canal Irrigation	3		Presently 50 per cent memberships of Kevdi PIM Project are women; similarly three women are members of the Community Institutions at Issar PIM

Source: Mandvi cluster office, Netrang, 2007.

and home-workers. The personal narrative conversations were chronologically free-flowing and open-ended, and linger over details of the periods spent in study areas. The information from the field is backed with the literature review and secondary data to inquire the two important questions and to validate the evidences. I have also shared the first draft of this report with the research participants and federation leaders and have tried to incorporate their perspectives on new challenges to what overtake as "farmwomen". As shown in Table 26.2, out of 53 GW in the Mandvi cluster nine were managed by the 73 female members till 2007.

Table 26.2: Group-well Irrigation User in the Mandvi Cluster

Name of the village	*No. of GWs*	*No. of women members*	*Area under cultivation in acres*	*Year of start*
Badtal	1	7	20	2003-04
Sathvav	1	8	15	2005-06
Bhatkhai	2	13 6	16 12	2005-06 2006-07
Devgiri	2	12 7	14 16	2002-03 2003-04
Bedada	2	6 8	20 18	2002-03 2003-04
Ghantoli	1	6	14	2006-07
Total	**9**	**73**	**145**	**2003-07**

Source: *Mandvi Cluster office Netrang 2007.*

Unlike the academics (Malcom Harper, 2002), NGOs and the development practitioners and professionals in India and elsewhere, AKRSP (India) has also emphasized the promotion of federations. Group conversation with the federation's committee members and SHG women of MJS are being explored to show the effectiveness of collaboration in shaping their performances in irrigation and technology management, and production relations *per se* male's control over cheap female family labour, but also reshape women's and men's motivation to invest their efforts.

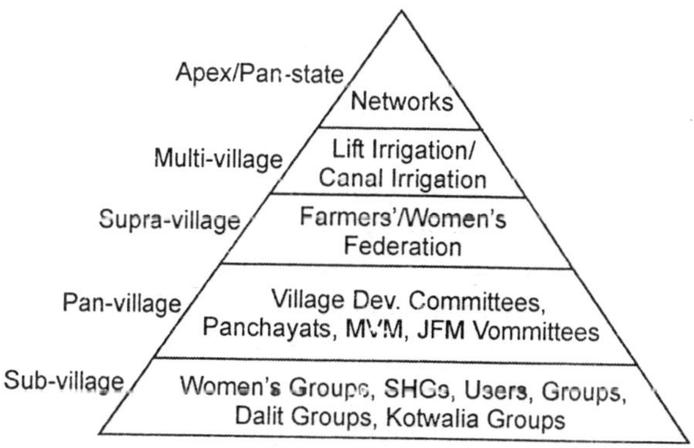

Fig. 26.1: Modified after Oza, 2007 on an Illustration of Five Tier Linear Federation and Network Structure of AKRSP (India)

In the year 2007 a total of 1586 SHGs with 26100 women members having Rs. 15.7 million savings under 622 Mahila Vikash Manadals (MVM) of 21 Women's Federations and cooperatives at the apex were divided around 400 villages (LFA based data 2007).

The Case of Badtal Village

My conversation with Kamuben, one of the para-worker and a well known leader of Jagruti Mahila Manch, Mandvi Block of Netrang District took place on a February afternoon in 2007. Kamuben aged about 55 years, dropped out of the 9th level of schooling. Well known for her activism she works as a member of the *Arogya Samiti* (health-committee) as a member of the Social Security Committee she deals with the widow and land issues. She has coordinated several *padyatras* (foot march) and mass movements[9] for awareness on "sustainable agriculture" and gender sensitization for the Mandavi Women's Federation. She lives with daughter and son in Godan *phalia* (hamlet) of Badtal village. Kamuben represents the post-structural socio-economic reform generation of women for whom commercial farming and social activism offer realistic opportunity as a form of occupation and affirmation. Kamuben told to me her story how widowhood changed the mode of her feminine roles and agriculture has become an important part of her way of life and not simply an occupation. Kamuben recalled how difficult and embarrassing it was to live in in-laws home. She said that:

> 'When my husband died, my brother-in-law took advantages of controlling joint family land and assets, without considering my children's future. I had returned from Rajasthan boarder where my husband was working as a teacher to in-law's home at Maritha located 5 kilometres away from my parents' home at Badtal. That time my elder son was five years old, daughter was 3 years and youngest son was only one year old. From early morning to bed I did house chores for in-laws and children without hesitation, but still I was living a life of great fear. My brother-in-law tortured me, all to show their control on me and my rights to properties.'

Kamuben's in-laws hold a patriarchal kinship structure. She left in-laws house within a year and her vulnerable situation obstructed her from equal access to the land productive. She stayed with her parents so that her children could grow up well. Kamuben failed to acquire her share of her land from in-laws home till her son grew up.

Badtal is a small village of just over sixteen hundred inhabitants, divided among three *phalias* with 365 households. Kamuben lives in Godan *phalia* consist of around 450 inhabitants divided among 115 households. The area receives 646 mm of rainfall, 72 per cent of which occurs during the three monsoon months. As derived from the group discussions that land is extremely limited, with an average of 1.2 hectares per household. Agriculture is the main occupation but most of the men in the village migrate to nearby towns of Surat and Ankaleswar to work in diamond polishing and textile industries, while females work in neighbouring villages as agricultural labourers for *Patel* families (the dominant caste who have left the land for more productive engagement).[10]

Kamuben used to help her father by giving Rs. 250 as the pension money she used to draw for her x-husband. She said that altogether those worked in her favour. After ten years of her stay in parent's home, realizing her struggle, her father then extended four acres of rain-fed land as a contingency. Her situation made her father to think about the future of other two married daughters' future as well. He then divided his ancestral parental property equally among her two brothers and two sisters. Kamuben said that this is not normal, if father have more land, and brother did agreed may provide a portion to their married daughter. Only when her son grew up did she petition for land from her in-law through Grampanchayat meeting. In the first meeting her brother-in-law agreed to provide two acres of land but that was rain-fed third time when Kamuben demanded for a piece of irrigated land (in the Panchayat), her brother-in-law exchanged an acre of land located near *Nala* (small tributary). Her second son is cultivating that land since five years. Since landownership was the primary source of collateral for credit, well-ownership closely followed contours of control over land. Only relatively wealthy people were able to mobilize capital to drill tube well.

The story abound effort to obtain access to ground water by the partners of Kamuben's groups. Kamuben recalled how these households did not have access to agricultural input, technology, loan or marketing arrangements and even their participation was ineffective in the Gramsabha. In Badtala, often *dalit* and poor tribal were deprived to have a well of their own and unable to mobilize collective efforts. Kamuben said that 'Jyamatiben from AKRSP (India) inspired us to attend meeting held in 2000 at AKRSP (India office). That was the beginning of the formation of new SHG and the Jagruti Mahila Manch at the apex in 2000. 15 women from our village attended that meeting. Back home we formed 3 female SHGs with 35 members. We worked on watershed project and set up bio-gas plants in the village.' During the years 2002-2006, SHG members of Badtal, Sathvab and Bhatkhai village Mahila Vikash Mandals on the basis of their excellent performances on micro-finance management have selected through the women's federation for financial grant for group-well under the Tribal Sub-plan scheme. Kamuben said that 'We collected information from MJS at Mandvi that group of farm men acquiring well irrigation project. In the year 2002, seven members with 11 acres of land came together and have shown our interest and assured the committee and extension workers to provide common land labour and cash whatever possible with us.'

The GW[11] committee of MJS then arranged an exposure visit for the new members to the previously managed GW scheme. Leaders of MVM organized 9 meetings. In those meetings members formed User Groups, finalized their priorities, and delegated responsibilities to the Extension Volunteer, construction committee, arrangement of material, labour, transportation etc. The land (36 sq. feet) identified for GW would be a group property and no rent whatsoever would be charged by the owner. Before going further into the Kamuben's perception I would brief in the next section on how the leadership quality among women like Kamuben has brought major advances in women's livelihoods through.

Mahila Jagruti Sangathan, Mandvi

The research participants informed that the Mahila Jagruti Sangathan (MJS) at Mandvi evolved in 2000 and registered in 2005. In 2007 MJS constituted 15-member committee with 1500 women members divided among 35 villages (Photo. 26.1).

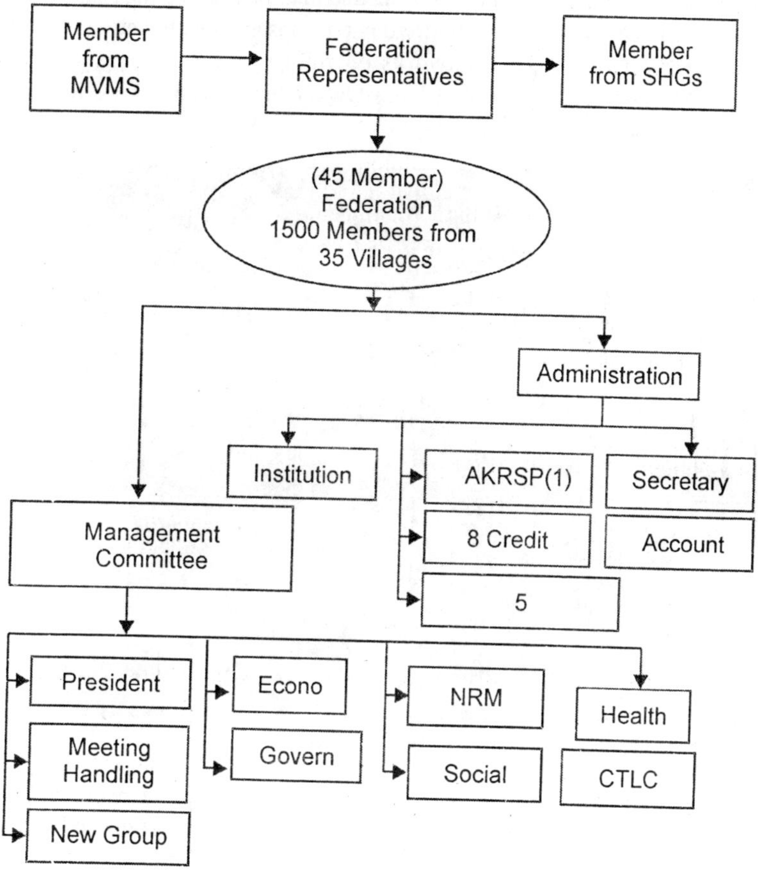

Source: Developed by the author in consultation with women members and staffs at Mandvi.

Fig. 26.2: Structure and Management System of the Mahila Jagruti Sangathan, 2007

Unlike other Women Federations in Gujarat, it has begun to promote and strengthen the SHG of the poor, provide a sense of solidarity and access to services as per the need and aspiration to enhance effective choices. SHG women members who are able to mobilize "social capital" through the micro-organization for MJS catalyze the new livelihood options and are strengthening existing livelihoods through developing forward and backward linkages with governance, government schemes, bank and market. Staging the issue of land right for women Kamuben and her groups are able to actively participate with the NGO Networks on right to information and Working Group for Women and Land Ownership (WGWLO, APR 2006; Velayudhan, 2008).[12]

Photo. 26.1: Members at Federation Meeting decision making process

Kamuben recalled how for the whole day Bhatiphalia people grappled with the appropriate site for the source of abundant ground water were following the drilling machine. The bore machine checked six sites, but failed due to the difference of opinion between men and women about the hard rock layer beneath the surface. At last when water burst out everybody present there were cheered up and dancing with happiness. Kamuben said that:

> 'Since generations we know the sky showers water, but underneath our land we have another source, which would back our equal standing with the big farmers cultivating crop during summer season. We inspired and speed the digging and cleaning work. Meanwhile three female members splited out and joined the male group in thought of relaxation from

> the hard rock head load. You know, unfortunately, the group-well dug by male SHG found long depth rock than ours.'

After 17 ft depth two blasts were used to dig another five feet. They worked out to network pipeline for 6 *kundis* (small square tank where pumped water can be stored and distributed directly to the field). Eleven acres of rain-fed land have now got irrigated water for another crop. Besides the *kharif* (rainy season) crop such as rice, *tuar* (lentil) and *makai* (maize), now these families grow onion, *mugfadi* (groundnut), *ganhu* (wheat) and *sherdi* (sugarcane). Kamuben is able to harvest two quintal of sugarcane and two quintal of lentil in her two hectares of land. She shared that not only gross cropped area has been doubled but also productivity is increased three-fold. Kamuben is able to produce 75 tonnes (six trucks) of sugarcane though the net profit in 2006 for first year was only Rs. 15,000, while in the second year she would get a net profit of Rs. 62,000, as the investment per acre of sugarcane would only be one-fourth than the first year. Each acre of groundnut provides them with Rs. 8,000 net profit.

Actually, Kamuben bore a loss of another 15 tonnes of sugarcane when the member on whose land the well was dug did not allow water at a crucial moment. Not only that, three other families cultivating groundnut were sharing the scarce water. Negotiation for the use of scarce water rule was violated by the groups. Groups have discussed such issues at the Federation office and seek expenses from AKRSP (India) for blasting morom (rock) till another seven feet depth in May 2007.

The group's struggle to maintain access to ground water throughout the year made me to trace the average depth of the wells around. Kamuben's group gave me a rough estimation of tube-well. There were 10 bore wells and (five run with diesel machine and five with electric motor), five open wells (run with either electric or diesel) on the left side and five bore wells and three open wells on the right side of the main road across, which provides water for approximately 30 hectares of land around. Unsurprisingly, while other machines are either five or ten HP, efficient to pump water to distant land, Kamuben's group has a 5 HP machine. The highest depth of well was forty feet, but the average depth varies from 25 ft to 32 ft each, deeper than Kamuben's GW though.

Kamuben argues that drying up well water during summer even in a drought period was not the phenomena. Only with the advent of new pumping technology, deeper excavation of ground and heavy extraction of water by pump depleted the water level and brought conflict around the surface water distribution. High yield varieties crop required more fertilizer and sufficient water so small and marginal farmers lacking investment get lower productivity. Those who have access to more land, have access to credit and construct deep tube-wells with submersible pumps to use and sell water to their neighbours put big farms in advantaged position to benefit from the wider water scarcity. In the Tenth Five-Year Plan in India increasing the user charges for water used in irrigation and industry implied a commodification of water and promotion of participation of private sector. It is the rural small and marginal farmers lack access to credit and a well have to pay a proportionately higher price for buying water as compared to those with more agricultural production. The phenomenon appears "tragedy" for those who have adversely affected from water pricing. Over cultivation to generate revenue and profit has lead to an over exploitation of ground water resource, and drying up the river catchment those supported surface water. It is likely to lead in time to deepen Kamuben's group like many rural poor and tribal women to dig deeper and deeper and cover longer distance for water those who provide water to the household survival and farm activities. Shiva (1998) refers to this as a 'patriarchal water order' where water is treated as an economic goods like any other commodity and rights to water are defined through rights to capital and political power. Such a privatization of water has clear gender and class implications. Water as a commodity is being assigned market value leverage men's decision on price fixing as they control the market. This would imply that social costs of damage are externalized and therefore, performances of women in the household economy and their knowledge in the management of water are excluded (*Ibid.*). The consequences of water privatization, therefore, have a disempowering impact as their right to water under Millennium Development Goals (2002) is undermined. Subsequently their equal participation in the community resources and contribution of knowledge to manage them are ignored.

Kamuben has decided to divide her land equally between her

son and daughter. When I asked if such a land distribution is a normative practice in the tribal family? Kamuben said to me that 'you know, a widow excluded from cultivated land of in-law and living with children to grew up can only explain to you how crucial agricultural productivity and hence control over a piece of land can be for her daughter'.[13]

Although there are no deeply entrenched taboos and restrictions ascribed on Vasava tribal widows, being debarred from a normal life, they generally find themselves isolated and at the receiving end. Research in male farming system in India and elsewhere indicates, women wanting to farm by themselves, are in a minority[14]—as the definition of a 'male farming' system and often 'taboos' (Agarwal, 1994 a and b) implies—facing deep rooted sexist cultural and developmental norms. Given the command over land Kamuben's performances ensure her better livelihood, food and social security.

She says, 'I hardly believe conventional ownership over land practices or living with the mercy of someone would be of any help. Me and my daughter do most of the strenuous labour in our field then tell me why men only legitimize their rights to land holding? Yet, my daughter and son-in-law have performed a more responsive role in farm activity; my son takes care of milching cows and buffalos and is a member of community dairy board.'

In the process of Kamuben's engagement with the Women's Federation by AKRSP (India), and within her life style she has found herself self-lived and enlightened with women in her community and around and other found Kamuben as a down-to-earth farmer and social activist. Kamuben's conviction for equal distribution of land between daughter and son is also supported by the fact that women took over the traditional male role of farm manager bringing an egalitarian distribution of resources. She hopes that higher chance of survival of agricultural productivity is more frequently tied to women farm heads. Hence, she wishes faster the registration of land under joint ownership process would add value to household productivity and women's interest. The next story is on female GW managers of Sathvav village.

The Case of Satvav Village

Sathvav village consists of seven hamlets. Ambaben aged about 28 is one of the SHG members and GW managers who belong to Bhati phalia. Bhati phalia consisting of 2000 inhabitants divided among 300 households. The main castes of the villages are *Vasava* and *Harijan*. With the inception of 3 female SHGs in the year 2001 with a total of 36 members Bhati phalia women took over the old Gram Vikash Mandal and set up their own Mahila Vikas Mandal. Presently, MVM has its own savings of Rs. 10,000. Of the three SHGs, Gayatri Juth had savings of Rs. 40,000. On the basis of the excellent performance on microcredit management Women Federation and MVM selected Gayatri SHG for financial provision from AKRSP (India) to set up their Group Well. In the year 2005, eight members with 8 acres of land around have started digging GW by having own contribution of Rs. 500 besides family labour contribution till the completion of the GW. At the depth of nine feet workers found thick layer of hard rock. With the financial help by AKRSP (India) they used eight time blast to dig another 17 ft which cost Rs. 17,500. Ten truck loads of stones cleared out from the well, which were sold to the *Vanias* for Rs. 6000. At the first stage five water distributaries points were set up with 800ft length pipe networking. Ambaben said that: 'women engaged in digging have to come after completing their household chores, such as cooking, cleaning, serving the child and old and animals at home, while this was not the case for men. Our finger paralysed and shoulders ached, and then once we returned home we had to clean all dirty clothes and serve dinner.'

Ground Water users avail water free of cost for cattle rearing and charges Rs. 5 per hour for irrigation from own members but Rs. 25 for the rest of the farmers at the village. The net saving in one season was Rs. 1500. Out of their savings SHG invested Rs. 25,000 to take four acres of land on lease for five years nearby their GW. The ownership of land not only facilitated them to get a loan of Rs. 25,000 from Bank of Baroda but also land was being developed for regular cultivation with higher productivity than earlier. They decided to pay back the bank loan within six months because they would not have to pay interest to the bank. SHG members used their revolving fund for growing vegetables. Four members cultivated grass sufficient for their livestock and continue the supply of four litres of milk to the milk society in this 42 degree centigrade summer heat. They have started purchasing cattle feed and other items in bulk. In case of jowar, green

Photo. 26.2: Members cleaning the rocks after blasting

Photo. 26.3: Members using irrigation water through small outlet network

lentil, and from groundnuts the productivity increased two-fold. Furthermore, GW members learned lesson from the first year of experiment.

SHG members expressed concerns at the incidence of violence against poor women and ignorance towards their children at villages by their drunkard men. This has not only prevented for being a part of the SHG, these women were irregular in minimum savings and attending meetings. Members have started intervening on some vexing gender related issues perceived at the community. Women have stopped liquor making and beating women and put forward this issue in the Gramsabha for a public decision. Little hesitance though Sarpanch came forward. Awareness on women atrocities is now increasing. The members also work as watchdogs for the cleanliness of the village road and sewerage, and have repaired all the tube wells for adequate drinking water. The capacity for autonomous action in the face of constricting social sanctions and structural inequalities has transformed through maintaining good governance at the local level.

The Case of Bhatkhai Village

Bhatkhai village at Mandvi Block consists of 105 households. Of the 71 women SHG members 46 have adopted bio-gas. Miraben, the leader of the GW members said that in the year 2007 three SHGs have linked with bank for microfinance and four SHG with Vadi (horticulture) project. In the year 2006, six members with 8 acres of land from Vandana SHG have constructed the GW with five distributaries constitute 1,560 feet of pipeline network. Of the Rs. 3655 earned by selling the surplus water, Champaben has alone spent one-third besides her share of water for irrigating three acres of sugarcane field. Champaben said that sugar factory at Surat would pay Rs. 1,200 per tonne for 150 tonnes of sugarcane. Group members have also planted good variety of mangos besides vegetable and maize.

Women have started spending more time nearby GW to perform much of the domestic work such as bathing, cleaning, cow and buffalo feeding. The pattern of investment has now changed. Champaben said that earlier the money borrowed from revolving fund was spent for school fees or daughter's illness, and now she has borrowed

Photo. 26.4: Sathvav GW members are celebrating their pride as able to grow sugarcane and vegetables

Rs. 5,000 for buying a Motorbike for her husband and Rs. 3,000 for cultivating vegetable. Miraben the finance secretary said that the profit from selling the water would be spent to buy a Gharghanti (machine for grinding wheat flour and Jawar) to grind flour. They would charge Re. 1 more per kg from villagers which will be still demanding as villagers would not have to walk long distance in the hot summer and muddy rainy season. The next section reflects the implications of such best practices on women's access to microfinance, equitable-based gender relations in intra-community resources and productivity.

Microfinance and Change in Agricultural Productivity

Exclusion in the present case of reference are a lack of land ownership, knowledge of technical knowhow and lack of microfinance services might have been a primary explanation for the late adoption of GW irrigation. SHG women who fall below a minimum landholding threshold for collateral can avail smaller amount of credit from revolving fund. However, this opportunity was not available to the marginalized non-SHG members. Experience from AKRSP (India)'s women GW managers shows that transforming the management status of SHGs and land resources members at Badtal, Sathav and Bhtkhai have now access to bank on regular basis. Officers of Tribal-sub-Plan, AKRSP (India) service credit agents, Para-workers and agricultural extension officer have now intensifying their input supply services, and training on diversified crop practices and information on new technologies towards women farmers. Gradually, with this knowledge and credit women are able to transform crop farming from a semi-subsistence mode to a commercial mode with an increase in productivity though they found the availability of water in summer is still insufficient.

Such a cost effective water irrigation management process has impact on out-migration caused from the push factors. Women and their family members are able to engage in their small agricultural plot and land on leased throughout the year. They have now started hiring wage labours from villages around. Women have now greater prestige in the family and community and have greater control over how their income would be spent. Women are able to take decision on household assets creation, food security and their children's well-being and even reduce the wasteful consumption and expenditure. Tribal men of these households too concede that transformation of land under joint ownership would be better for the households, as there is greater chance of women, unlike men, will lose the land and assets because of drinking or debt or will increase interest among women on farming and productivity. Instead of waiting for a state-enforced land reform and transformation of ownership to the women headed households. Hence, enabling environment (supply) created by AKRSP (India), MJS federation and WGWLO network contributed to mobilize women's interest (demand) to manage input services, access to credit, (microfinance institution and bank) and technology and exercise their rights to some extent on community governance.

Women's Transformation Roles

The most significant impact of such an inclusive movement has been the empowerment of women in terms of transformation of water management right to women, land right to widow and capacity building of women (technical and microfinance training) to undertake activities independently or collectively in the public domains. The success of the transformative role of women can clearly be attributed to the institutional arrangement at the community level (formation of SHGs, water user and management groups, *pani samities*); the leadership initiatives by Kamuben, Ambaben and Miraben like cases and the role played by them instrumental for designing and implementing a gender participatory intervention in the factor of production (land, water, labour, capital, inputs, credit and markets); the women's sense of ownership of local water resources, infrastructure and technology through operation and maintenance; influencing the local water selling prices; the democratic and transparent process of distribution of water among the group members and community; conflict resolution; creation of new assets and additional savings which would provide alternative livelihood choices and above all, the strong group solidarity. Women Federations are making use of the WGWLO-NGO's network to acquire ownership certificate on land. The unique thing about the GW users is that they are members belonging to the different SHGs and MVM further collectivising their capability to manage finance and creating assets with insurance for various farm and non-farm livelihoods.

Traditionally, women wanting to farm by themselves are those who are widowed, single, divorced or destitute, but that is a minority. However, in the project intervened area women working alone and directly involved in implementation and supervision of construction have set their own priorities for emergency supply and maintenance. The PIM associations under AKRSP (India)'s intervention elsewhere have learnt the benefit of integrating women's participations. Another spontaneous trend observed among the research participants in Mandvi was that women farmers increasingly engaged in collective action and decision-making. In rural Gujarat, unlike 11 federations under AKRSP (India), Self Employed Women's Association and its ambitious "Women Water and Work Campaign (SEWA, 2003)", Utthan, Unnati, Kutch Mahila Vikash Sangathan, Sadhana, NM Sadguru Water and Development Foundation and many other where collective action is a strategy that women have successfully adopted to come together as a group, network and work with the parent organization. Individual their voices are subdued, though collectively they are able to undertake activities, foot-march, tractor-march, rallies and campaign along with men or independently for issues influencing land and water policies thereby demonstrating an increased sense of ownership and confidence with and within. What clearly emerges is that through women's self-mobilize actions good governance for sustaining local resource management and engendering the governmentalism of the state, market and bankers to some extent are possible. Within the household women still tend to take on the burden of all household responsibilities so that they are not prevented by their male counter from going out to participate in the SHGs, construction sites and new workplaces. While an active participation in irrigation, machine maintenance, accounting and crop production places women in an advantageous bargaining position within the household and at community, they have potential to produce more significant impacts. Next section concludes the paper by offering some ideas on the ways forward in enhancing women's participation in agriculture and water management.

Conclusion

Experience from AKRSP (India) emphasizing women's self-mobilized participation shows that the time has come for water sector laws to operationalize the human rights to water, thereby, integrating social and economic challenges (land-ownership, microfinance, assets creation and participation in local governance) relevant to centre-stage women's participation in water management. Although the water sector reform promote community participation in which women and other marginalized groups may have ample opportunity to participate the complex gendered relationships within which they depended heavily on the former for employment as agricultural labourers or as sharecroppers or to pay the commodity cost of water. Importantly, men at the communities and projects are beginning to realise the opportunity costs of women's labour and collective actions and this made them ready to share spaces previously denied.

Staging women's performance through personal narrative and oral account at various sites of struggle (Mohanty, 1991; Berg and Acharya, 2004) was an attempt to dismantle the complexity of gender relations in the water management. This would imply that a gender perspective on agricultural productivity and water management has potential to address gender inequalities and injustices in societies on ownership of land and economic opportunities for women. Major advances by building women inclusive local institutions as this study suggest is that it has able increase of functional space for sharing experiences on access and leverage cohesiveness and effective choices, and struggle social life, capacity building, diminishing role of moneylender and marginalization of women in the privatization approach to water (Shiva, 1998; Dutta, 2003; SEWA, 2003; Acharya, 2003; Panda, 2006). Women's participation in the new form of village institutions/decision-making processes redefining gender roles and structural conditions have transformatory potential (Mukhopadhyay 2004) to dealing with new financial, livelihood, market and social services. Integration of new forms of women's organization into water management skills comes out to be the major reason to improve their ability to make strategic choices and increase their self-esteem as they shape the 'waterscape' and agricultural productivities. The significance of women's oral account highlights that SHG women farmers' association has gone beyond mere instrumentalism by being with first addressing questions of joint ownership to land, assets and active participation of farmwomen in income and resources those foster women's agency and their well-being. As the question is not one of number or fine sounding intervention strategies of a location alone, impressive as they may sounds. Instead this study has argued, while acknowledging the growth that has been made, there are still miles to go and gap to close before women and marginalized communities can have equal access to membership in various schemes, avail the services, technology, loan and insurance needs and participate in the decision making process of institutions and at household units.

If gender equality in the land and water management is a political project it is essential to cater the experience of women into the canal irrigation and revival of river basin management projects, and to demonstrate at the *Gramsabha* (village meeting) or *grampanchayat* (local governance). PIM training commenced with an objective of improving gender equality and poverty reduction would demonstrate such best practices through training and exposure visit while building capacity of new generations of trainees and stakeholders. Mainstreaming gender equality in the agricultural and ailed training, agri-business, rotation schedule and cropping calendar and organizing meetings with different stakeholders are prerequisite for profitable farming and sustaining the functional spaces.

This would imply that prospect for effective participation of women in land and water management hinge on how the transformation of gender relations call for an egalitarian mode of production practice within which tribal women in Gujarat would be seen as farmers and could exercise their full human potential and negotiate their identity(ies). Furthermore, there is a need to bring all gender mainstreaming efforts made within the SHG-federations on common platform and being with first addressing questions of women's agency and their well-being. Agricultural policy-makers including irrigation agencies and NGO-practitioners institutionalising with the objective of increasing participation would bring about a convergence among varieties of political and institutional constituencies with a common understanding of farmwomen's control over water and on farm irrigation management. Consideration of women farmers' perceptions can contribute significantly to an understanding of how struggles over meanings and discourses, and how truth claims are made in gendered spaces. Their conversations with me have enabled them to evolve from an aggregate of individual issues into a cohesive collective, in which they can look at themselves and their gendered spaces in new ways. Counter narrative of thinking and making sense of those realities developed a positive self-image, recognize their strengths, and explode sexual misconceptions, are and should, therefore, assume a much more central place in attempts to mainstream gender in workspace culture. Gender and Water Alliance Network, Geographical discourses on the negotiating power of individual and collective space, place and gender would work as a strong influential advocacy for articulating demand and carving out rights and insisting on participation of women in policy on water resource governance so as to make a significant difference to the agricultural productivity and lives of women, their family and communities in India.

Acknowledgements

I am grateful to the staffs at Netrang and Mandvi field office of AKRSP (India) for their discussion and facilitating the contact with the women

members. I especially thank Kamuben, Ambikaben and Miraben and group members of MJS, Mandvi for sharing their time and experiences with me. I am grateful to Apoorva Oza, CEO, AKRSP (India) for his insights and discussion. My sincere thanks goes to the participants and the Convener of the National Seminar on "Population, Development and Environment: Issues and Challenges" of BHU held on 18-20 February, 2010.
Note: An abstract version of this paper was published in the SaciWATERs conference in December 17-20, 2008.

ENDNOTES

1. See SOPPECOM. 1998. 'Society for Peoples' Participation in 'Ecosystem Management', Recommendation for State Irrigation Acts 1998. National Water Policy (2002), Government of India. The Odisha Pani Panchayat Act, 2002.
2. See Shah, Anil, Apoorva Oza, and R. Parthasarathy, 2006. 'Impact of Participatory Irrigation Management on Growth and Equity', in Workshop Proceedings by Development Support Centre, Ahmedabad. And see Mukherjee, A. and Shilpa, V., 2002. 'Canal Irrigation Management by Tribal Communities: Case study of AKRSP (India) supported PIM societies of South Gujarat', IWMI Tata and Ford Foundation.
3. See Shah, A. and Yagnik, J., 2007. 'Estimates of BPL-households in Rural Gujarat: Measurement, Spatial Pattern and Policy Imperatives', Gujarat Institute of Development Research, Ahmedabad.
4. 31 Canal Irrigation, 148 lift irrigation and 955 check dams and irrigation tanks and 1108 bori-bunds, 186 Group Well along with 8038 water efficient irrigation technologies were used to improve the land and agricultural productivity so far. 6392 household RRWHS and 97 community RRWHS, 1334 hand pump and 168 drinking well were constructed in collaboration with community people under various schemes (Water and Sanitation Management Organization, PRAVA an NGO network on drinking Water Issues) for securing adequate drinking water and reduce women's hard-work for fetching water from distance sources (APR, 2007). The state government had formed a 'Planning Group for PIM Training in Gujarat' in 1998 which was headed by Chairperson of DSC and former CEO of AKRSP (India) Anil C. Shah along with WALMI and AKRSP (India) and is leading the PIM training since then. www.akdn.org/india
5. Spivak, Gayatri Chakravorty, 1987. Can the subaltern speak? in Williams, P. and Chrisman, L. (eds), 66-111. Prentice Hall, London;

Nussbaum, M.C., 2000. Women and human development: the capability approach. Kali for Women, New Delhi.

6. Attempt at gender mainstreaming would entail visibility of the differing roles and responsibilities of both women and men in the water irrigation management programme and policies. It would create policies that institutionalize the equitable participation of women and men. It would entail establishing budgeting processes that ensure necessary water infrastructure.
7. See World Bank's Gender Equality Action Plan 2007-10 addressed that 'The global community must renew its attention to women's economic empowerment and increase investments in women...Increased women labour force participation and earnings are associated with reduced poverty and faster growth; women will benefit from their economic empowerment, but so too will men, children and society as a whole...'. (p. 2).
8. Transformatory potential is about person's sense of agency; a person believes that they control their lives that make difference their lives are within the power to influence. So empowerment is about the ability to make, to teach and to participate in the wider decision-making process which has very important influence on the way you can lead your life (Acharya, 2004, Kabeer, 2005).
9. A five-day long mass awareness campaign on "Sustainable Agriculture" was held at Mandvi Cluster of Surat District. The message of Sustainable Agriculture has been delivered through different communication materials, viz. Padayatra, Skit and Audio-Visual to around 5,500 farmers of 13 PIM villages. This awareness campaign was culminated in a form of large gathering at Isar village in a ceremony participated by 3,000 farmers, of which, 75 per cent were women. Renowned development activists and academicians associated with the promotion of "Low External Input Sustainable Agriculture" and "Organic Farming" has addressed the issues. The contribution of Padayatra team towards the success of the event was remarkable, especially the women in the sense, since they had to set aside their household responsibilities for greater social and environmental cause. Around 20 per cent of the total cost of the awareness campaign was borne by different Canal Irrigation Societies and member institutions of Mahila Manch.
10. Of the 115 households, 15 households are landless out of which 5 SHG members have got one buffalo each on loan from AKRSP (India). 15 have got Biogas and 25 have producing Vermin compost for their agricultural field and few also sell it. Chaudhry is the dominant caste, but people take different *dharma*. One primary school is available at the village, but to continue high school and college children have to

travel Ghantoli Ashram or Sarkoi or Mandvi which are five kilometres by auto rickshaw (which is not affordable) or by bicycle. Such a situation compels most of the girls to drop out from higher education. There are 12 hand pumps and four wells available for drinking water which is also used for bathing and animal drinking in Badtal.

11. A study done by Agarwal and Patel, N. in 1998 on the Group well in Moskut and Sanjavan suggests that GW members of AKRSP (India) had an exposure visit in 1995 to Gawali village. Study done by Koppen *et al.*, 2001 documented that the option to deepen the well and use it for irrigation purposes by installing a mechanized pump was discussed independently in the male GVM and in the MVM in Sagbara. With some hesitant from men women have got charged to manage the irrigation scheme. Then Jambar group women offered a visit to Hazaribag in Bihar to experience how women group successfully manage a mechanized pump scheme with the support of the NGO PRADAN.
12. See APR 2006 and Velayudhan, M. 2007. 'Women's Right to land: Voices from Grassroots Movement and Working Women's Alliance from Gujarat', AKRSP (India)-Ford Foundation.
13. Legally though daughter has equal claim over land and even daughter play a key witness role on family land and property distribution among the brothers still as a daughter or daughter-in-law her say on her need is little considered. Lack of in-depth research on the kinship-support system, social and security such as life insurance, benefits or pension and on the changing status of widows for example Kamuben's life story are need rigorous analysis.
14. Widows, single, divorced and whose husband engaged in factory work or non-farm activities, or women who have inherit land rights in their own name face many problems. As in a highly skewed gendered organization of farming, agricultural growth requires not only more gendered-balanced agricultural external support but also, and certainly more difficult, profound challenging of the economic, political and cultural gender discrimination intrinsic to male monopolization of production factors, including women's labour.

REFERENCES

Acharya, J. (2003). Women's well-being and gendered practices of labour and workspace in traditional craft productions in Odisha, India. Gender *Technology and Development*, 7(3), 333-357.

Agarwal, A. and Sivaramkrishnan, K. (eds.) (2001). *Social Nature: Resources, Representation and Rule in India*. New Delhi: Oxford University Press.

Agarwal, A. and Clark, G. (1999). Enchantment and dischantment: the role of community in natural resource conservation. *World Development*, 27(4), 629-49.

Agarwal, B. (1994). *A Field of One's Own: Gender and Land Rights in South Asia*. Cambridge University Press.

Agarwal, B. (2003). *Gender Inequality, Cooperation, and Environmental Sustainability*. In J.M. Baland, S. Bowels. and P. Bardhan (eds.), New York: Russell Sage.

Berg, N.G. and Acharya, J. (2004). The place of personal narratives in research on self-employment: the case of craftswomen in Odisha, India. In J. Acharya (ed.), Norwegian University of Science and Technology Press.

Dubash, N.K. (2002). *Tubewell Capitalism: Groundwater Development and Agrarian Change in Gujarat*. Oxford University Press, p. 300.

Hardiman, D. (1998). Well irrigation in Gujarat: systems of use, hierarchies of control. *EPW*, 33 (25), 1533-44.

Kelkar, G. (2005). Development effectiveness through gender mainstreaming: gender equality and poverty reduction in South Asia. *EPW*, October 29, 4690-4699.

Koppen, B. Van., Nagar, R.K. and Vasavada, S. (2001). *Gender and Irrigation in India: The Women's Irrigation Group of Jambar, South Gujarat.* IWMI-AKRSP(I), Working Paper 10, p. 9.

Mohanty, C.T., Russo, A. and Torres, L. (eds.) (1991). *Cartographies of Struggle: Third World Women and the Politics of Feminism.* Bloomington: Indiana University Press.

Mukhopadhayay, M. (2004). Mainstreaming gender or "streaming" gender away: feminists marooned in the development business. *IDS Bulletin*, 13(2), 95-103.

Oza, A. (2007). *Irrigation and Water Resources*. Indian Infrastructure.

Panda, S.M. (2006). *Women's Collective Action and Sustainable Water Management: Case of SEWA's Water Campaign in Gujarat, India.* CAPRI Working Paper No. 16, Washington.

Sen, A. (2000). *Development as Freedom*. New York: Alfred A. Knopf.

SEWA (2003). *Women's Struggle for Water*. Ahmedabad.

Shiva, V. (1998). Women's water rights. *Waterlines*, 17(I), 9-11.

Shah, T. (1993). *Groundwater Markets and Irrigation Development: Political Economy and Political Policy*. Bombay: Oxford University Press.

United Nations (1997). *Mainstreaming the Gender Perspective into all Programmes and Policies of the United Nations' System. Economic and Social Council,* Geneva.

Swaminathan, M.S. (2005). The sons of the soil. *The Hindu*, April 19.

Index